T0383208

Fluorescing World of Plant Secreting Cells

Fluorescing World of Plant Secreting Cells

Victoria V. Roshchina

Laboratory of Microspectral Analysis of Cells and Cellular Systems
Russian Academy of Sciences Institute of Cell Biophysics
Pushchino, Moscow Region
Russia

SCIENCE PUBLISHERS
An imprint of Edenbridge Ltd., British Isles.
Post Office Box 699
Enfield, New Hampshire 03748
United States of America

Website: *http://www.scipub.net*

sales@scipub.net (marketing department)
editor@scipub.net (editorial department)
info@scipub.net (for all other enquiries)

Library of Congress Cataloging-in-Publication Data

Roshchina, V.V. (Viktoriia Vladimirovna)
Fluorescing world of plant secreting cells/Victoria V. Roshchina.
p. cm.
Includes bibliographical references.
ISBN 978-1-57808-515-6 (hardcover)
1. Plant luminescence. 2. Plant cells and tissues. 3. Plants--Secretion.
4. Cytofluorometry. I. Title.

QK844.R67 2007
571.7'92--dc22

2007036631

ISBN 978-1-57808-515-6

Published by Science Publishers, Enfield, NH, USA
An Imprint of Edenbridge Ltd.
Printed in India

Preface

Autofluorescence, e.g. fluorescence of intact living cells induced by ultra-violet or violet light and seen under a luminescence microscope, has been of interest for investigators from the beginning of the 21st century. The study of this phenomenon has the potential to observe cellular life without invasive methods.

Autofluorescence of intact plant secretory cells induced by ultra-violet or violet light is clearly seen in the visible region of the spectra under a luminescence microscope. Light emission of secretory cells depends on the chemical composition of their secretions. Identification of intact secretory cells filled with products of secondary metabolism may be based on their fluorescence characteristics. Autofluorescence of intact secretory cell of plants, which contain various fluorescent compounds, is considered as a possible biosensor for cellular monitoring. This phenomenon could be recommended for: 1. diagnostics of secretory cells among non-secretory ones; 2. express-analysis of the content of secretory cells at norm and under the influence of various factors; 3. analysis of cell-cell interactions. The filling of cells with a secretion and their removal are easily observed under various types of luminescence microscopes – from confocal microscopes to different microspectrofluorimeters. Cellular fluorescence can serve as an indicator of the cell state *in vivo*. The development of earlier diagnostics based on the autofluorescence may be an area of remote monitoring of agricultural crops and yields of medicinal plants as well as the remote sensing of environmental stress. Moreover, some fluorescing intact secretory systems appear to be appropriate cellular models for cellular biology. In some cases, the autofluorescence of isolated natural secretory components could be recommended as fluorescent markers *in vivo* or as fluorescent dyes and probes for a histochemical staining of many living cells due to binding with certain cellular compartments.

This book summarizes information on autofluorescence of plant secretory cells as a phenomenon and the possibilities of the practical use of light emission by cell biologists, biophysists, biochemists, botanists and ecologists. The spectrum of its readers may include lecturers, post-graduate students, students and others interested in biological light emission.

Introduction

The term "autofluorescence" is used for a luminescence of naturally occurring molecules of intact cells in a visible region of the spectra induced by ultra-violet or violet light (Taylor and Salmon, 1989; Haugland, 2000). Sometimes it is determined as fluorescence of natural compounds within an organism. Luminescence is the light emission by a molecule, which is excited by shorter light wavelengths (in this case called fluorescence) or spontaneously emitted by an organism (bioluminescence or chemoluminescence). Fluorescence is an emission of light by a substance immediately after it has absorbed some radiant energy falling on it. Usually the short-lived emission disappears just after the actinic light is switched off, unlike phosphorescence, which is long-lived and a stable process, lasting after the end of the exciting irradiation. The process of fluorescence is quite different from phosphorescence and bioluminescence. The term fluorescence is used when the interval between the act of excitation and emission of radiation is very small (10^{-8}-10^{-3} sec).

The modern concept of the fluorescence is considered as an electron transfer from a single level orbit to a lower level. When a compound absorbs the energy of actinic light, some of its electrons are temporarily excited to a higher state of energy than normal. As the electrons return to their normal state, they lose energy, which is emitted by the substance in the form of light emission at a characteristic wavelength. Fluorescence has a longer wavelength than the absorbed radiation which is due to the emitted photons, that are less energetic than the absorbed ones.

The mystery of the phenomenon "fluorescence" is so far unclear. This emission is observed for both inorganic and organic substances, and depends on many conditions. Many living cells – microbials, plants and animals – demonstrate similar phenomenon of fluorescence. Moreover, one could presume that the phenomenon is peculiar to objects in space as a whole: light emission (or may be a part of the light) from Earth as seen by astronauts (cosmonauts) which may belong to fluorescence excited by UV-radiation of the Sun. Organic and inorganic components of our planet may give various colour pictures due to their fluorescence. Similar pictures could be seen for other planets and stars. Like modern techniques (fluorimeters, luminescence microscopes and microfluorimeters) for measurement of the fluorescence, our eye sees an induced emission from the Earth's surface. Ancient Greeks said that the word "cosmos" means "world" and "human"; i.e. keeping in mind a complexity of an unknown heavenly space and an unknown human world. The term also may be applied to any living organism – "the plant cosmos" or "the microorganism cosmos" because each of them is very complex and mysterious too.

The history of luminescence for microscopic native objects dates back to 1838. The first observations were of David Brewster, a Scottish preacher, who experimented with optics. Naturally occurring luminescence was first described 20 years later by George G. Stokes in his monograph "On the Change of Refrangiblity of Light" as the light emanating from biological objects such as paper, bones, horn, ivory, leather, cork and cotton. A possible origin for the word "fluorescence" can also be traced to Stokes. In 1904 August Kohler of Jena made the first UV microscope (with ultra-violet light excitation) and photographic system, and later, in 1911-1913, Kemstadt and Lehmann built the first fully functional fluorescence microscope. This marked the beginning of the studies of living cell fluorescence, especially for pigmented objects. Tswett (1911) was one of the first to observe autofluorescence of plants under a fluorescence microscope. The widefield reflected light fluorescence microscope has also been a fundamental tool for the examination of fluorescently labelled cells and tissues since the introduction of the dichromatic mirror in the 1940s. The studies were mainly devoted to the cellular analysis with applications of natural and synthetic fluorophores as histochemical dyes and fluorescent probes as well as autofluorescence or self-fluorescence of the living cells. Later Brumberg (1956) constructed a modern type of fluorescence microscope, which has become the base for modern microspectrofluorimeters. This permitted to register the fluorescence spectra of microscopic objects.

Ellinger (1940) summed up the the first observations in biology by fluorescent microscopy. The method was applied in botany (Goryunova, 1952; Alexandrov and Sveshnikova, 1956), microbiology (Meisel and Gutkina, 1961) and later has been used for all types of living cells (Barenboim et al., 1969; Aubin, 1979; Benson et al., 1979; Host, 1991; Rost, 1995; Wang and Hermann, 1996; Reigosa Roger and Weiss, 2001). A new epoch in the study of autofluorescence began with the appearance of the first microspectrofluorimeter constructed by B. Chance who registered the first fluorescent spectra of intact living cells and showed the participation of NAD(P)H in blue emission and flavins – in green fluorescence (Chance and Thorell, 1959). There was also another microscopic technique that allows good observation (Brumberg, 1959) with the examination of images by imaging detectors (White et al., 1991) and even the registration of the fluorescence spectra of the objects if photomultiplier tubes are used (van Gijzel, 1967; Karnaukhov, 1972; Kohen et al., 1981; Candy, 1985). In the 60-70's of the 20th century a series of modern microspectrofluorimeters led to new information received by van Gijzel (1967-1970) and Karnaukhov et al., (1968-1978). The autofluorescence studies of animals and plants have been analyzed in some monographs, in particular "Luminescent spectral analysis of cells" (Karnaukhov, 1978) and "Spectral analysis in cell-level monitoring of environmental state" (Karnaukhov, 2001). The development of registered microspectrofluorimetry (van Gijzel, 1967; 1971; Karnaukhov, 1972; Karnaukhov et al., 1981; 1982; 1983; 1985; 1987; Palewitz et al., 1981) leads to a solution of the problem via an understanding of the connection of light emission of individual natural substances with fluorescence of separated cellular compartments. In this period, the autofluorescence measurements on animal cells took place in flow cytometry with laser excitation, but in connection with the correction of the fluorescent probe determination (Willingham and Pasan, 1978; Alberti et al., 1987; Szöllosi et al., 1987). Since 1990, some specialists in physics (see book edited by Pawley, 1990) were the first to recommend Laser-Scanning Confocal Microscopy for the study of the fluorescence of biological objects, in particular autofluorescence of animal and plant cells. Fluorescence imaging of living cells became a separate method of cellular studies (Whitaker, 1995; Ruzin, 1999). Besides this, there has been an attempt to apply the optical fiber technique (Thompson, 1991) for the measurement of cellular autofluorescence, in particular bacteria (Saxena et al., 2002) and human skin (Zeng et al., 1995).

Today we are aware of autofluorescence in many living cells – in microbials, plants and animals, excited by ultra-violet or violet irradiation and registered by the luminescent technique, especially with luminescence microscopy in various modifications. A reader can see them from the Fluorescence Image Family on Internet sites, where there are several similar photomicrographs (http://www.microscopyu.com/galleries/smz1500/index.html). This gallery examines the fluorescence microscopy of both cells and tissues with a wide spectrum of fluorescent probes. Stereomicroscopy fluorescence images of living organisms are especially attractive – from fungi to insects which is an addition to traditional fluorescence specimens, such as stained thin sections, cell culture mounts, and autofluorescence in plant tissues. Fluorescence microscopy with super-resolved optical systems also start to be included in cell analysis (Egner and Hall, 2005). Luminescence microscopy was used for biological science along with usual fluorimetry.

Notwithstanding all the above-mentioned achievements, we know little about the contribution of emitted compounds in whole autofluorescence of living organisms as yet. This fluorescent world only half-opens its doors for researchers permitting the discovery of red emission of chlorophyll and its participation in photosynthesis. But, beside chlorophyll, many components of living cells can fluoresce. A modern look at plant fluorescence has been presented in some reviews (Buschmann et al., 2000; Buschmann and Lichtenthaler, 1998). Among plant cells, as it has been established recently (Roshchina et al., 1997 a, 1998 a; Roshchina, 2003), secretory cells contain various fluorescent products, mainly secondary metabolites, but their autofluorescence in intact structures is not described completely in literature. At the end of the 20th century, the main attention was paid to fluorescence of sea animals due to fluorescent proteins found, and was used as fluorescent probes for the study of other organisms by using methods of genetic engineering (Matz et al., 1999; Hanson and Köhler, 2001; Labas et al., 2003; Shaner et al., 2004; Habuchi et al., 2005). The discovery of fluorescent proteins (see review Tsien, 2004) have been given a new direction in the problem of autifluorescence, related to the directed mutagenesis of fluorescent proteins. This added to the assortment of tools and created an avenue for scientists to probe the dynamics of living cells in culture.

Recently new aspects of plant autofluorescence became of interest to fundamental investigations. Plant secretory cells, which are clearly seen under a luminescence microscope, have been shown to lighten all colours of the visible spectrum due to the presence of various fluorescent secondary products (Roshchina et al., 1997a; 1998a; Roshchina, 2003). Fluorescent substances are released and concentrated in the cell wall, vacuole and extracellular space. Modern scientists may convert to new Leeuwenhoekes*, when they observe various colours of fluorescing secretory plant cells under a luminescence microscope and its modifications – from microspectrofluorimeters to laser scanning confocal microscopy. A bright luminescing picture is observed for specialized secretory cells. The fluorescence may be used for the identification of the compounds in intact secretory cells if there is an appropriate optical system of registration that enables the fluorimetric analysis of the cell *in vivo*.

A historical record shows that the UV-excited light emission of secretory cells under a luminescence microscope was first demonstrated on pollen (Ruhland and Wetzel, 1924; Berger, 1934; Asbeck, 1955). Later, using the luminescence microscopy, the fluorescence of secretory cells has been found in leaves (Ascensao and Pais, 1987; Curtis and Lersten, 1990), roots (Kuzovkina et al., 1975; Eilert et al., 1986), in pistil of flowers (Kendrick and Knox, 1981). Microspores are single secretory cells themselves, and the interest in the fluorescence of pollen grains was maintained by botanists and geologists (van Gijzel, 1971; Willemse, 1971; Driessen et al., 1989). The phenomenon of the autofluorescence of the above-mentioned cells is used for pollen analysis in botany (Willemse, 1971; Driessen et al., 1989), in geology (van Gijzel, 1971), meteorology (Satterwhite, 1990), and microphotography of fluorescing secretory structures (Curtis and Lersten, 1990; Zobel and March, 1993). The studies, which are related to the character and nature of the fluorescence from individual secretory cells and perspectives of the application of the phenomenon to the practice, were done in the 90's of the 20th century (Roshchina et al., 1994; 1995; 1996; 1997a, b; 1998a-d; Roshchina and Melnikova, 1995; 1996; 1998a, b; 1999). The fluorescence spectra of more than 100 plant species with various secretory cells are characterized in special papers (Roshchina et al., 1995; 1996; 1997a, b; 1998a; Roshchina and Melnikova, 1995; 1996; 1999).

*Leeuwenhoek, Dutch researcher, and Hooke from England were first, who saw various living cells under microscope in 17th century.

So far the phenomenon of bright luminescence emitted from plant secreting cells was not analyzed in literature, except for our review (Roshchina, 2003), and it is necessary to do so. The aim of this book is to show the resources of autofluorescence, peculiar to secretory cells, as a biosensor and a bioindicator reaction for various types of diagnostics - from cellular monitoring of environment to the study of cellular development.

Acknowledgement

The author is grateful to colleagues of the Laboratory of Microspectral Analysis of Cells and Cellular Systems for their cooperation and help in the experimental studies. I especially wish to thank Dr. Valerii N. Karnaukhov, Head of the Laboratory of Microspectral Analysis of Cells and Cellular Systems and Vice-Director of the Institute of Cell Biophysics at the Russian Academy of Sciences for his constant support and Assistance with the work. I am also grateful to the contributors for the experiments conducted, Drs. Eugenia V. Melnikova, Valerii A. Yashin, Rita Ya. Gordon, Larisa A. Sergievich, Alexandra V. Yashina, Vladimir I. Kulakov and Nadezhda K. Prizova as well as to the computor assistants Ludmila I. Mit'kovskaya, Larisa F. Kun'eva and Andrey Rodionov. In addition, I am thankful to my other collegues, who also participated and contributed in my publications and the support by valuable plant materials: from other laboratories of our Institute such as Professors Victor I. Popov and Vladimir I. Novoselov and from Institutes of Russian Academy of Sciences such as professor Lidia V. Kovaleva and Dr. Inna Kuzovkina (Institute of Plant Physiology RAS, Moscow), Professor Boris N. Golovkin (Central Botanical Garden RAS, Moscow), Drs. Nikolai A. Spiridonov (Center for Drugs Evaluation and Research U.S. Food and Drug Administration, Bethesda, USA), Dimitrii A. Konovalov (Pyatigosk Pharmaceutical Academy, Russia).

Contents

CHAPTER 1 Autofluorescence of Secretory Cells as a Phenomenon

Fluorescence of plant cells, which are clearly seen under a luminescence microscope, includes all colours of the visible spectrum. It is not only the red emission of chlorophyll located in chloroplasts and the blue or blue-green luminescence of the cell wall (Kasten, 1981). A bright luminescent picture is also observed for specialized secretory cells of leaves, stems, flowers and roots (Eilert et al., 1986; Ascensao and Pais, 1987; Curtis and Lersten, 1990; Zobel and March, 1993; Roshchina et al., 1995; 1997a, b; 1998a) and unicellular secreting cells as microspores served for breeding (Berger, 1934; Asbeck, 1955). Fluorescence of nectar was first seen by Frey-Wyssling and Agthe (1950).

Specialized secretory cells accumulate the substances, and the excretory function of such cells prevails (Roshchina and Roshchina, 1989, Roshchina and Roshchina, 1993). Plant secretory structures were usually studied with an electron microscope (Vasilyev, 1977; Fahn, 1979) which does not allow the physiological activity of a secretory process *in vivo* to be investigated. Fluorescent analysis may overcome this problem. Secretory cells contain secondary substances, such as phenols, flavins, quinones, alkaloids, polyacetylenes, coumarins, terpenoids and others (Roshchina and Roshchina, 1989, Roshchina and Roshchina, 1993), which can fluoresce under ultra-violet radiation. Moreover, it has been established that many secretory cells, such as idioblasts and complex glands are not seen on the plant surface without the excitation by UV-light or staining with artificial fluorescent dyes (Roshchina et al., 1997a; 1998a; Roshchina, 2003). The fluorescence may be used for the indentification of compounds in intact secretory cells if there is an appropriate optical system of registration that enables the fluorimetric analysis of the cell *in vivo*.

1.1 SECRETORY CELLS

Secretory cells – specialized structures filled with secretory products (S) that are located in secretory vesicles or reservoirs as seen below in Fig. 1.1:

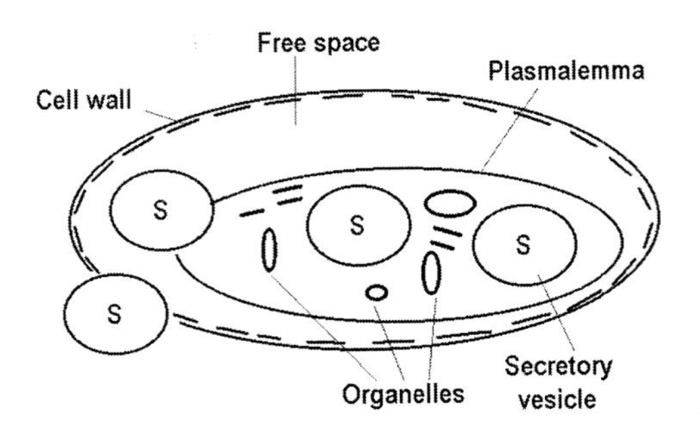

Fig. 1.1 Scheme of secreting cell

Often the secretion moves the cellular organelles to a cell wall (the organelles may be reduced and even disappear), or may be accumulated in the free space between plasmatic membrane plasmalemma and the cell wall, and then be excreted out of the cell.

Main plant secretory structures can be distinguished in the following groups:

Secretory cells of vegetative organs (leaves, shoots, roots)	Secretory cells of generative organs (petals of flowers, nectaries, pistils, glandular cells)	Microspores for vegetative and generative breeding	Secreting cells of non-differentiating tissue

The terms and images of some secretory structures can be found in the Glossary (See Appendix 1). In detail one should see as sources (Essau, 1965; 1977; Stanley and Linskens, 1974; Fahn, 1979: Roshchina and Roshchina, 1993; Werker, 2000).

Among the studied secretory structures were intact single cells such as idioblasts, hairs, generative (pollen or male gametophyte) and vegetative (mainly for species of the family Equisetaceae) microspores, and cells of multicellular secretory structures of Gymnosperms and Angiosperms (glands, glandular trichomes, secretory hairs, and others). Secretory cells are seen as both unicellular cellular structures (unicellular hairs, unicellular spores or microspores that are needed for plant breeding, single secretory cells among non-secretory ones, etc) or as a part of multicellular secretory structures (multicellular hairs, multicellular glands, resin ducts, nectaries, laticifers, etc). These unicellular and multicellular structures may play roles in plant defence against pests and parasites, attract insect-pollinators and others. (Fahn, 1979; Roshchina and Roshchina, 1993). Electron microscopy and electron scanning microscopy permit analysis of the secretory structures only after fixation (Vasilyev, 1977; Fahn, 1979) as seen in Fig. 1.2. Among non-invasive methods are optical coherence microscopy (combination of confocal microscope and optical coherence tomography) for plant optically nontransparent secretory cells seen in scattering near-infrared light (Roshchina et al., 2007c) and any modification of luminescence microscopy (see Section 1.2) for fluorescing cells.

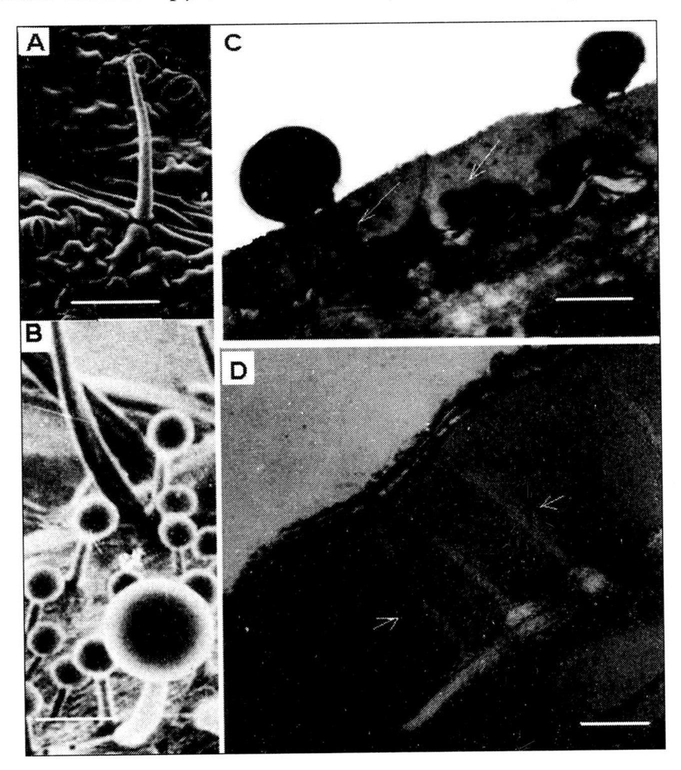

Fig. 1.2 The photograph of the surface of secretory cells and organs done with scanning electron (A and B) or usual electron (C and D) microscopy. Adopted from Roshchina et al., 1998c and unpublished data. A. – secretory hair of *Achilea millefolium*, bar 0.1 mm; B. – secretory hairs of the carnivorous tissue of *Drosera capensis*, bar 0.1 mm; C. – pollen surface of *Papaver somniferum*, bar 0.1 μm and D. – Slice through the pollen surface of *Betula verrucosa*, bar 0.1 μm. The microchannels with secretion are shown with arrows.

Unicellular structures may be secretory hairs, simple (ordinary) cells and microspores. Microspores of plants are also unicellular structures with a multilayer cover, which prevent damage of the cell. Among the microspores two types are distiguished: vegetative (in spore-breeding plants, belonging to mosses, ferns and horsetails) and generative (pollen of seed-breeding plants, belonging to Gymnosperm and Angiosperm plants). These are considered secretory unicellular systems. Pollen grains and their external cover are postulated as simple secretory systems (Gimenez-Martin et al., 1969). Microspores such as pollen (Fig. 1.2) and vegetative microspores of Cryptogam species have a complex secretory structure: 1) multilayer cover with an outer layer exine (rigid product of the cellular secretion during the anther development) which polymerized material (sporopollenin) brightly glowing due to phenolic or carotenoid residues; 2) microchannels in exine filled with liquid secretion, consisting of fluorescent pigments, lipids, proteins, etc (Roshchina et al., 1998a; Roshchina, 2003, 2005b). Some of the visible secretory organs can fluoresce *in vivo* under a luminescence microscope (excitation with UV – or violet light).

Distinct mechanisms operate in different secreting systems at the cellular level (Dunant, 1994). Among them are 1) free diffusion through the plasma membrane; 2) exocytosis resulting from fusion of a secretory granule with the plasma membrane; 3) fleeting release from a granule through a transient pore without full fusion; 4) release through a specialized plasmalemmal molecule such as the mediatophore. The mode of the release depends on the nature of the substances to be evoked. Lipophilic substances such as NO or steroid hormones can freely diffuse across biological membranes without any specific releasing machinery. The rate of the process is regulated by the rate of their synthesis. Usually the processes and regulation are relatively slow–from seconds to a few minutes. Hydrophylic substances and water-insoluble particles need other modes of the secretion. These are full fusions via the secretory vesicles or secretory granules by exocytosis or the rupture of the plasmatic membrane. The secretion of the large granules is supposed to connect with the formation of a fusion pore between the interior of a granule and the extracellular space. Proteins, usually being synthesized far from the site of the secretion, participate in a complex pathway to the plasmatic membrane. Some compounds such as neurotransmitters acetylcholine and biogenic amines, which are formed and then stored in high concentrations locally in secretory vesicles, and act at a short distance, operate in a very rapid time course up to 0.2 ms. Together with the hormone or mediators, there is a liberation of associated proteins or enzymes, ATP, ions and other compounds. Neurotransmitters released in *Torpedo* electric organ and plants have pointed toward a plasmalemmal protein, called mediatophore (Dunant, 1994; Roshchina, 2001a).

Functions of the secretory products, mainly secondary metabolites, in secretory structures consist of the following (Roshchina and Roshchina, 1989; 1993; Roshchina, 1999a, b):

1. Secretory components play the role of the store compounds and are used if it is necessary for its life;
2. Secretory components play a defensive role, when organisms need to be protected against pests;
3. Secretions released from secretory cells may serve as a medium for transmitting the chemosignal from one cell to another.

In multicellular tissues of leaves, stems, roots and seeds, fluoresced secretions accumulated within the cell and in the extracellular space between plasmalemma and the cell wall (Fig. 1.1) and excretions are mainly in a liquid state, apart from the deposition of crystals on the surface. Secretory structures fluoresce which allows us to see them on the tissue surface without fixation. Moreover, it has been established that many secretory cells, for instance idioblasts and complex glands are not seen on the plant surface without the excitation by UV-light or staining with artificial fluorescent dyes. Fluorescence under ultra-violet light was used for the microphotography of secretory cells (Curtis and Lersten, 1990; Zobel and March, 1993) and could be observed under a luminescence microscope and measured by microspectrofluorimetry. The fluorescent spectra were fixed *in vivo* from individual cells both of non-generative and generative tissues: including leaves, flowers, buds, anthers, pistils and pollen grains (Roshchina et al., 1996; 1997a, b; 1998a; Roshchina and Melnikova, 1999).

1.2 TECHNIQUE FOR THE STUDY OF THE SECRETORY CELL'S AUTOFLUORESCENCE

The phenomenon of autofluorescence may be observed under a usual luminescence microscope or confocal microscope, and be fixed by the application of special microspectroflurimeters (Fig. 1.3). Principles of the technique based on modification of a luminescence microscope, in which one can see the fluorescing object in a whole field of view.

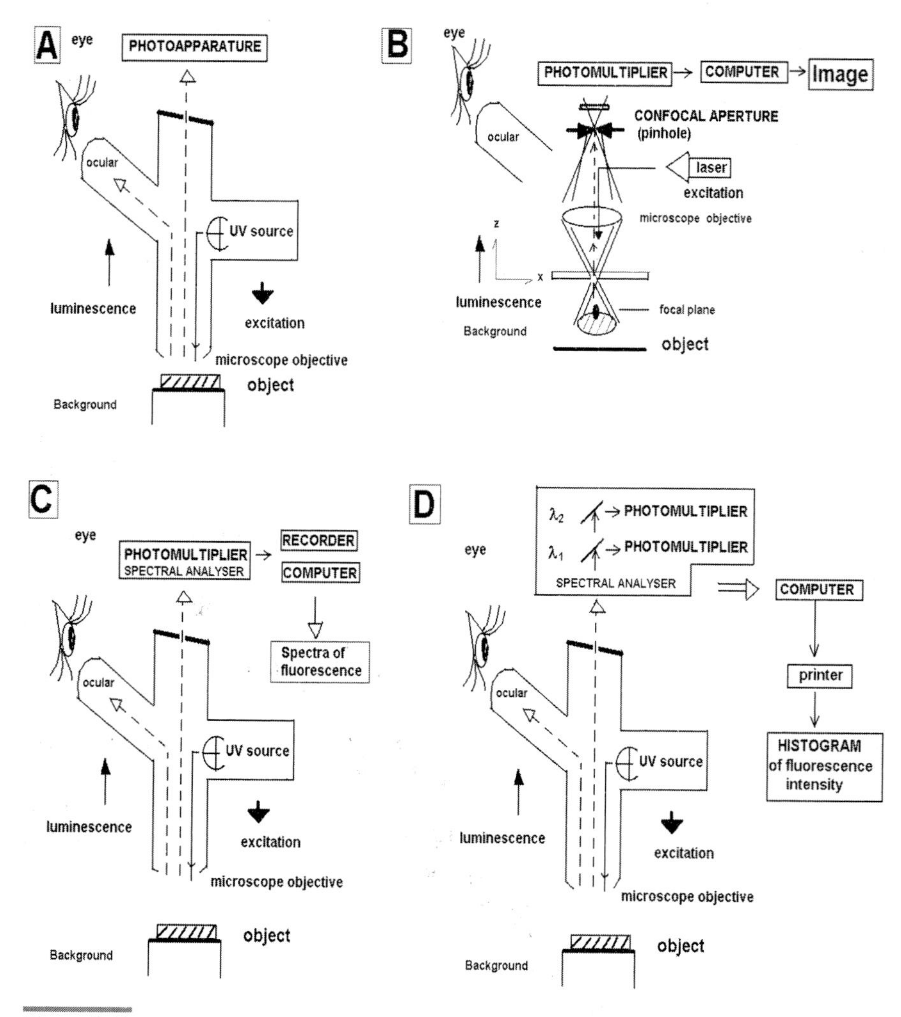

Fig. 1.3 Technique for the measurement of plant autofluorescence. Adopted from Roshchina (2003). A. Luminescence microscopy; B. Confocal microscopy; C. Microspectrofluorimetry with registration of the spectra; D. Double-beam microspectrofluorimetry with registration of histograms.

Luminescence microscopy. Luminescence microscopes are able to photograph samples. Fluorescence from intact secreting plant cells, which are induced by ultra-violet light 360-380 nm, was observed and photographed on a high-sensitive photofilm under a fluorescence microscope used for aeroshooting from an aeroplane (Roshchina and

Melnikova, 1995; 1999; Melnikova et al., 1997; Roshchina et al., 1998a) or on photofilm Gold Kodak 400 (Roshchina and Roshchina, 2003; Roshchina, 2003). Photographs, which are represented in colour Figures (Fig. 1 from Appendix 2), demonstrated the examples of light emission which is seen under a luminescence microscope.

Fluorescing secretory cells are clearly seen among non-secretory or secretory cells with other non-fluorescing products (Roshchina, 2003). Bright blue luminescence is observed for secretory cells of woody bud scales in birch *Betula verrucosa*, blue-green – for secretory hairs of potato *Solanum tuberosum*, glandular surface of pistil stigma of *Campanula persicifolia*. Microspores served for plant breeding are also secretory unicellular systems (Gimenez-Martin et al., 1969; Roshchina and Roshchina, 1989; Roshchina and Roshchina, 1993; 2003). Generative microspores from seed-breeding plants (called pollen, male gametophyte) fluoresce in different sections of the spectrum depending on the composition (Roshchina et al., 1998a; Roshchina and Melnikova, 1999). The blue-green emission is from pollen grains of birch *Betula verrucosa*. Spore –breeding plants have vegetative microspores, which have chloroplasts, and fluoresce (Colour Fig. 3) within the range of blue to red depending on their state. (Roshchina et al., 2002; 2003a; 2004). Root secretory cells are usually represented as secretory hairs and single cells known as idioblasts. Idioblasts of *Ruta graveolens* root fluoresce in the orange region of the spectrum.

Confocal microscopy. Unlike a usual luminescence microscope, in confocal microscopy before being caught by a photomultiplier, the fluorescent beam from the sample studied passes through a confocal aperture called pinhole (Fig. 1.3 B). Changing the diameter of the aperture a pinhole limits the scattered light from the parts of the object outside the focal plane and contrasts it with the image (Pawley ed., 1990; Pawley and Pawley, 2006). Construction of a confocal microscope enables the observation of cellular structures by the regulation of the depth of an object slide. Laser scanning confocal microscopy (LSCM) may produce images of high quality from fluorescing cells. The images of plant secreting cells, which excrete allelochemicals, or plant cells, which serve as acceptors of allelochemicals, may be changed in allelopathic chemical relations that register by this technique.

The advantages of the technique are 1) 3-channel simultaneous detection that allows one to see images excited by three different wavelengths of light from a laser and to receive a common complicated interference image of the object; 2) a possibility of having the increased depth penetration for receiving 20 visual slices (optical sections) or the complete volume (the information must be also quantitatively extracted); 3) interchangeable filters; 4) Graphical User Interface and production of accurate computer models as well as mathematical analysis; 5) pattern analysis of the structure. Confocal microscopy offers several advantages over conventional optical microscopy, including controllable depth of field, the elimination of image degrading out-of-focus information, and the ability to collect serial optical sections from thick specimens. The key to the confocal approach is the use of spatial filtering to eliminate out-of-focus light or flare in specimens that are thicker than the plane of focus. A laser beam of sufficiently high power can burn the animal skin, but, as will be shown in Chapters 2 and 6, the short time of observation does not prevent the normal development of some plant cells. There has been a tremendous increase in the popularity of confocal microscopy in recent years, due to the relative ease with which extremely high-quality images can be obtained from specimens prepared for conventional optical microscopy, and because of its large number of applications in many areas of current research interest.

Cells of algae and non-secretory cells of some higher plants were studied by the method (Cheng and Summers, 1990; White et al., 1996). Confocal imaging of secreting plant cells was studied for pollen and vegetative microspores analysis (Salih et al., 1997; Roshchina et al., 2004). Colour images can also seen in colour Figs. 2 and 3 (Appendix 2). Crystals of oil secretions, fluorescing in green are clearly seen within the secretory hair. Colour demonstrates the stack of optical slices of secretory leaf hair in *Solidago virgaurea* L. (Fig. 2 in Appendix 2). Scanning of the object along the Z-coordinate (see image of slice on section 3) with an interval 1.0 μm one can see the slices of the microspore (Colour Fig. 6). The slices can be collected with the help of a special computer programme for LSCM 510 and reconstruction of the separate fragment of the cell surface may be received.

One can also see the fluorescent drop of oil (Colour Fig. 2 Sections 7-13). The drop of oil secretion contains fluorescing green, yellow and red colour layers that shows the various lipofilic inclusion in the secretory structures. Dried oil is crystallized on the surface of the secretory hair and seen as green-fluorescing bodies. Smoke-like secretion, fluorescing in the green region, is also seen along the multicellular trihome, and the emission is concentrated at the end of the hair.

Green-fluorescing crystals of the secretions are also observed on the surfaces of the leaf and the trichome itself. Chloroplasts of the cell are also observed as red-fluorescing spots. Cell walls fluoresce in green. Deep in the hair and surface of the leaf, red fluorescing cell walls can be seen, which is possible due to the concentration of fluorescing oil in the extracelluar space between the cell wall and the plasmalemma. Secretory hairs of the medicinal plant *Solidago virgaurea* L. contains a lot of oil, which include fluorescent terpenes.

In Appendix 2 (colour Fig. 3) colour images of the *Equisetum arvense* cell (luminescing organelles have been concentrated, perhaps, around non-visible nucleus) can be seen. When light from all three channels excites the fluorescence of crystalline individual compounds such as flavonoids quercetin and rutin or pigments of plant cells - azulene, chlorophyll and carotenoids fluoresce in different regions of the spectra: in yellow and red or blue, red and yellow-orange, respectively (Appendix 2, colour Fig. 21). It enables one to compare the light emission of the substances within cellular structures.

Microspectrofluorimetry. The recording microspectrofluorimetry is also applied in cellular biology (Fig. 1.3 C,D). Microspectrofluorimeters have been first used for the study of fossil pollen (van Gijzel, 1967), microalgae and higher plants (Karnaukhov, 1972; 1978). A modified technique different from the one mentioned above and was also used for the study of vacuolar fluorescence in onion stomata guard cells (Zeiger and Hepler, 1979; Palevitz et al., 1981) in connection with the effects of blue light and of vacuolar movement. Weissenböck et al., (1987) also measured the fluorescence spectra of the object with a new variant of microspectrofluorimeter. The commercial apparatus, which consists of two variants of microspectrofluorimeters (Fig. 1.3, C,D) made at the Institute of Cell Biophysics RAS has been patented in different countries (Karnaukhov et al., 1981, 1982; 1983, 1985, 1987) and produced by the Institute of Biological Technique (Pushchino).

Microspectrofluorimetric technique is one of the non-invasive methods used for a cellular diagnostics (Karnaukhov, 1978; 2001).

Luminescence of microobjects, excited by short wave radiation of an arc lamp and after spectral decomposition is registered by detectors-photomultiplier(s). Microspectrofluorimeters, having a detector with optical probes of various diameters up to 2 μm (the changed areas or probeholes composed with the system of mirrors) have been constructed at the Institute of Cell Biophysics of the Russian Academy of Sciences (Karnaukhov et al., 1981, 1982; 1983; 1985; 1987). This technique may register the fluorescence spectra or measure the fluorescence intensities at two separate wavelengths. Microspectrofluorimeters can receive a magnitude fluorescence image of a certain area of the specimen that appears on a spherical mirror. Unlike electron microscopy (Fahn, 1979), this method allows the investigation of physiological activity of a secretory process *in vivo*. The emission data may be written in the form of a fluorescence spectra with the help of an XY-recorder (can be attached to a computer). By using such microspectrofluorimeters, luminescence is registered from individual cells and even from a cell wall, large organelles and secretions in periplasmic space (the space between the plasmalemma and the cell wall) as well as from drops of secretions evacuated out of a secretory cell on the cellular surface. The fluorescence spectra were also measured by other microspectrofluorimenric techniques (Leitz MPV-SP microspectrophotometer), in particular for the analysis of the flavonoid accumulation in Fabaceae at the nodule formation (Mathesius et al., 1998). The fluorescence spectra of secretory cells were registered recently with the microspectrofluorimetric technique (Roshchina et al., 1995; 1996, 1997a, b; 1998a; 2002; Roshchina and Melnikova, 1995; 1996; 1999; 2001). The examples of the fluorescence spectra of some secretory cells can be seen in Fig. 1.4. Microspectrofluorimetry has been also used for the analysis of secondary products formed in leaves treated with herbicides (Hjorth et al., 2006).

Various modifications of microspectrofluorimeters may not only register the fluorescence spectra, but also measure the fluorescence intensities at two separate wavelengths in the form of histograms related to the fluorescence intensity at λ_1 or at λ_2 as well as to the ratio of their values (See scheme in Fig. 1.3). A special programme "Microfluor" makes it possible to obtain the distribution histograms of the fluorescence intensities and to perform statistical treatment of the data, using Student t-test (Karnaukhov et al., 1987). By this mode a histogram comparison of the state of different secretory cells was made, especially of the microspores (Roshchina et al., 2002; 2003a,b).

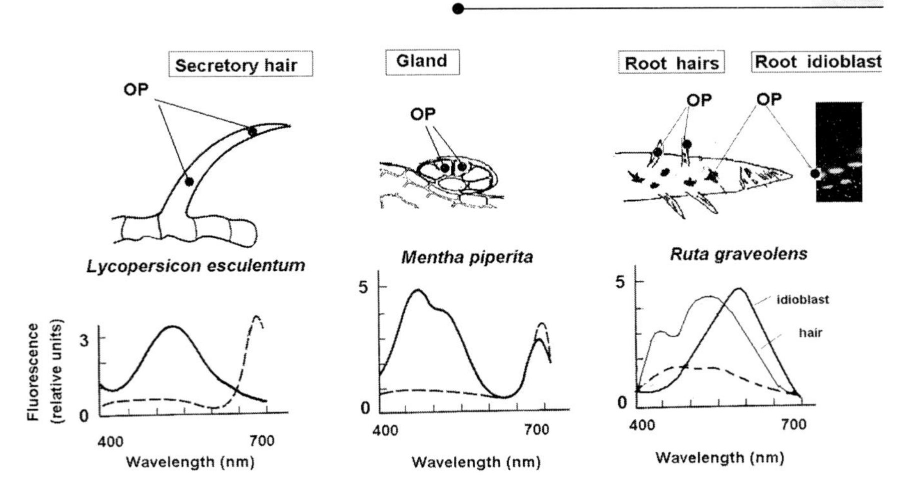

Fig. 1.4 The fluorescence spectra of secretory cells measured by microspectro-fluorimetry. Unbroken line - secretory cells; broken line – non-secretory cells. The position of optical probe (OP) is shown for leaf secretory leaf hairs of *Lycopersicon,* leaf gland of *Mentha* and on the surface of root with idioblasts (single secretory cells from *Ruta graveolens.*

Standard fluorimetry. Beside the microscopy-related technique, the usual standard fluorimeter or MPF fluorescence spectrophotometer with laser 360-380 nm excitation (also in a combination with liquid nitrogen kryostation) was used for the study of the fluorescence spectra on 1-3 mm – film (layer) of pollen at room temperature and at the temperature of liquid nitrogen (Butkhuzi et al., 2002). Due to the apparatus, more peaks in the fluorescence spectra of the samples have been observed in comparison with the spectra of pollen from the same species that was registered by microspectrofluorimetry (Roshchina et al., 1997a, b; Melnikova et al., 1997).

1.3 CHARACTERISTICS OF FLUORESCING SECRETORY CELLS

The observed fluorescence of intact secretory cells is the sum of emissions of several different groups of substances, both excreted out or accumulated within the cell and linked on the cellular surface (Fig. 1.5). The contribution of the secretions in the fluorescence will be considered first.

The contribution of internal content of the secretory cell to the total fluorescence also relates to the transparency of epidermal cells. The actinic

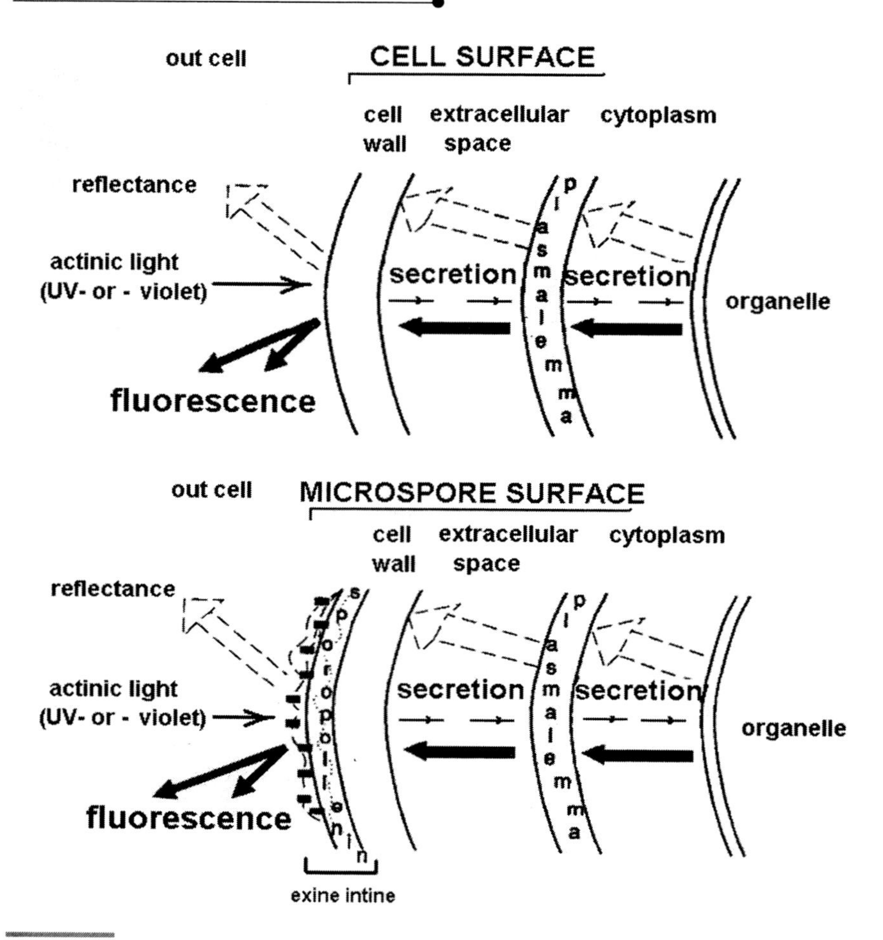

Fig. 1.5 Scheme of the possible fluxes of UV-radiation (excitation) and fluorescence in plant secretory cell as a whole (upper side) and, in particular in microspores such as pollen and spores of Cryptogams (Below)

ultraviolet light easily penetrates through them and excites the luminescence of the secretion located both in periplasmic space and within a cell, as well as the luminescence of cellular organelles.

1.3.1 Fluorescence of Secretory Cell

Estimating the fluorescence spectra of secretory cells, where abundant secretion is seen, the main fluorescence detected is defined by the composition of secretion and differs from surrounding non-secretory cells

(Fig. 1.4). As will be shown in Chapter 2, cells of hairs, glands, and nectars have maxima in the blue, blue-green and yellow-orange regions of the spectra, unlike non-secretory cells of the leaf, stem or parts of the flower, which either have no maximum at 680 nm, peculiar to chlorophyll or have a small one (Roshchina et al., 1997a; 1998a). Due to this characteristic it is easy to distinguish secretory cells from non-secretory ones (For example, see Colour Fig. 1). Moreover, floral and extrafloral nectaries differ in their nectar composition which is seen from the different fluorescence spectra (Roshchina et al., 1998a). Surrounded non-secretory cells have no maxima in the spectral region, which is shorter, than 650 nm. Chlorophyll-less cells of root tissue (Roshchina, 2005b) also demonstrated differences between secretory and non-secretory cells. Idioblasts, which are secretory cells, fluoresce in the orange region of the fluorescence spectra whereas surrounding non-secretory cells fluoresce at shorter wavelengths. According to Vogelmann, (1993), the contribution of the reflected light from any plant cells is not more than 10%.

What are the main signs that the visible fluorescence of secretory cells actually belongs to a secretion? First of all, the evacuation of the secretion from the secretory cells leads to the quenching of the luminescence within the secretory cell, as can be seen for leaf secretory hair of the common nettle of *Urtica dioica* (Fig. 1.6), tomato *Lycopersicon esculentum*, and calendula *Calendula officinalis* (Roshchina and Melnikova, 1995; 1999). The empty space within the secretory hair (Fig. 1.6) is observed as a dark space. Often drops of lipophilic secretion may stay at the end of the secreting hair and can fluoresce, appearing as lightening against the dark background of the empty hair (Roshchina and Melnikova, 1995). Only cell walls, when they include phenols, have a weak emission, which is non-measured or measured maximally as less than 7-10% of the total emission of the fluorescing secretory cell (Roshchina et al., 1998a). Therefore, about 90% of the emission of the secretory cell is due to its secretion. The second evidence of the main contribution of the secretion to light emission of the secretory cell is an analogous fluorescence of the secretion released on the surface such as liquids or crystals of some alkaloids, terpenes and phenols (Roshchina et al., 1997a, b; 1998a; Roshchina and Melnikova, 1999).

For microspores such as pollen or vegetative microspores the brightest fluorescence is observed when the microspores are dry. The contribution to light emission belongs to both rigid polymer secretion of exine and slightly released secretion of the microchannels (Roshchina et al., 1996;

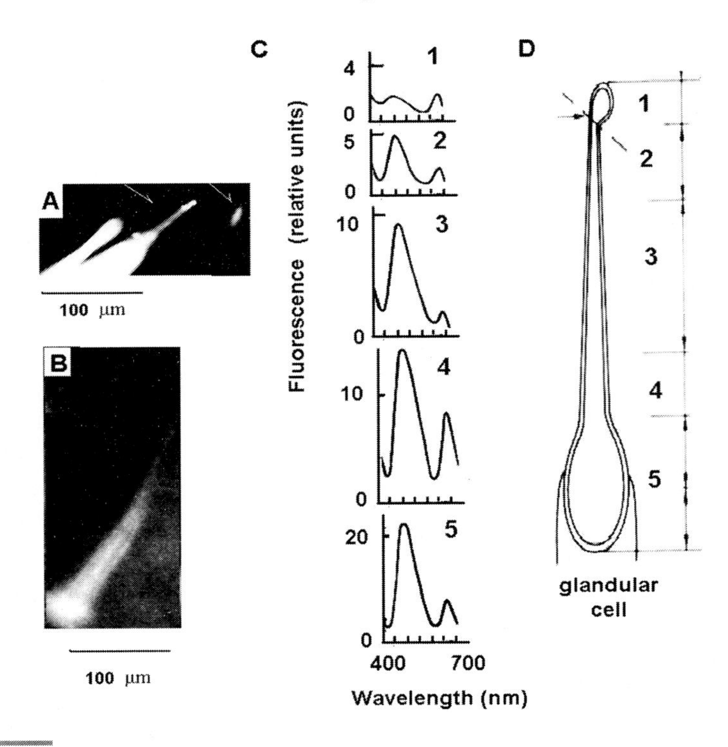

Fig. 1.6 The photographs of fluorescing secretory stinging hair of *Urtica dioica* (A, B) and its fluorescence spectra (C) measured along the trichome as shown on scheme (D). The fluorescing drop of the secretion is clearly seen at the tip of the empty hair (A). Adopted from Roshchina et al., (1997a; 1998a).

1997b; Roshchina and Melnikova, 1996). Water quenched the luminescence of viable pollen, but not of non-viable pollen (Roshchina et al., 1996; 1997b). The liquid secretion is released at moistening and can fluoresce out of a cell, but more often it is weak or is completely quenched by water. Rigid polymer material of exine includes pigments (excreted earlier during the development of microspore) such as carotenoids or phenols (Stanley and Linskens, 1974). The extraction of these pigments of *Petunia* pollen by 80% acetone and then by benzene or chloroform leads to no signs in the green-red region, with only the weak blue fluorescence being seen (Fig. 1.7).

Although the fluorescence of the non-secretory products is not large in comparison with the secretory ones (only 7-10%) because the cellular fluorescence intensity drops dramatically when the secretion is released

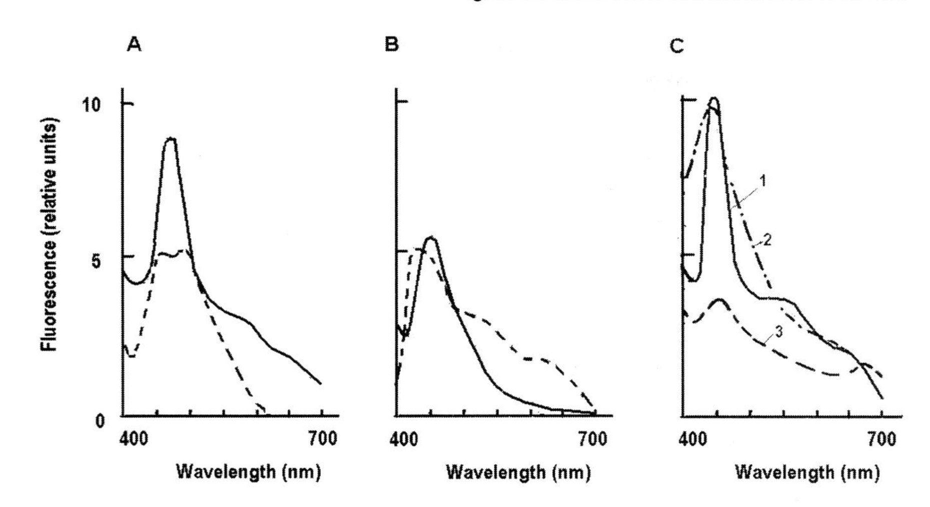

Fig. 1.7 The fluorescence spectra of *Petunia hybrida* pollen. A. Clone without anthocyanins, B. Clone with anthocyanins and without azulenes; C. Clone with azulenes and without anthocyanins. A and B. unbroken lines - intact pollen; broken lines - water extracts from pollen (1: 50 w/v); C. intact pollen before any treatment (1); or after the removal of the pigments with solvents for 3 h at first with the 80% acetone (1:50 w/w) (2) and then with benzene (1:50 w/w) (3).

from the cell (Roshchina et al., 1998a), these also need to be considered apart from the secretory products themselves. Among excreted metabolites are, those which have participated in energetic reactions and in the formation of cellular wall cover as well as components linked or excreted by cellular membranes which also may contribute to the visible fluorescence.

The possible effects of a reflection that are measured by fiber optics and the spectral radiation gradient can be of special interest for the researcher (Vogelmann, 1993). Light may be reflected from either the cuticular surface or within the mesophyll of some tissues from the numerous air/cell interfaces. Absolute values for surface reflectance are estimated at 675 nm, and 13% of light is reflected from the upper surface of leaves (Vogelmann, 1993). 140 μm of the leaf tissue absorbs 90% of the blue and red light, and in a lesser degree the green light. The contribution of emitted light in a reflection has not been estimated yet.

1.3.2 Contribution of Individual Components in the Cellular Fluorescence

The fluorescence of secretory cells includes the emission from 1. structural cellular elements such as the cell wall and cellular compartments with primary metabolites - proteins and nucleic acids, and 2. secretion-containing compartments filled with secondary metabolites - such as vacuoles and secretory vesicles (intracellular secretion) and extracellular space (extracellular secretion). As a rule, the fluorescence of primary metabolites from secretory plant cells in visible parts of the spectrum is much lower, than the emission of secondary metabolites. Therefore, our main attention will be to the emission of secondary metabolites, which are concentrated in the secretory structures. Besides, the contributions of the structural elements and primary metabolites of a cell will also be considered

1.3.2.1 Secondary metabolites

Secondary metabolites such as phenols, terpenoids, alkaloids and their derivatives are usually present in the secretory cells. As analyzed in Table 1.1, most studied secretory cells have one or several maxima in the fluorescence spectra: 1 for generative microspores or pollen of some plant species), 2 for generative microspores, vegetative microspores, glandular cells and trichomes on leaves and stems, 3 or 4 for generative microspores, vegetative microspores, glandular cells and trichomes on leaves and stems (Roshchina et al., 1996; 1997b; 1998a; 2002, Roshchina 2003). It depends on the chemical composition of their secretions and excretions. Multicomponent mixtures in secretions, however, demonstrate luminescent characteristics of prevailing substances. Table 1.1 shows the fluorescence maxima of known secondary metabolites in secretory cells, which can determine the light emission of the secretions in intact secretory cells from various plant species in comparison with their excretions. The details of the fluorescence of secondary metabolites can be seen in Chapter 3, Section 3.2.

In the blue region of the spectra alkaloids, phenols, terpenes and some aliphatic compounds fluoresce (Table 1.1). The glands and glandular hairs of terpenoid-rich species contain substances, fluorescing at—420-480 nm - terpenes such as monoterpenes (monoterpene alcohol menthol), sesquiterpenes (azulenes), and phenols (flavonoids, aromatic acids). The fluorescence scales of the substances which are included in the secretions of secretory hair and other secretory cells may also be useful for the pharmacological analysis of medicinal plants (Roshchina et al., 2000 a, b). This permits the quick control of the accumulation of active matter without long biochemical procedures. Microspectrofluorimetry enables one to analyze both the content of intact nectary cells and the small drops of the secretions of floral and extrafloral nectaries on *Passiflora coerulea* or *Impatiens balsamina* (Roshchina et al., 1998a).

In the green-yellow and yellow spectral region coumarins and some flavonoids fluoresce (Table 1.1). This is observed in the secretory cells of leaf buds during the development of their secretory structures and at the accumulation of the secretory products which appear by the microfluorescent technique (Roshchina and Melnikova, 1995, 1999; Roshchina et al., 1998a). The above-mentioned secretory cells may be recommended for cellular control of the development in nature. In the orange and red region of the spectra the fluorescence of some alkaloids, anthocyanins and azulenes is seen (Roshchina et al., 1998a; Roshchina and Melnikova, 1999; Roshchina, 2001b, d). Secretory cells of roots contain alkaloids, while anthocyanins and azulenes are found in generative (pollen, male gametophyte) and vegetative microspores (mainly in the family Equisetaceae). Terpenoid-containing cells often include azulenes which also contribute to the emission when they are linked with cellulose (Roshchina et al., 1995). The contribution of chlorophyll at 675-680 nm is seen only for chlorophyll-containing cells. The medium also plays a certain role in the orange-red emission of intact cells. It can be illustrated in the model system - root cells of *Ruta graveolens* which have several secretory structures, filling with lipophilic secretion enriched acridone alkaloids (Eilert et al., 1986), and brightly fluorescing with maximum 595-600 nm among surrounded non-secretory cells with blue (maximum 480 nm) fluorescence (Roshchina, 2001d; 2002). Lipophilic crystals of rutacridone, representative of acridone alkaloids, and its solutions in non-fluorescing immersion oil also fluoresced orange with maximum 600 nm. Similar orange emission was observed only for the root extract by chloroform or, in a lesser degree, by 100% ethanol, but not for extracts by water or acetone.

Special attention is focussed on microspores. Microspores are usually covered by a multi-layer cell wall where the outer layer, know as exine, consists of polymeric material sporopollenin. Sporopollenin components of pollen are mainly fluorescent because phenols (maxima at 440-480 nm), carotenoids (maximum at 500-560 nm), and azulenes (maxima at 440-460 nm, and for azulenes linked with cellulose – at 620-640 nm) are present here. The exine of pollen grains also contains anthocyanins (maxima at 450-470 and 600-640 nm), carotenoid bodies (maxima at 500-560 nm), and other fluorescent products (Roshchina et al., 1997a, b, c; 1998a). Composition of exine of vegetative microspores such as of *Equisetum arvense* has not been studied well. It is known that microspores contain fluorescent components, such as flavonoids, quercetin and kaempferol, various alkaloids, like carotenoids, chlorophyll (Plant Resources of Russia and Surrounded Countries, 1996), and recently azulenes have also been found here (Roshchina et al., 2002). The maxima of the main known components found in secretory cells are presented in Table 1.1. Among the compounds are lipophilic and hydrophilic as well as a mixture of the compounds. Sometimes crystals of the substances are seen on the cell surfaces.

1.3.2.2 Structural components and primary metabolites

Plant tissues may fluoresce in all visible spectral regions (Stober and Lichtenthaler, 1992; 1993; Lang et al., 1994). The structural components and primary metabolites (amino acids, proteins, pyridine nucleotides, nucleic acids, etc) are usually weak fluorescent, except components of the cell wall in some cases.

Cellulose and metabolites in the cell wall. The light emission of the cell wall is of special interest. Pure cellulose in the cell wall demonstrates very weak emission, but the fluorescence is strengthened when phenolic residues (maximum at 440-460 nm) are included (Roshchina et al., 1998a). Aromatic acids are found in isolated cell walls for example ferulic acid fluorescing in green (Lichtenthaler and Schweiger, 1998).

Among the products secreted through the cell wall is hydrogen peroxide, which induces the accumulation of insoluble fluorescent material, on the surfaces of the cell walls. When polymeric material containing phenolic components (cinnamic acid derivatives), accumulates in the cell walls, H_2O_2 takes part in formation of phenolic cross-links, the fluorescence of the surface increases and becomes bright greenish-yellow. The peroxidase-catalyzed polymerization of the pollen sporopollenin also needs H_2O_2 and the pattern of light emission during the process of pollen development (Roshchina et al., 1998a) may be related with the peroxidation reactions. The fluorescence of the secondary metabolites of the secretion, which may be released out of extracellular space, has been described above in Section 1.3.2.1.

Table 1.1 *The fluorescence maxima of main substances occurred in plant secretory cells (SC) and excretions (E)*

Class of substances	*Occurrence*			*Representative*	λ_{ex}/λ_f *(fluorophore)*
	In plant taxa	In organs	In secretory structures		
Alkaloids					
Acridone type	Rutaceae	Roots	SC	Rutacridone and other acridone alkaloids	380/590-595 (acridone)
Isoquinoline type	*Berberis vulgaris*	Leaves, roots, stems and fruits	SC	Berberine, chelerythrine	380/510, 540 (isoquinolinium)
	Chelidonium majus L.	Stems and roots	Latex and laticifers	Berberine ,chelerythrine	380/510, 540 (isoqunolinium)
Tropolonic type	*Crocus autumnalis L.*	Bulbs	SC	Colchicine	360/435 (lactone)
Anthocyanins	*Petunia hybrida, Papaver orientale.*	Flower	Pollen	Petunidin	360/Shoulder 510, 555, 570-585, shoulder 610
Azulenes (Derivatives of ses-quiterpene lactones)	*Artemisia, Achillea, Matricaria,* *Equisetum arvense*	Flowers, leaves and stems Microspores	Oil glands, trichomes, pollen Vegetative microspores	Azulenes of pollen bee-collected, Chamazulene 1,3-chloroazulene Guaiazulene	360-380/420, 620, 725 360-380/410, 430, 725 360/430 360/400, 433 (lactone)
Carotenoids	Many families	Flowers	Pollen cover	Carotenes, xanthophylls	360/520-560
Coumarins (Dicoumarins)	Many families. *Aesculus hippocastanum*	woody plant buds	SC and PE	coumaric acid esculetin 7-hydroxy- and 5-7-hydroxycoumarins	360/405-427, 360/475, (524) 300/479
Cytokinins	Many families	All parts of plants	E	Kinetin	380/410, 430 (adenine)
Flavonoids	Many families	Buds of woody plants	SC or glands	Galangin Kaemferol Quercetin	365/447-461 365/445-450 365/440, (584)

Table Contd.

Table Contd.

Furanocoumarins	*Heracleum sibiricum.*, *Psoralea corylifolia*	Leaves, stems, flowers, and roots	SC, glands, E	4-Methylpsoralen	360/440-420 (furanocoumarin)
C_6-C_3 Hydrocinnamic acids and their esters	Many families	Buds	E, SC	Caffeic acid, Cinnamic acid Ferulic acid	365/450, 360/405-427 350/440 (phenolic ring)
Indole derivatives	Many families *Urtica dioica L.*	Leaves, stems, flowers and fruits	E, stinging trichomes	Serotonin	360/410-420 (indole)
Monoterpenes and their alcohols	Rutaceae *Mentha piperita* Asteraceae Many species Labiatae *Salvia*	Flowers, leaves and stems	SC, glands, E	Menthol Camphor Camphor derivatives	360-380/415-420 310-313/404-413
Polyacetylenes	*Cicuta virosa* *Artemisia capillaris*	Roots, Leaves, roots and flowers	SC, secretory reservoirs E	Cicutotoxin Capilline	360-380/580 (three triplet bonds) 360/408, 430 (two triplet bonds)
Sesquiterpene lactones	Genus *Artemisia*, *Gaillardia pulchella*	Flower	E	Artemisine Gaillardine	360/395, 430 (lactone) 360/415 (lactone)
Tannins	Many species	Leaves, stems, fruits and roots	Idioblasts	Valoneaic acid	360-380/500(polyphenol)

λ_{ex} and λ_f - wavelengths of excitation light and fluorescence. Sources: Wolfbeis, 1985; Roshchina et al.,1998a; Roshchina, 2003, and Roshchina, unpublished data

Pyridine nucleotides and flavins. The higher fluorescence of the cell wall in other plants is, probably, due to surface NAD(P)H (maximum at 460-470 nm) or flavins (maximum at 520-540 nm) redox reactions, taking place on the plant surface. The plasma membrane also produces NAD(P)H and dependent products, therefore the pyridine nucleotides fluoresce.

The light emission of nuclei, microsomes, mitochondria and other cell compartments, except chloroplasts, is weak if at all. It may be caused by the production of NAD(P)H and flavins. In cells of animals and plants the fluorescence originates in discrete cytoplasmic vesicle-like regions and is absent in nucleus (Aubin, 1979; Roshchina et al., 2002). Most fluorescence appears to arise from intracellular nicotine amide adenine dinucleotide or NAD(P)H, flavin and flavin coenzymes (Aubin, 1979). NADPH and flavins also may contribute in the blue and green emission of chloroplasts.

Pigments. Pigments of some secretory cells contribute to the cell fluorescence. They are found both in cellular cover, for instance in exine of microspores, and, mainly in vacuoles and secretory vesicles, within the secreting cell. Chlorophyll located in chloroplasts may contribute to the total cellular luminescence of chlorophyll-containing secretory cells, but its maximum at 680 nm differs from the red emission of the other secretions (< 640 nm). Azulenes fluorescing with maxima at 620-640 nm are found in exine of pollen from many species (Roshchina et al., 1995) and vegetative microspores of *Equisetum arvense* (Roshchina et al., 2002) as well as in isolated chloroplasts of *Trifolium repens* and *Pisum sativum* (Roshchina, 1999a). The pigments were purified from the surface tissue of needles of *Picea excelsa* var. *blue* (Roshchina, 1999a). The absorbance and fluorescence spectra of the blue pigments are given in Chapter 3. Carotenoids found both in chloroplasts of all plants and in exine of some microspores fluoresce with maxima at 530-560 nm (Roshchina et al., 1997b; 1998a; Roshchina, 2003).

Proteins. Most cellular structural proteins in water media have a weak or lack any fluorescence in visible spectrum, except porphyrin-containing proteins and some fluorescent proteins in sea animal organisms. Although crystals of albumin and some other proteins emit in blue (at 430-460 nm), but the emission disappears after water or water solution addition (Roshchina et al., 2003c,d).

The red emission related to porphyrines (maxima 625, 628 or 690 nm) such as cytochromes, peroxidase, catalase, haemoglobin and their relative compounds in plants has been described by Karnaukhov (1978). Weak fluorescence with maxima 465-470 nm is peculiar to cytochrome C_{553} from algae *Chlorella* (Roshchina and Kukushkin, 1984). In plants and photosynthesizing bacteria fluorescence was related not only to pigments chlorophyll and phycobillins, but also to chlorophyll-a-proteins, chromoproteids and phycobilliproteins.

Our information about fluorescing proteins as a whole is not large. The exception may be only for some specialized proteins, peculiar to some organisms, like the known green fluorescent protein (GFP) from a variety of coelenterates, both hydrozoa such as *Aequorea*, *Obelia*, and *Phialidium*, and anthozoa such as *Renilla* (Tsien, 1998; Miyawaki and Tsien, 2000). Derivatives of green fluorescent proteins have been found by Labas et al., (2003) in the Russian Academy of Sciences Institutes of Bioorganic Chemistry and Biochemistry. Some of the animal fluorescent proteins are included into plants by methods of genetic engineering in the experiments. Fluorescent proteins also expressed by bacteria *Pseudomonas fluorescens* in the plant rhizhosphere, in particular in tomato seedlings (Bloemberg et al., 2000).

Nucleic acids. Nucleic acids fluoresce in the ultra-violet spectral region-maximal emission for DNA at 330-335 nm and for amino acids of proteins – at 282-348 nm as described by Wolbeis (1985). Quantum yield of the DNA emission is lowest ~ 10^{-4}. However, crystals of DNA and, in a lesser degree, RNA, also emit in blue (at 430-460 nm), but the emission disappears after water or water solution addition (Roshchina et al., 2003c).

1.3.2.3 Correlation in the fluorescence between secretions and the secretory cell

The correlation between the fluorescence of intact secretory cells and prevailing components in their secretions is shown in some examples of Table 1.2. In many cases, the fluorescence of the secretions occurs with maxima in shorter wavelengths, than the emission of secretory cells. The lightening of the cell wall in blue and chlorophyll in the red spectral regions is also seen due to maxima 460-480 nm and 680 nm, relatively.

The fluorescence of some photosynthesizing cells includes maximum 680 nm related to chlorophyll. The emission spectra of chlorophyll fluorescence of green leaves, taken from the whole surface of many cells at room temperature, show two maxima near 685 nm and 735 nm (Gitelson et al., 1998).

Table 1.2 The fluorescence maxima of intact secretory plant cells, their secretions and prevailing components of the secretions. The emission excitation by 360-380 nm.

Plant species	*Wavelength (nm)*		
	Secreting cells	*Secretion*	*Individual fluorescenting component in secretion*
	Terpenoid (oil) -	**secreting**	**structures**
Asteraceae *Achilea millefolium* L.	floral hair 465, 500, 550	oil 460, 410	Monoterpenes 420 sesquiterpenes 380, 410
Matricaria chamomilla L.	floral cells 460, 550, 650	oil 440, 460	sesquiterpenes 380, 410
	leaf cells 460, 550, 675, 450, 680	oil 460	sesquiterpenes 380, 410
Labiatae *Salvia splendens Sello ex Nees*	floral glands 465, 550, 460	oil 410	monoterpenes 420 (menthol)
Pinaceae *Picea excelsa*	resin ducts of needles 430, 460, 680	resin 430	terpenoids 410, 420, 430
	Alkaloid-and amine	**- containing**	**structures**
Papaveraceae *Chelidonium majus* L.	laticifer 480, 560, 680	latex 560	Berberine, chelerythrine 520
Solanaceae *Lycopersicon esculentum* L.	leaf secretory hair 540, 680	oil 410	Terpenoids, alkaloids 410-420, 460
	Phenol (quinone) -	**containing**	**structures**
Betulaceae *Betula verrucosa*	secretory cell of bud scale 495-498, 520, 680	exudate 460-470	flavonoids 460-470
Hippocastanaceae *Aesculus hippocastanum* L.	secretory cell of bud scale 480, 680	exudate 460	flavonoids 460-480
Guttiferae *Hypericum perforanum* L.	leaf glands 460, 680 leaf glands of flowers 460, 440, 680	exudate 440	quinones 460

• The sources: Wolfbeis 1986; Roshchina and Melnikova 1995 and unpublished data. Exct –Excretion, Edt – Exudate, and WL – water leachate

1.3.3 Light Emission of Different Cellular Compartments and Organisms, which Could Live on Plants

A special aspect of the autofluorescence problem is an emission of different cellular compartments and organelles. Moreover, some organisms, which live on plants such as pests or neighbours in biocenosis, also fluoresce, and the observer may see the emission.

1.3.3.1 Cellular compartments

Cellular compartments such as the cell wall, extracellular space, plasmalemma and cytoplasm with organelles contribute in different degrees in the fluorescence of secreting cells. Fig. 1.8 and colour Fig. 4 (Appendix 2) shows their different luminescence excited by UV-light.

Cell wall. The cell wall of intact cell fluoresce weakly in blue. In the emission spectra of isolated cell walls from *Spinacea oleracea* (Lichtenthaler and Schweiger, 1998), there is only one maximum in blue (450-465 nm). The cell wall emission depends on the substance(s) which impregnates the cellulose – mainly in blue (at 450-465 nm) or in green (at 500-520 nm). Pure cellulose cell wall has a weak emission as was demonstrated on the seed parachutes from *Taraxacum officinale* (Roshchina et al., 1998a) whereas dry paper filters from pure commercial cellulose fluoresce in blue. Colour LSCM images of various cells wall in stomata are shown green or green-yellow fluorescent in Colour Fig. 4 (Appendix 2).

According to Lichtenthaler and Schweiger (1998), leaves of green plants, when excited by UV-light 315-390 nm, exhibit a blue-green emission with a maxima 440-455, 466, and 474 nm, besides a red and far-red chlorophyll fluorescence. The authors have shown that, depending on the plant species, the presence of covalently bound ferulic acid (in a lesser degree p-coumaric, caffeic, chlorogenic and sinapic acids may contribute) in isolated cellular walls from non-soluble leaf residue of the methanol extraction. In this spectral region many compounds may fluoresce such as other aromatic (cinnamic) acids, flavonols, stilbenes and coumarins. Some alkaloids also appear to be present. The participation of similar compounds in the vacuolar emission of intact cells is not excluded. If the cell wall contains other phenolic compounds the green or blue-green or yellow emission is seen. In intact onion guard cells, the green emission with maximum at 520 nm, which was excited by 436 nm light, decreased, but under excitation by 365 nm light the cell wall fluorescence became blue with maximum 470 nm (Zeiger and Hepler, 1979; Zeiger, 1980; 1981; 1983). Hemicellulose and pectin emit weakly in yellow-green.

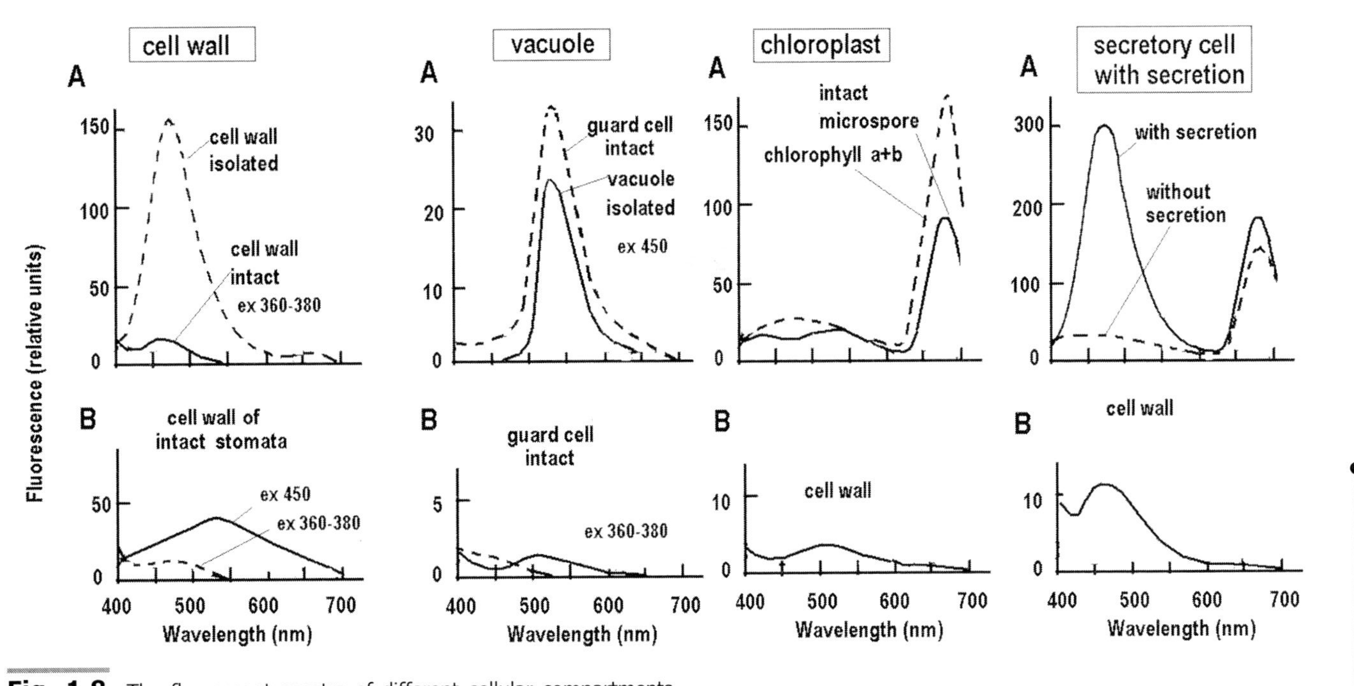

Fig. 1.8 The fluorescent spectra of different cellular compartments.
Cell wall – A. intact cell wall of *Spinacea oleracea* leaf and cell walls isolated from these objects, excitement 360-380 nm. B. – Cell wall of stomata on leaf of *Solidago canadensis* L., excitement 360-380 nm or 450 nm.
Vacuole – A. intact guard cells of stomata on bulb scale *Allium cepa* and isolated vacuoles from this object excited by blue light 450 nm. B – intact guard cells of stomata on bulb scale *Allium cepa* excited by UV-light 360-380 nm or 450 nm.
Chloroplast. – A – intact secreting microspore of *Equisetum arvense* L. (unbroken line) and the ethanolic chlorophyll a+b fraction isolated from the microspores (broken line). B – cell wall of the microspore. Excitement 360-380 nm.
Secretion in secretory cell. A – secretory hair of *Lycopersicon esculentum* Mill. with secretion (unbroken line) and without secretion (broken line), B. – Cell wall of the secretory hair. Excitement 360-380 nm.

Table 1.3 demonstrates the lower fluorescence intensity of cell walls from *Allium cepa* in comparison with other parts of the cell. The fluorescence of the cell wall also depends on the excited light wavelength: when it was excited by blue or violet lights the emission maximum shifted to longer wavelengths. For instance, the cell walls of stomata guard cells in the leaf of *Solidago canadensis* L. fluoresce weakly in blue, when they were radiated by UV-light 360-380 nm, whereas at the excitation by violet 420 nm or blue 440 nm, green-yellow fluorescence with maxima 530-535 nm was seen in Fig. 1.8 (see also colour Fig. 4 in Appendix 2). Bright luminescence is especially peculiar to stomata chink formed by two guard cells.

Plasmalemma. The plasma membrane produces NAD(P)H and dependent products, therefore the pyridine nucleotides fluoresce in blue at 460 nm. Perhaps, other components could fluoresce, for instance, plasma membrane of cyanobacteria *Anacystis nidulans* contains a small fluorescent pool of both carotenoids and protochlorophyllide - chlorophyllide (Peschek et al., 1989). The contribution of plasmalemma in the summed cellular blue fluorescence is also possible due to the NADH formation and other redox reactions (Dahse et al., 1989). More complex protochlorophyll participation in the red emission was described for some blue-green bacteria (Peschek et al., 1989) and may be assumed for other chlorophyll-producing organisms.

Cytoplasmic structures. The contribution of internal content of the secretory cell to the total fluorescence also relates to the transparency of epidermal cells. The actinic ultra-violet light easily penetrates through them and excites the luminescence of the secretion located both in periplasmic space and within a cell, as well as luminescence of cellular organelles. The fluorescence of the secondary metabolites of the secretion has been described above. The light emission of microsomes, mitochondria, chloroplasts may be caused by the production of NAD(P)H and flavins. The fluorescence of carotenoids mainly, chlorophyll, prevails in a total luminescence of the chloroplasts (Vogelmann, 1993). Chlorophyll may greatly contribute to total cellular luminescence. In some cases, it is an indicator. For instance, immature pollen without completely formed sporopollenin of exine (in anthers) is transparent for ultra-violet light and the 680 nm fluorescence maximum of chlorophyll is seen clearly (Roshchina et al., 1997b). Chloroplasts were found directly in spermia of pollen from *Lupinus luteus* (Ruhland and Wetzel, 1924).

Vacuole. Vacuole is the organelle, in which wastes may be evacuated, and in this case it is called "intracellular sac of the secretions". Some of the components may be in crystalline form. The vacuolar autofluorescence was first analyzed by several authors (Palewitz et al., 1981; Weissenböck et al., 1987) which showed the green emission of the organelle in various objects irradiated by a blue light. Living cells of the guard cell of *Allium cepa* L. and *A. vineale* L. fluoresce in green or yellow-green when exposed to blue 436 nm light (Zeiger and Hepler, 1979; Palevitz et al., 1981). The maximum vacuolar fluorescence is 520-525 nm (Zeiger and Hepler, 1979), but the chlorophyll red fluorescence also contributed. Green fluorescence was assumed to belong to flavins and flavoproteins. As demonstrated in Fig. 1.8, the vacuole luminescence studied on guard cells of stomata on *Allium cepa* scale emits, depending on the light wavelength excitation, in green with maximum 500 (excitation by 360 –380 nm) or 520-530 nm (excitation by 450 nm). As seen from the data received with optical probes of different sizes (2 mm and 10 mm) by double-channel microspectrofluorimeter (Table 1.3), the intensity of green fluorescence, peculiar to vacuole and secretions crystallized on the cellular surface, is several times higher, than in chloroplasts and the cell wall. There are data that anthocyanins of vacuoles fluoresce in the blue and red spectral regions.

Table 1.3 The fluorescence intensity of cells in *Allium cepa* bulb scale (excitation 420-436 nm). Optical probe means a diameter of the fluorescing area measured n = 100.

Cell part	*Green emission* $I_{520\text{-}530}$	*Red emission* $I_{640\text{-}680}$
Vacuole with crystal (optical probe 2 μm)	0.02 ± 0.005	0.02 ± 0.007
Vacuole with crystal (optical probe 10 μm)	0.07 ± 0.009	0.26 ± 0.02
Chloroplasts (optical probe 2 μm)	0.01 ± 0.001	0.11 ± 0.006
Chloroplasts (optical probe 10 μm)	0.01 ± 0.002	0.26 ± 0.02
Cell wall (optical probe 2 μm)	0.01 ± 0.002	0.01 ± 0.004
Cell wall (optical probe 10 μm)	0.03 ± 0.004	0.2 ± 0.008
Green-fluorescing crystal druses (optical probe 2 μm)	0.06 ± 0.004	0.02 ± 0.005
Green-fluorescing crystal druses (optical probe 10 μm)	0.28 ± 0.09	0.39 ± 0.08

Chloroplasts. As for chloroplasts, as seen in Fig. 1.8, their emission differs from cell wall fluorescence and shows, depending on the cell, one maximum in the blue-green spectral part and large characteristic maximum 680-685 nm in red. This picture is similar when observed for isolated chlorophyll a+b fraction purified by thin-layer chromatography. Table 1.3 shows that chloroplasts lying separately and chloroplasts surrounded vacuole fluoresce with the same intensity.

Secretory products in extracellular spaces and within cells. Unlike the above-mentioned cellular compartments and organelles, in specialized secretory cells, the release of the secretion out of the cell induces the quenching of the cellular fluorescence (Fig. 1.8) as seen in the secretory hair of *Lycopersicon esculentum* Mill. Only cell walls maintain blue emission whereas the empty middle of the cell looks like a dark space. Extracellular space, when secretions are accumulated before evacuation out a cell, also fluoresce from blue to red, depending on the chemical nature of prevailing secretory components (see Chapter 2 and 3). The compounds may participate in the redox reactions due to a participation of excreted diaminooxidase (Federico and Angelini, 1986) and superoxide dismutase (Roshchina and Roshchina, 2003).

1.3.3.2 The autofluorescence of other organisms on plant surface

Other organisms coming in contact with plants can also fluoresce, it is espsecially significant in order to distinguish autofluorescence of plant cell from pest cell fluorescence. Although information is limited, we have tried to analyze this problem. As shown in Table 1.4, various organisms have an autofluorescence. Colour examples of similar emission for bacteria *Pseudomonas aeruginosa* and ticks are in Colour Fig. 5 (Appendix 2).

Table 1.4 The autofluorescence fluorescence of other organisms excited with UV-light

Object	*Colour of fluorescence*	*Fluorophore*
Bacteria	Blue or blue-green (for most species if any emission)	NAD(P)H, flavins
Photosynthesizing (cyanobacteria Rhodopseudomonas, etc).	Blue, green and orange-red	NAD(P)H, flavins, bacterio-chlorophyll, phycobillins, phycoerythrins
Fungi	Blue or blue-green	NADH, flavins, flavoproteins
Spores	Blue or blue-green	NADH, flavins, flavoproteins
Hypha	Blue or blue-green	NADH, flavins, flavoproteins
Animals		
Insects	Blue or blue-green	NADH, flavins, flavoproteins
Ticks	Blue or blue-green	NADH, flavins, flavoproteins
Spiders and spider web	Blue or blue-green	NADH, flavins, flavoproteins
Sea organisms (medusa, etc)	Green, yellow or red	Green, yellow or red fluorescent proteins

Bacteria fluoresce mainly in blue, in particular as *Pseudomonas fluorescens* (Angell et al., 1993) or in orange-red as photosynthesizing bacteria, cyanobacteria, representatives of Rhodophyta and others (Ahmad et al., 1999), or have no fluorescence at all. Orange-red colour of the emission may belong to cytochromes or pigment phycoerythrin, in particular for cyanobacterium *Beggiatoa* (Ahmad et al., 1999) as well as bacteriochlorophyll in a lot of photosynthesizing bacteria. In many cyanobacteria such as *Aphanizomenon flos-aquae* pigments phycocyanin and allophycocyanin demonstrate blue emission. Many fungi also emit in blue (Mann, 1983). Cytoplasmic autofluorescence of an arbuscular mycorrhizal fungus *Gigaspora gigantea* in plants was observed by Sejalon-Delmas et al., (1998). The fluorescence of bacteria and fungi may be related to flavins, such as riboflavin, although porphirins and vitamins can also contribute (Meisel and Pomoshchinikova, 1952; Meisel and Gutkina, 1961; Konstantinova–Schlesinger ed, 1961), although there are various other fluorophores. Bright blue-fluoresced microorganisms are also related with NAD(P)H emission. Flavins and flavoproteins from *Azotobacter vinelandii* and *Escherichia coli* have a marked fluorescence around 500 nm, but some of them have a blue-shifted maximum (Veeger et al., 1980). This emission, in particular for 3-methylflavin, differs in water and glycerol, shifting in the last case to a longer wavelength. The difference was also there for room and 77°K temperatures. In photosynthesizing organisms, chlorophyll fluoresce with maxima at 678-685 nm, phycoerythrine – at 572 nm, phycocyanins – 637-660 nm (Karnaukhov, 1978).

Fungi or insects such as ticks and spiders, fluoresce, mainly in blue or blue-green (Table 1.4). Blue autofluorescence of many animal cells is related to NADH whereas green emission – to flavins (Chance and Thorell, 1959; Aubin et al., 1979; Benson et al., 1979; Rost, 1995). The emission of fungal hypha and mycelium are much higher and in a shorter wavelength region, than plant cells which makes it possible to differentiate fungal infection from fluorescing secretory plant cells. Insects fluoresce in all parts of their body, and it may be observed separately on the plant cell surface. On the spider's abdomen, several silk glands that hold viscous liquids fluoresce in bright blue which pass through minute tubes called spinnerets; normally a spider has three pairs. Upon being drawn from a droplet, the liquid protein structurally rearranges to polymerize into solid silk threads. Many spiders have blue autofluorescence which like the emission of many animal cells is related to NAD(P)H whereas green emission – to flavins (Chance and Thorell, 1959; Aubin et al., 1979; Benson et al., 1979).

New aspects of the analyzed problem have been made by genetic engineering that constructed vectors from the genome of animals and include them into plant cells (Stewart, 2006). This is a case with green fluorescent proteins. A variety of marine organisms are known to fluoresce under the actinic light. The lightening of sea and ocean species was seen many times and has been described for the small medusa *Aequorea victoria* by the naturalist Forskal at the end of 18th century. Although the luminescence picture may belong to both chemiluminescence and secondary excited fluorescence, two centuries later fluorescent protein equarin was discovered in the medusa, and in 60's of 20th century – another green fluorescent protein (GFP). Our information about fluorescing proteins is not large. The exception may be only for some specialized proteins, peculiar to some organisms, like the known green fluorescent protein (GFP) from a variety of coelenterates, both hydrozoa such as *Aequorea*, *Obelia*, and *Phialidium*, and anthozoa such as *Renilla* (Tsien, 1998; Miyawaki and Tsien, 2000). Derivatives of green fluorescent proteins have been found in some laboratories (Matz et al., 1999; Labas et al., 2003) as in the Russian Academy of Sciences Institute of Bioorganic Chemistry and Biochemistry. GFP from the jellyfish *Aequorea victoria* is especially used as a fluorescent probe or Ca^{2+} -indicator, which in water solution emits with maximum 508 nm. The protein has the highest fluorescent yield 0.8 that made it perspective probe. The chromophore is a p-hydroxybenzylideneimidazolinone. In a base of its formation are reactions of cyclization of the conformated amino acid residues (of serine or thyrosine), then dehydration, and, at last, aerial oxidation. Only at this stage of molecular oxidation can the chromophore fluoresce, and atmospheric oxygen is required for the emission (Tsien, 1998). Green fluorescent protein (GFP) can be genetically concatenated to many other proteins, and the resulting fusion proteins are usually fluorescent and often preserve the biochemical functions and cellular localization of the partner proteins (Miyawaki and Tsien, 2000). GFP fusion has major advantages over previous techniques for fluorescent labelling of proteins by covalent reaction with small molecule dyes. Among fluorescent proteins are those with the emission in the red spectral region, such as proteins derived from *Discosoma* sp. red fluorescent protein (Shaner et al., 2004). Today GFP with molecular mass 28 kDa may serve as genetic markers for the study of embryonic development and development of tumours (Labas et al., 2002, 2003). Genetic engineering permits one to include gene coding the protein into various organisms, where GFP is expressed in different cellular compartments and tissues. Similar colour proteins gave several images with excitation ranging from 480 to 560 nm.

Components of higher animal cells can also fluoresce if excited by light 350-380 nm. For example human skin emits in blue-green with maximum 470 nm (Zeng et al., 1995), muscle – in blue with maxima 410 and 480 nm (Ha et al., 1999), while bovine cornea – in ultraviolet (Uma et al., 1994).

Transgenic plants inoculated with bacterial or fungal pathogens may show accumulation of autofluorescent compounds like resistant tomato clone inoculated with bacterial gene from *Pseudomonas syringae* (Tan et al., 1999). Non-transgenic plant clones had no visiable fluorescence.

Besides above-mentioned emission in living organisms, it keeps in mind the fluorescence of non-organic crystals, which could be included in living cells, for example like ZnSe crystals fluoresced in blue (460-490 nm) at the excitation laser beam 353 nm (Agaltzov *et al.*, 1997). This question is not studied yet.

Conclusion

Autofluorescence excited by ultraviolet, violet or blue light is peculiar to many plant secretions that show the phenomenon in intact secretory cells. Any modification of luminescence microscope – from a simple technique to microspectrofluorimetry and confocal microscopy allows one to see it. The fluorescence of secretory cells is well distinguished among non-secretory cells that there is no necessity to stain the cells with artificial fluorescent dyes. The main fluorescence of secretory cells is related to secondary metabolites concentrated in extracellular space, vacuoles, secretory vesicles as well as some amount of the fluorescent secretions released out. Non-invasive observation of the secretory cells, based on their autofluorescence, enables one to analyze the vital state of the structures.

CHAPTER 2

Autofluorescence of Specialized Secretory Cells

Secretory cells are present as both unicellular cellular structures (unicellular hairs, unicellular spores or microspores which are needed for plant breeding, single secretory cells among non-secretory ones, etc) or as a part of multicellular secretory structures (multicellular hairs, multicellular glands, resin ducts, nectaries, laticifers, etc). These structures may play roles in plant defence against pests and parasites, in attracting insect-pollinators and others (Fahn, 1979; Roshchina and Roshchina, 1993). Some of the visible secretory organs can fluoresce *in vivo* under a luminescence microscope (excitation with ultra-violet or violet light).

Fluorescence of secretory cells may vary depending on their anatomy and taxa, as shown in Table 2.1. Below we shall consider the fluorescence and the fluorescence spectra of secretory structures in various organs of different taxa as well as concentrate on a more detailed consideration of most known secretory cells. All cells are able to secrete in different degrees, but this ability may vary in various plant taxa. Some systematic groups of plant species have special secretory cells, where the secretory function prevails. In plants there are external and internal secretory tissues. External tissues include glands, glandular hairs, reservoirs of the secretions, hydathodes and nectaries (Roshchina, V.D. and Roshchina, V.V. 1989; Roshchina, VV. and Roshchina, V.D., 1993). Internal secretory tissues are represented as resin ducts, laticifers and idioblasts. Idioblasts, single secretory cells, are scattered among non-secretory cells and differ in their characteristics. They may accumulate various compounds: calcium oxalate in the form of single crystals, druses or raphids, terpenes, slimes, tannins, oils, etc. The light emission of various taxonomic plant groups – from spore-breeding to seed-breeding ones will also be analyzed.

Below the fluorescence of external and internal secretory structures will be discussed. The analysis will begin with examples from representatives of more ancient taxonomical groups such as spore-bearing plants and end with those of seed-bearing plants. In Table 2.1 the data for more than 150 species are summarized, and the most significant features of the secretory structures which vary in their anatomy will be considered as examples in separate sections of the chapter.

Table 2.1 The fluorescence maxima of secreting plant cells in various taxa

Plant species (Latin names)	*Plant species (English names)*	*Secreting cells and secretions*	*Fluorescence maxima, nm*
		Fungi	
Agaricaceae *Agaricus campestris* (Fr.) Quel	True mushroom	Sporangium Spore Secretion of sporangium	550 450 510
		Mosses	
Polytrichaceae *Polytrichum piliriferum* L.	Haircap-moss, common hair moss	Sporangium Spore Secretory hair of sporangium	500 530 540-545
Sphagnaceae *Sphagnum sp.* (Dill.) Ehrh.	Bog moss, peat moss	Sporangium	520-530
		Horsetails	
Equisetaceae *Equisetum arvense* L.	Field horsetail	Entrance of hydathode in thallus Epithem Drop of secretion from hydathode Vegatative microspore Antheridium Spermatozoid Archegonium Egg cell	530, shoulder 475 460-470, 680 500 465, 540, 680 460, 540, 680 500, 560 520 shoulder 500, 585
		Ferns	
Polypodiaceae *Dryopteris filix-mas* (L.) Schott/	Shield fern	soruces	475, 530, 680
		Seed-breeding plants	
Acanthaceae *Beloperone guttata* Brandengel	Beloperone	Pollen	450, 405, 565, 670

Table Contd.

Aceraceae *Acer campestre* L.	Hedge maple	Pollen	440
Acer negundo L.	Box elder, ash-leaved maple	Pollen	475
Alstroemeriaceae *Alstroemeria aurantiaca* D. Don	Alstroemeria	Pollen Stigma of pistil Secretory cells of the flower petal Nectary	475, 550 465, 530, 680 500 520, 620
Amaryllidaceae *Clivia sp.* *Eucharis grandiflora* Planch et Lindem.	 Kaffir lily Amazon lily	 Pollen Pollen Stigma of pistil Secretory cells of the flower petal Secretory cells of the crown (stamen with anther)	 shoulder 475, 540 460-475 500 460 or 500, 530 or 475, shoulder 550 450, 500, 680
Hippeastrum hybridum	Knight's star,	Pollen Stigma of pistil Secretion of seedling	480 465, 525 535
Hymenocallis speciosa Regel	Spider-lily	Pollen	525, 670
Narcissus pseùdonarcissus L.	Common daffodil	Pollen	470, 675
Zephyranthes sp.	Zephyr lily	Pollen	490, 520
Arecaceae *Cocos nucifera* L.	Coconut palm	Secretory cell of fruit pulp	500, 560
Asclepiadaceae *Hoya carnosa* (L.) R.Br.	Common wax plant	Pollen Floral nectar	480, 550 495, 610
Asteraceae *Achilea millefolium* L.	Common yarrow	Pollen Floral glands Secretory cell of petal Leaf gland or secretory hair Secretory cell of leaf Secretory cell of root	480 465, 475, 500, 550 565 450, 680 475, shoulder 540 465, 545 460

Table Contd.

Arctium tomentosum Mill	Cotton burdock	Secretory cell of leaf upper side Secretion (crystal on the leaf surface) Secretory cell of leaf lower side Secretion (crystal on the leaf surface) Secretory hair of flower Nectary in red sports of petals Anther Pollen Pistil	440 or 450, (500) 550 440 440 460 465 440 465, 530 465-470 465, 550
Artemisia vulgaris L.	Mugwort, worm wood	Pollen Ligulate floral glands Leaf glands Oil	500 465, 500, 550 450, 680 460
Calendula officinalis L.	Pot marigold	Pollen Secretory non-capitate hair of tubular flower Secretory capitate hair of ligulate flower Secretory cell of sepal from ligulate flower Secretory non-capitate hair of ligulate flower Secretory capitate hair of sepal Secretory non-capitate hair of sepal Secretory of calix sepal Secretory capitate hair of leaf Secretory non-capitate hair of leaf	460, 520, 580 430, shoulder 500 475, 500 445, shoulder 530 450, 550 445, 550 440 or 440, 550, 680 435-440 430 560 or 585, 520, 600 or 500, 605
Cineraria hybrida Hort.	Silver ground = dusty miller	Pollen Secretory cell of leaf Secretory hair of leaf Secretion (crystal)	Shoulder 460, 545 439, 475, shoulder 530 475, 530, shoulder 600 500, shoulder 600
Cirsium arvense (L.) Scop	Canada thistle, creeping thistle	Secretory cells of tubular flower	595

Table Contd.

Echinops globifer (sphaerocephalus) L.	Common globe thistle	Secretory cells of petal from red tubular flower Secretory cells of petal from green tubular flower Pollen of red tubular flower Pollen of green of tubular flower Secretory cells of leaf Secretory hair of leaf	450, 680 460, 680 450 470, 680 465, 550, 680 475, 550, 680
Gaillardia pulchella Foug.	Gaillardia	Secretory cells of tubular flower Pollen Glands of ligulate flower Glands of tubular flower Glands of leaf Secretion on the leaf surface (crystal) Secretory cell of root	480 or 550 and 585 475 485, 625 465, 550, 680 475, 680 475 500
Gerbera jamesonii (Bolus) Hook.	Gerbera	Pollen	450, 510
Matricaria chamomilla L.	German camomile, wild camomile	Pollen Glands of ligulate flower Gland of leaf Oil	435, 465, 570, 620 460, 550, 650 460, 550, 675 or 450, 680 440, 460
Santolina chamaecyparissus L.	Cypress lavender cotton	Secretory hair of leaf Secretory cell of leaf	475, 560 or 500, shoulder 600 475 or 500, shoulder 550
Tagetes erecta L. (yellow)	Aztek marigold, African marigold	Pollen Gland of ligulate flower Gland of tubular flower Secretory cell of root	480, 625 500 480, 625 565, shoulder 630
Tagetes patula L. (orange)	French marigold, spreading marigold	Pollen Tubular flower Ligulate flower Hairs of tubular flower Gland of ligulate flower Gland of tubular flower	475 550 525, 610 525 485, 500, 625 480, 625
Tanacetum vulgare L.	Common tansy	Pollen Floral gland	490, 600 490, shoulder 550

Table Contd.

Taraxacum officinale Wigg.	Common dandelion	Pollen Laticifer Latex Secretory hair of leaf	520 550, 680 465 465
Tussilago farfara L.	Common coltsfoot	Pollen	460, 520, 675
Balsaminaceae *Impatiens balsamina* L.	Garden balsam	Extrafloral nectar Extrafloral nectary	465 465, 520
Begoniaceae *Begonia rex* Putrey.*	Begonia, assam king begonia, elephant's ear	Pollen	520
Berberidaceae *Berberis vulgaris* L.	European barberry	Fruit peel Pollen Stigma of pistil Leaf secretory cell Flower secretory cell	475, 520, 680 465 (510), 540, 640 505 465 535
Betulaceae *Alnus* sp. *Betula verrucosa* Ehrh.	Alder European white birch, weeping birch	Pollen Pollen Exudate of bud scale Secretory hair of bud scale Secretory cell of bud scale	520 440 460-470 500, 520 480-495, 520, 680
Brassicaceae (Cruciferae) *Capsella bursa-pastoris* L.	Shepherd purse	Secretory cell of leaf Secretory hair of leaf	500-520, 680 500-520, 680
Cactaceae *Echinocereus pentalophus* Rumpl	Echinocereus	Anther Pollen	475 550
Epiphyllum hybridum	Leaf cactus	Pollen Stigma of pistil Floral nectar	540 460, 525 no
Gymnocalycium castellanosii Brekby.	Gymnocalycium	Anther Pollen	No 525
Gymnocalycium zegarra Card.	Gymnocalycium	Anther Pollen	- 525
Lobivia jajoiana Backb.	Lobivia	Anther Pollen	485 550
Mammillaria dioica K. Brand.	Mammillaria	Anther Pollen	530 555-560

Table Contd.

Mammillaria sheldoni Bod.	Mammillaria	Anther Pollen	480, 545 560
Turbinicarpus lophophoroides (Werd) Buxb. et Backb.	Turbinicarpus	Anther Pollen	No 565
Campanulaceae *Campanula grandiflora* L.	Bell-flower, blue-bell, balloonflower	Pollen Stigma of pistil Pistil	520 475, shoulder 520, 680
Campanula persicifolia L.	Peach-leaved bellflower	Pollen Stigma of pistil Pistil	530-560 475, 560 475, 530
Chenopodiaceae *Atriplex patula* L	Common orach(e), fat hen	Leaf salt gland Glandular cell	465-475, 545, 680 500, 680
Chenopodium album L.	lamf's quarters goosefoot	Leaf salt gland	440, 530
Chenopodium rubrum L.	Goosefoot	Leaf salt gland Glandular cell of flower Pollen	475, 565 475, 565 475, 560
Convolvulaceae *Calystegia sepium* (L.) R. Br.	Hedge glorybind	Pollen	550
Convolvulus arvensis L.	European glorybind	Pollen Stigma of pistil Pistil Floral nectary Floral nectar Secretory hair on pistil Secretion of the leaf hair	480, shoulder 600 510, shoulder 600 565 580 580 475, 585 465, 540
Cornaceae *Swida alba* Opiz.		Anther Pollen Stigma of pistil Secretory cell of petal in flower Secretory cell of ovule sac Secretory hair of stem Secretory cell of leaf Secretory hair of leaf	465 450-465 440 465 440, 500 450, 530 425-440 450, 550
Crassulaceae *Bryophyllum daigremontianum* Hamet et Perr	Bryophyllum	Pollen Floral nectar	475 460

Table Contd.

Sedum hybrida L.	Stonecrop	Secretory cell of leaf Secretory cell of root	465 or 465, shoulder 550 475, 550
Cucurbitaceae *Cucumis sativus* L.	Cucumber	Pollen Stigma of pistil Secretory cell of flower petal Secretory hair of the flower petal	475 475, 525, 680 550-565, 680 475
Cucurbita pepo L	Gourd pumpkin	Pollen Stigma of pistil	530-540 450, 550, 680
Cupressaceae *Thuja occidentalis* L.	American arbor vitae	Resin duct of leaf Resin Secretory cell of seed cover	450, 535 460 475
Dipsacaceae *Knautia arvensis* Coult.	Field scabious	Pollen Leaf glandular hair	475, 515 475, 515
Droseraceae *Drosera capensis* L.	Sundew, dew plant	Slime Leaf granular hair	545 560, 680
Euphorbiaceae *Euphorbia viminalis* L.	Leafy euphorbia	Latex Laticifer	465 540, 680
Fabaceae *Medicago falcata* L.	Sickle alfalfa	Pollen	480, 550, 620
Medicago sativa L.	Alfalfa, medick	Pollen	460, shoulder 520
Melilotus albus Medik.	White sweet clover, Bokhara clover	Pollen	480, 520
Melilotus officinalis or officinale (L.) Pall	Yellow sweet clover	Pollen	480, 500-510
Trifolium pratense L.	Red clover	Pollen	500, 675
Trifolium repens L.	White clover	Pollen	515, 675
Geraniaceae *Geranium pratense* L.	Meadow geranium	Pollen	430
Pelargonium graveolens L' Herit.	Stock's-bill	Glandular hair of leaf Glandular cell of leaf Gland of leaf Glandular hair of flower Secretion from petal in flower	Shoulder 465, 550 or shoulder 500, 570 490, 565 500, 555-560 shoulder 490-500, 560 500, 560 or 550

Table Contd.

Gesneriaceae *Saintpolia ionantha* Wendl.	Common African violet	Gland of petal in flower	655-660, (680)
Gramineae *Alopecurus pratensis* L.	Meadow foxtail	Pollen	500, 680
Dactylis glomerata L.	Orchad grass	Anther Pollen Secretory cell of flower	500 500 (540) or 475, 520 475, 530
Guttiferae *Hypericum perforatum* L.	Common St. John's-wort	Pollen Floral glands on petal Leaf glands Exudate of leaf gland	460, 550 460, 680 440, 460, 680 440
Hippocastanaceae *Aesculus hippocastanum* L.	Common horse chestnut	Pollen Secretory cell of bud scale Secretory hair of bud scale	470, 680 480, 680 500
Hydrophyllaceae *Phacelia tanacetifolia* Benth.	Tansy phacelia	Pollen	430
Iridaceae *Gladiolus* sp.	Gladiolus, coneflag	Pollen	470
Crocus vernalis L.	Common crocus	Pollen	550
Funkia sp. (Hosta).	Plantain lily	Pollen	470
Lamiaceae (Labiatae) *Lamium album* L.	White dead nettle	Secretory cells of flower petal Secretory cells of leaf	475, 560-570 465, 560
Lamium maculatum L.	Spotted dead nettle	Secretory cells of anther Secretory hair of petal in flower Secretory cell of petal in flower Secretion (crystal) on the surface of petal	500, shoulder 550 475, 540 475, 540 465, 540
Leonurus cardiaca L.	Motherwort	Secretory cells of leaf lower side Secretory cells of leaf upper side Secretory hair of flower Secretion (crystal) in the hair	465, 680 475, 680 465, 600 or 465 530, 600

Table Contd.

Mentha piperita L.	Peppermint	Secretory cell of leaf Secretory hair of leaf	460, 530-540, 670 or 450, 525, 605 450, 595, 680
Salvia splendens Sello ex Nees	Scarlet sage	Floral glands Oil	465, 550 410
Lentibulariaceae *Utricularia* sp.	Bladderwort	Leaf trap glands Slime	560, 680 545
Liliaceae *Chlorophytum commosum* (Thumb.) Jackue. var. *variegatum*	Chlorophytum	Secretory cell of root	470, 550
Hemerocallis fulva L.	Forrest's day lily, yellow day lily	Pollen Stigma of pistil	550, shoulder 620 470
Tulipa sp.	Tulip	Pollen	470
Malvaceae *Hibiscus rosa-sinensis* L.	Chinese hibiscus	Secretory hair on pistil Anther Pollen	440 475, 560 475, 560
Hibiscus syriacus L.	Sharon rose	Secretory hair on pistil Anther Pollen	440 475, 560 470-475, 560
Lavatera thuringiaca L.	Thuringian mallow	Anther Pollen Secretory cell of petal in flower Secretory surface of pistil stigma Secretory cell of sepal in calix Secretory cell of leaf upper side Secretory cell of leaf lower side	490 475 475 475 475, (680) 475-480 475
Malva verticillata L.	Mallow	Anther Pollen Secretory cell of petal in flower Secretory surface of pistil stigma Secretory cell of sepal in calix	475 475 475 540 485

Table Contd.

Onagraceae *Chamerion angustifolium* Holub.	Fireweed, great willow herb	Pollen	460 or 500
Epilobium hirsutum L.	Willow-herb	Pollen Secretory cell of petal in flower Secretory surface of pistil stigma Secretory cell of leaf upper side	465 500 475 465
Fuchsia boliviensis Carr.	Fuchsia	Pollen Stigma of pistil	500 455-460
Papaveraceae *Chelidonium majus* L.	Greater celandine	Pollen Latex Laticifer	540 560 480, 560, 680
Eschscholtsia californica Cham.	California poppy	Pollen	no
Papaver orientale L.	Oriental poppy	Pollen	470
Passifloraceae *Passiflora coerulea* L.	Blue-crown passionflower	Pollen Stigma of pistil Extrafloral nectar Extrafloral nectary Floral nectary Floral nectar	475, 570 475 475, 550 465-475, shoulder 515 475, 550 480,545
Pinaceae *Abies sibirica* L.	Siberian fir	Resin duct of needle Resin of needle Resinous cell of cone Secretory hair of cone Resin of cone	465, 600 550 465, 500, 600 500 580
Larix decidua L.	Common larch, European larch	Pollen	460 (675)
Larix sibirica Ledeb.	Siberian larch	Resin duct of needle Resin of needle	525 465
Picea abies Karst.	Norway spruce, common spruce	Resin duct of needle Resin of needle Resinous cell of cone Resinous hair of cone Resin of cone	525, 600 525, 600 475, 575 475, 525, shoulder 600 575

Table Contd.

Picea excelsa (L.) Karst.	Spruce fir, Norway spruce	Resin duct of needles Resin of needle Resinous cell of cone Resinous cell in scale of cone Resin of cone Secretory cell of 3-10 d old seedling Cell in root meristem of seedling Resin of seedling	430 550 460 445-450 470 500, 580 485, shoulder 585 450, 550
Picea pungens Engelm.	Colorado spruce, Blue spruce	Resin duct of needles Resin	440, shoulder 610 550
Pinus sibirica Du Tour.	Cedar pine	Resin duct of needle Resin of needle Resinous cell of cone Resin of cone	465, 475 465 475 500, 550
Pinus sylvestris L.	Scotch pine	Resin duct of needle Resin of needle Resinous cell of needle Resinous cell of cone Resin of cone	450, 525-566 475-490 or 530 485-490, shoulder 570 475, 550 530
Plantaginaceae *Plantago major* L.	Ripple-seed plantain, great plantain	Pollen	540
Portulaccaceae *Portulaca hybrida*	Purslane	Pollen	480, 520
Ranunculaceae *Delphinium consolida* L.	Forking larkspur, field larkspur	Pollen	480, (560)
Rosaceae *Agrimonia eupatoria* L.	Agrimony	Glandular hair of leaf and stem	465
Cerasus vulgaris Mill.	Cherry	Pollen	500, 550
Crataegus oxyacantha L.	English hawthorn, European hawthorn	Pollen	480
Filipendula ulmaria (L.) Maxim	European meadow sweet	Pollen	480, 550
Geum rivale L.	Water avens	Glandular cell of petal	460, 550, 680
Geum urbanum L.	Common avens	Pollen	475-500

Table Contd.

Malus domestica Borkh.	Apple tree	Pollen	540
Rosa canina L.	Dog rose, wild rose	Pollen Floral secretory cell	465, shoulder 500 485
Rosa duvarica(ta) or duhurica Pall.	Duhurian rose	Pollen	500
Rosa rugosa Thunb.	Ramanas rose	Pollen Secretory cell of petal	500 500-520
Rubus idaeus L.	Red raspberry, European raspberry	Pollen	430
Rubus odoratus L.	Fragrant thimbleberry, purple-flowering raspberry	Secretory hairs of leaves, flowers and stems	420-430 (weak intensity)
Spiraea sp.	Spirea	Pollen	550
Rubiaceae *Gallium boreale (is)* L.	Northern bedstraw	Pollen	475
Gallium verum L.	Yellow bedstraw	Pollen	475
Rutaceae *Citrus cinensiss* (L.) Osbeck	Sweet orange	Secretory cell in peel of fruit	450, 560
Ruta graveolens L.	Rue	Pollen Secretory glandular cell of leaf upper side Secretory cell of root (idioblast) Secretory hair of root Secretory cell of root tip	Shoulder 475, 550 565 585-590 480, 575 500, 590
Salicaceae *Populus balsamifera* L.	Southern poplar	Pollen Exudate of bud scale Secretory cell of bud scale Secretory hair of bud scale	470, 515 470 500, 680 500
Salix virgata L.	Basket willow	Pollen	470, 515
Scrophulariaceae *Symphytum officinale* L.	Common comfrey, woundwort	Idioblast of leaf Secretory hair of leaf Slime cell of leaf Slime Secretory cell of petal from flower Stigma of pistil Water extract from leaf	535-556, 680 510 535-540 535 530, shoulder 465 530 440, 460

Table Contd.

Solanaceae *Capsicum annuum* L.	Red pepper, Cayenne pepper	Secretory cell of fruit peel Secretory cell of fruit pericarp Secretory cell of the seed stalk within fruit	465, 530 465, 550 525, 550 475, 550
Lycopersicon esculentum Mill.	Common tomato	Pollen Stigma of pistil Secretory hairs of petal Oil of flower Secretory hair of leaf	475, shoulder 575 460, 525 460, shoulder 520 410 460 or 465-470, shoulder 520, 680
Nicotiana alata Link et Otto	Tobacco avion	Anther Pollen Secretory cell of petal upper side in flower Secretory cell of petal lower side in flower Secretory surface of pistil stigma Secretory cell of leaf upper side	475 475 475, 550 475, 550 465, shoulder 560, 680 510
Petunia hybrida (L.) Vilm.	Common petunia	Pollen Stigma of pistil	450, 525, 663 538
Solanum tuberosum L.	Potato	Secretory cell of leaf Secretory hair of leaf Secretory cell of petal in flower Pollen Stigma of pistil	465, 550 or 475, 540 475, 540 or 465, 525, 680 465 460 465
Saxifragaceae *Philadelphus grandiflorus* Willd.	Big (scentless) mock orange	Pollen Secretory cell of petal in flower	455, 510, 620 450
Tiliaceae *Tilia cordata* Mill.	Little leaf linden	Pollen	470
Umbelliferae *Heracleum sibiricum* L.	Siberian cow parsnip	Anther Pollen	450 480, 520
Urticaceae *Urtica dioica* L.	Big-stinging nettle, stinging nettle	Pollen Stinging hair Water leachate	445, (680) 440, 460-470, 538, 680 440
Urtica urens L.	Small nettle, dog nettle	Stinging hair	440, 470, 530, 680

Spore-breeding and seed-breeding plants differ in their life cycles. The first (club mosses, horsetails, and ferns) can include free living unicellular forms such as spores (diploid cells) and sessile multicellular bodies, on which special sexual organs with male and female gametes (haploid cells) are formed during their development. The attached gametes fuse each other and form a diploid cell (zygote), which develops from the embryonic stage to the adult plant with vegetative organs, producing spores. Spores are vegetative generation which germinate and form multicellular thallus, multicellular body. In this way the life cycle continues.

Among spore-bearing organisms, fungi are heterotrophs, e.g, unlike other plant taxa, they have no chlorophyll. Some of them, such as yeast, may develop by simple cellular division, forming two cells. Others have fruit bodies with sporangia filled with fluorescent spores, which germinate and produce hyphae, branched and cylindrical filaments (bright blue fluoresced).

2.1 SECRETORY CELLS IN SPORE-BREEDING PLANTS

Spore-breeding plants are fungi, mosses, ferns and horsetails. Fungi include unicellular (such as yeast) and multicellular (for example, fruit bodies in Basidiomycetes) forms. All parts of the organisms may release various metabolites, which are able to fluoresce. As for mosses, ferns and horsetails, the major unique feature of spore-bearing forms in their life cycle is the alteration between free-living haploid gametophyte generation and diploid sporophyte generation. The groups of plant species also have other common features: multicellular sex organs in the form of archegonia and antheridia, flagellate male gametes, meiosis at spore formation and aerial spore dispersal.

All spore-breeding organisms have no leaves and roots, and secreting cells are located on the above-ground part of the organism – fruit body (in fungi) or thallus (for mosses, horsetails and ferns) with the spore cases. In many of them, rhizomes, the underground part for anchoring to a substrate, mainly to a soil, are present. They have spores for breeding and secreting cells on multicellular generations, which are non-specialized or weak-specialized, and are able to release slime and other secretory components. The secretion occurs in different degrees, and this ability may vary in various plant taxa. The cells and spores also release the secretions and may fluoresce. If organisms have both the vegetative and sexual modes of breeding such as mosses, ferns and horsetails, for which they have special organs – gametophytes with male and female sexual structures known as antheridia and archegonia, respectively. The structures also may fluoresce depending on the products released.

2.1.1 Sporangia and Spores

Spores of spore-breeding species are formed and stored within the sporangia, which rupture after maturing and this leads to spreading of the spores. Figure 2.1 demonstrates the fluorescence spectra of the sporangia, spores and non-sporangial parts of various plants from fungi as true mushroom *Agaricus campestris* L., moss as *Sphagnum sp.* L., and shield fern *Dryopteris filix-mas* L. There are differences between slime-releasing cells of sporangium and spores of the plants as well as between species due to the different chemical composition of the secretions. Spores of fungi true mushroom often fluoresce in the blue region of the spectrum with maximum 450 nm whereas sporangium – in green-yellow at 520-550 nm. In chlorophyll-containing cells of the thallus of moss and fern we see a wide emission in the blue-yellow region 460-570 nm (and sometimes – maximum 680 nm, peculiar to chloroplasts) for secreting cells and only red emission for non-secreting cells enriched in chlorophyll. In the fern *Dryopteris filix-mas* the fluorescence spectra of soruses with spores has two maxima at 475 nm (possible emission of flavonoids) and 525 nm (possible emission of flavins) in the blue-yellow part of the spectrum whereas cells surround ing the soruses demonstrate the same maxima in a smoother spectra. Non-secretory cells fluoresce only with one maximum at 680 nm, peculiar to chlorophyll. Moreover, unicellular spores of horsetails, mosses and ferns, which serve for non-sexual breeding, can also fluoresce. Dry non-developing spores of the plants emit in blue. Fresh and fossil spores of peat moss *Sphagnum* sp. have a bluish green and (or) green emission with maxima 470-474 and redemission 618-621 nm (van Gijzel, 1971). The fluorescence spore intensity may decrease under UV-light exposure (negative fading) or increase (positive fading) as demonstrated by van Gijzel (1971) and Willemse (1971). The fluorescence colour of club-moss *Lycopodium* spores (family Lycopodiaceae) is similar for exine and exosporia, and changed from a yellow light to a greenish yellow under various treatments (van Gijzel, 1971).

Fluorescent components of the fern *Dryopteris filix-max* can be phenolic compounds such as filicin, albaspidin, aspidinin, aspidinol, desaspidin, flavaspidic acid, filicic acid and various flavonoids (Plant Resources of

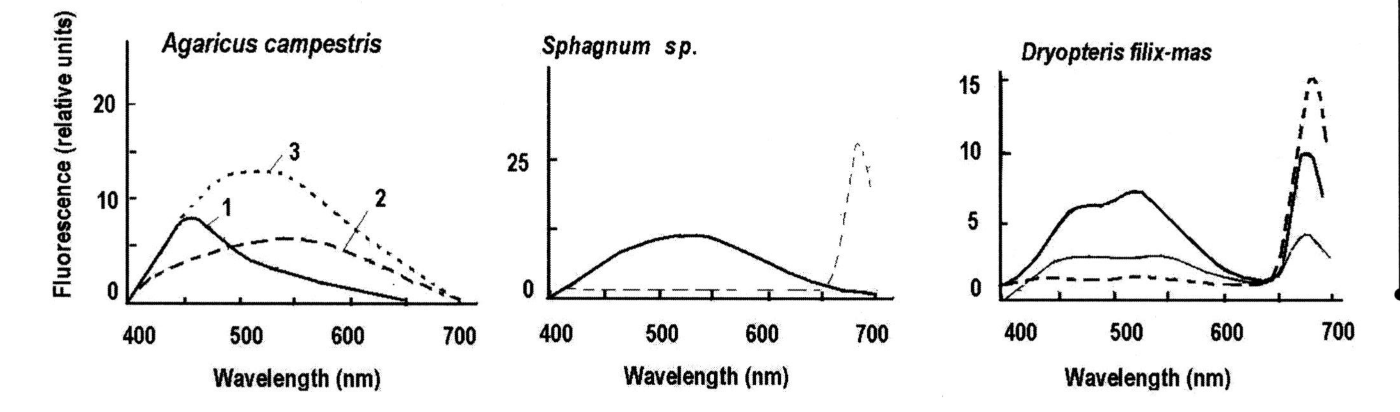

Fig. 2.1 The fluorescence spectra of the fungi true mushroom *Agaricus campestris* L., moss *Sphagnum sp.* L., and shield fern *Dryopteris filix-mas* L. Left-fungi true mushroom *Agaricus campestris* (1- blue-fluorescing spore, 2- yellow fluorescing sporangium, 3- secretion of sporangium; Middle - bog moss *Sphagnum sp.* unbroken line – secreting cell, broken line – non-secretory cell; Right-shield fern (male fern) *Dryopteris filix-mas* L., unbroken lines-soruses, broken line – non-secretory cell.

Russia and Surrounded Countries, 1996; Golovkin et al., 2001). In rhizomes of the species phloroglucides, flavaspidic acid and filicic acid are found (von Schantz, 1962a, b) , which can also fluoresce. The composition of fluorescent exine of vegetative microspores (Colour Fig. 6 Appendix 2) of spore-breeding plants such as of *Equisetum arvense* has not been studied enough. It is known that microspores contain fluorescent components, such as, flavonoids quercetin and kaempferol, various alkaloids, carotenoids, chlorophyll (Plant Resources of Russia and Surrounded Countries, 1996), and recently azulenes have been found here (Roshchina et al., 2002). The maxima of main components are in Table 1.1.

The phenomenon of autofluorescence from vegetative microspores of the spore-breeding plant *Equisetum arvense* has been studied by methods of laser-scanning confocal microscopy (LSCM) and microspectrofluorimetry during the development of the cells (Roshchina et al., 2002; 2004; Roshchina, 2004a). The spore is a single cell with a rigid cover, with an external layer called exine. It has special structures – elaters serves for anchoring with the soil (Colour Figs. 6, 7 Appendix 2). The microspores have demonstrated a difference in structures: blue-fluorescing cover and red-fluorescing chloroplasts (see colour Fig. 6 in Appendix 2). LCSM images are also given in colour Fig. 3 that show optical slices of the vegetative microspores in dry or wet states. The fluorescence spectra of the studied cells have also been measured by original microspectrofluorimeter (Colour Fig. 7). As shown in Colour Fig. 7, the middle part of the microspore has three maxima – 460 nm, 550 nm and 680 nm, the latter is peculiar to chlorophyll fluorescence. Thus, the presence of chlorophyll is seen in the dry microspores. The cover of the microspore has no maximum 680 nm, as well as in elaters. The red fluorescence of the microspores was, mainly, due to the presence of chlorophyll and azulenes (Roshchina et al., 2002; 2004; Roshchina, 2004a; 2005a). The character of the spectra and the colour of fluorescence may be changed during microspores germination. Unicellular microspores are recommended as natural probes of cellular viability and development (Roshchina, 2004a).

2.1.2 Gametophytes and Sexual Organs

Spore-bearing plants are mosses, horsetails, ferns, as well as fungi which today are in intermediate positions between the plant and microbial kingdoms. Among the mentioned taxonomic groups, mosses, horsetails and ferns have both vegetative and sexual modes of breeding. Most of them form vegetative microspores, which germinate, divide and then produce a multicellular thallus – gametophyte. Later, on the surface of the thallus sexual organs arise such as archegonia or female gametophyte, and antheridia or male gametophyte. Sometimes, both archegonia and antheridia arise on one and the same thallus, and this looks like a hermaphrodite. In other cases, male and female gametophytes are on separate thalli. Vegetative microspores and gametes of antheridia and archegonia are unicellular secretory systems. Moreover, non-gamete cells of the organs may excrete slime for attracting and recognizing the opposite sexual cell at breeding. Thus, examples of the cellular fluorescence should be considered.

Undeveloped microspores look like blue- lightening, while red fluorescence with maximum 680 is clearly seen in developed spores. When the spores germinate, they form a multicellular thallus (gametophyte), where sexual organs are formed. Sexual organs, such as the female, known as archegonia and the male known as antheridia, are located on the thallus, belonging to the female gametophyte or male gametophytes. Sometimes, the thallus may be a hermaphrodite, e.g. carries both the female and male organs. Sexual and non-sexual organs also differ in their fluorescence. It enables one to distinguish them on the thallus. Female and male gametophytes produce egg cells (female gametes) and spermatozoids (male gametes), relatively. Spermatozoids penetrate into the egg cell, and thus fertilization occurs.The egg cell and spermatozoids fluoresce in different regions of the spectra and can also be distinguished. Figure 2.2 demonstrates the fluorescence spectra of multicellular thallium (gametophyte) with male organs – antheridia and female organs – archegonia. Spermatogenous tissue and individual spermium fluoresce at 530-570 nm without chlorophyll maximum, whereas young and old gametophytes stored the same character of the spectra as unicellular microspore, only chlorophyll fluorescence intensity at 680 nm became much more -6 times (Roshchina et al., 2002; 2003b; 2004). The archegonia differ in the fluorescence between cells. The egg cell lightened in the blue (shoulder 500 nm) and orange (maximum 595 nm) regions, whereas surrounded cells – only in the red region (main maximum 680 nm belonging to chlorophyll) like thallium-gametophyte. Colour Fig. 8 (Appendix 2) show LSCM images of the sexual organs on the thallus. When the green-fluorescing spermatogenous sac with yellow-emitted spermatozoids, is ruptured, the male gametes are liberated into the

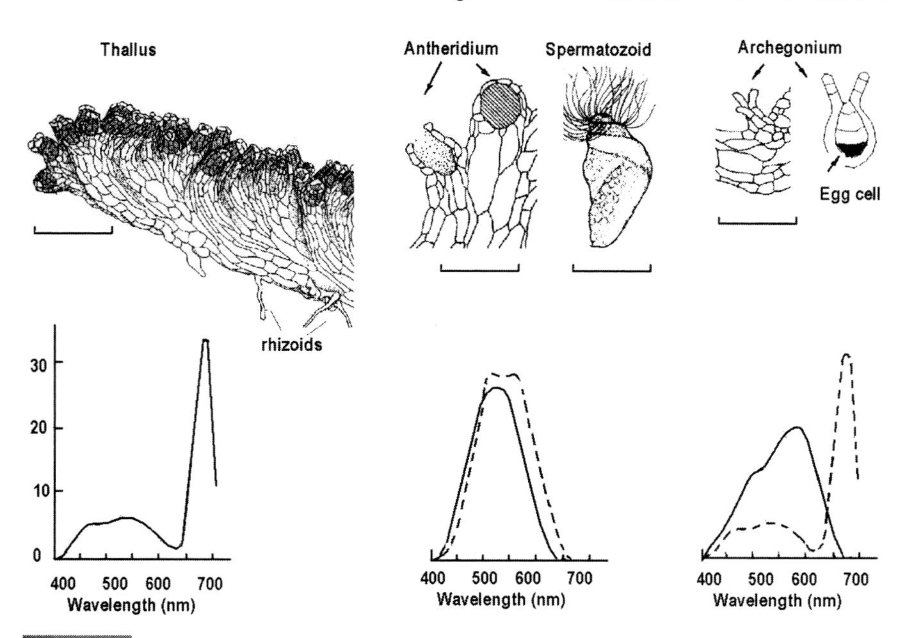

Fig. 2.2 The fluorescence spectra of gametophyte and sexual organs in horsetail *Equisetum arvense.* Left – thallus (rhizoids have no measured fluorescence); Middle – antheridium (broken line) and spermatozoid (unbroken line); Right – archegonium, where the egg cell (unbroken line) and the surrounded cells of bottle-like cover (broken line). Bar 1 mm for every structure on the picture, except spermatozoid (10 μm).

water environment in order to move to the archegonium within the egg cell. The archegonium looks like a bottle, is located within the egg cell and surrounded by red cells. Thus, the autofluorescence of the *Equisetum arvense* allows one to distinguish the stages of the development *in vivo* without any dyes. Enhanced red fluorescence serves as a marker for the development to start in microspores. Any scientist without special training can distinguish antheridia and archegonia on the thallium due to their differences in fluorescence.

2.1.3 Hydathodes and Slime-releasing Cells

Vegetative parts of spore-bearing plants have secretory structures, which may release water or water-soluble substances and are called hydathodes (Fahn, 1979; Roshchina and Roshchina, 1989; 1993). They also show significant fluorescence. Figure 2.3 shows one of the examples where the hydathodes of horsetail *Equisetum arvense* L. located on the vegetative part of plant contains not only water, but also fluorescent components. The drop of secretion evacuated from the orifice of hydathode has 480-490 nm fluorescence maximum, whereas cells of the orifice show two maxima, 470 and 530 nm, and cells of epithem show only one (beside 680 nm), this is also characteristic of some flavonoids maximum 465 nm (Table 2.1). Various fluorescing veins of the leaf have maxima at 465 and 505 nm or at 525 nm, which indicates the similarity between the vein content and that of the hydathode cells. On the contrary, non-secretory cells show marked red luminescence only at 680 nm, which is typical of chlorophyll. This species contains flavins (riboflavin), flavonoids (dihydroquercetin,

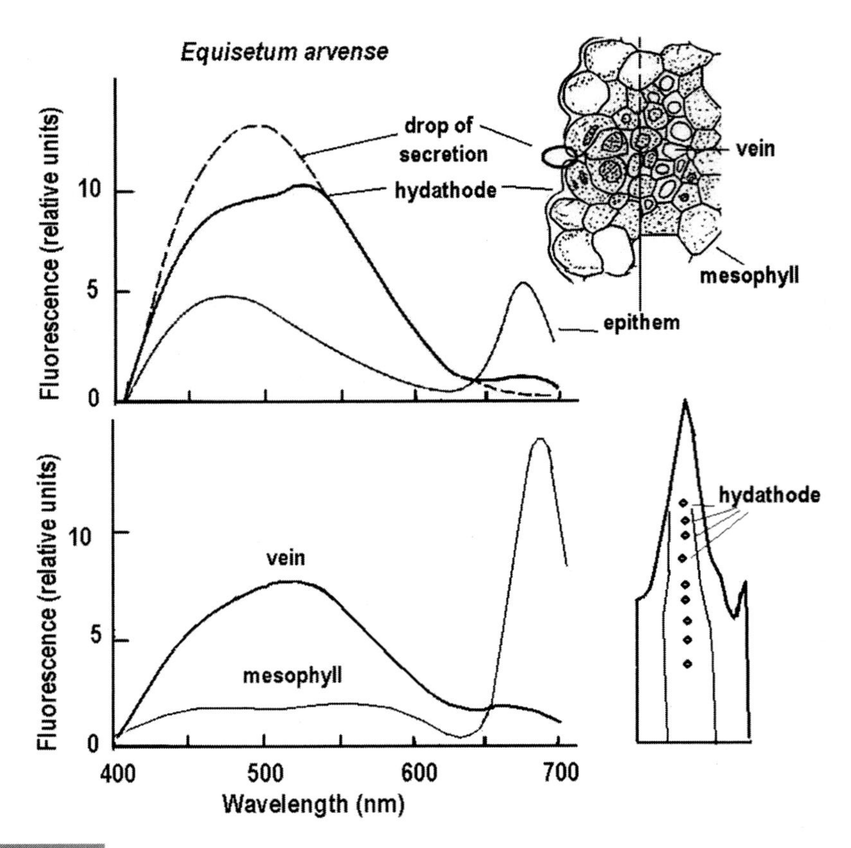

Fig. 2.3 The fluorescence spectra of the hydathode cells and surrounded cells from overground parts of *Equisetum arvense.* Adopted from Roshchina et al., (1998a). Right upper and lower pictures – schemes of the hydathode structure and position of hydathodes on the overground part fragment.

naringenin, etc), phenolcarbonic acids (n-coumaric, ferulic, caffeic and others) and other substances (Plant Resources of Russia and Surrounded Countries, 1996; Golovkin et al., 2001) which may fluoresce. Flavonoids and phenolcarbonic acids emit in blue (450-470) (althougth some of them, as seen in Chapter 3, have fluorescence in green and yellow-orange), while emission at 510-520 nm is characteristic of flavins. We do not know the exact composition of the hydathode secretions, but preliminary fluorimetric analysis gives some information.

Cells of thallus, gamethophytes, the sexual organs in the sporophyte stage of the development and overground parts of adult spore-breeding plant may release slime, which can contain fluorescent compounds. This mucilage is observed in spermatogenic tissue of antheridium and especially in the archegonium, when the cells of the channel surrrounding the egg cell are decomposed in order that spermatozoids can move along the channel for fertilization. Figure 2.4 shows the fluorescence spectra of cells on the surface of moss *Polytrichum piliferum* L. Slime cell fluoresces with maxima 475-480 nm, but pured slime - weakly (without maximum in visible region). The sporangium (pod boll) with or without the cap as well as spore emit weakly in green with maximum 530 nm and strongly in red with maximum 680 nm, which is peculiar to chlorophyll. The entrance of the sporangium with spores fluoresces with shoulder 450 nm and maximum 567 nm. The secretory hair on the sac with spores has a weak emission with maximum 540 nm.

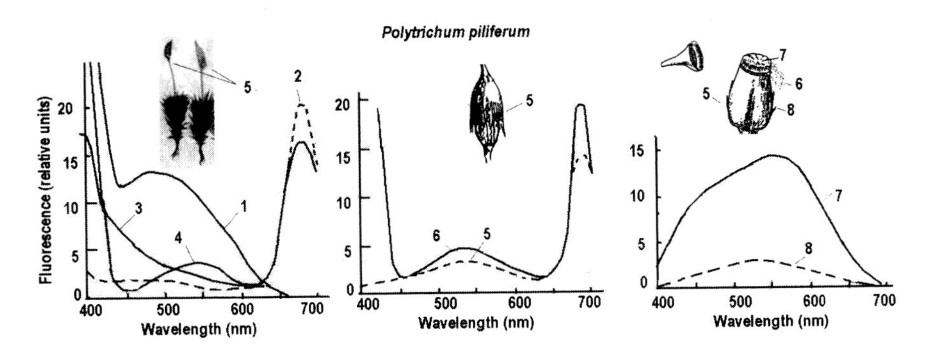

Fig. 2.4 The fluorescence spectra of moss *Polytrichum piliferum*. 1-slime-releasing cell, 2-non-secretory cell; 3-slime; 4-hair; 5-pod boll or sporangium with spores; 6-spore; 7-the entrance to pod boll (sporangium); 8-hair on the sporangium.

2.2 SECRETORY CELLS OF SEED-BREEDING PLANTS

Seed-breeding plants include Gymnosperms (which have no flowers) and Angiosperms (flowering species). These groups of plants have secretory structures in all parts (Fahn, 1979; Vasilyev, 1977; Roshchina and Roshchina, 1989; 1993). The structures in the reproductive and vegetative parts of the plants should be considered separately.

2.2.1 Reproductive Organs

Reproductive organs of seed-breeding plants are male and female gametophytes. Male gametophytes (microspores) or generative microspores are known as pollen for both Gymnosperms and Angiosperms. The female gametophyte, ovule, includes megaspore (egg cell), surrounding with the whole assemblage of other cells. Male gametes (spermia) are released from microspores and move to the ovule, the process ends by a fusion of one of spermia within the egg cell. And so the fertilization occurs. In Gymnosperms, egg cells are in female cones—macrosporangia while pollen grains (microspores) are in the microsporangia. Unlike Gymnosperms, female organs of Angiosperms include pistil, in which the ovule sac contains the egg cell. The surface of the pistil is a stigma with papillae which actively secrete for pollen anchoring and recognition. After fertilization and the embryo development, seeds are formed. Secretions of secretory cells in generative organs participate in plant breeding. The cells of spore- and seed-breeding plants differ in their structure and content. Table 2.1 shows that the generative structures also fluoresce under ultraviolet light. Several species can show mono-, di-, and three-component systems.

2.2.1.1 Reproductive organs of Gymnosperms

Reproductive organs of Gymnosperms are represented as microsporangia with microspores called pollen (male gametophyte with male gametes – spermia) and as macrosporangia with macrospores known as female gametophyte, which contain the female gamete. Secretory cells in the organs are mainly resin-produced, and their secretion – resin consists of various terpenoids, often mixed with phenols (Roshchina and Roshchina, 1989; 1993). The resin can fluoresce in UV-light of a luminescent microscope. Especially marked luminescence is observed for conifer plants such as the genera *Pinus*, *Abies*, *Picea*, *Larix* and others. Figure 2.5 shows the example of the fluorescence emitted by secretory cells of microsporangia with microspores (pollen grains) and secretions of a mature structure called cone taken from *Pinus sylvestris*, the conifer plant in Gymnosperms.

The lightening is peculiar to secretions consisting of resins and oils, whose main components are terpenoids. All components of resin in the microsporangium and pollen fluoresce in blue at 460-480 nm (pollen grain fluoresce in blue, while pollen wings – bubles – so weak in order to be measured). But secretory cells also have emission in the green-yellow spectral region whereas non-secretory cells have a weak fluorescence in

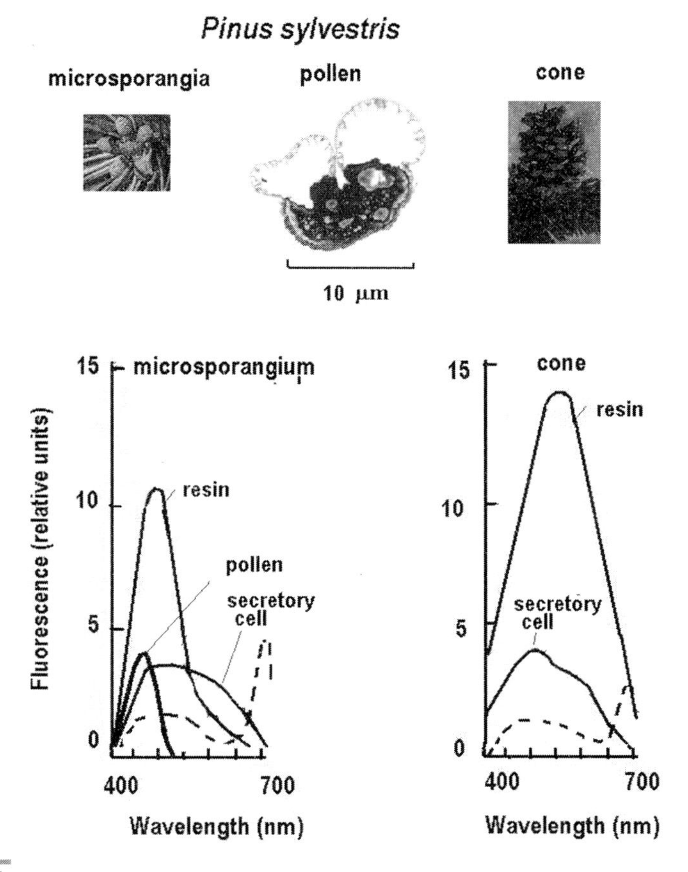

Fig. 2.5 The fluorescence spectra of the secretion (resin), secretory cells (with resin) of microsporangium, pollen (microspore) and the mature cone from Gymnosperm plant *Pinus sylvestris.* Broken line – non-secretory cell.

blue-green and strong emission in red with maximum 680 nm, peculiar to chlorophyll. One can compare the light emission from the microsporangium, cone and needle surface (Table 2.1). In all cases, the contribution of terpenoids such as α- and β- pinenes, terpeneol, sesquiterpene lactones is in a blue spectral region from 420 to 470 nm that is peculiar to the resin released. Green and yellow components in the fluorescence were observed for secretory cells, where other compounds may be present, for example carotenoids (emission 530-540 nm), flavins (emission 510-520 nm) and flavonoids (emission 460-480 and 550-570 nm). Pollen grains weakly fluoresce in the short blue spectral region, where only terpenes emit.

Figure 2.6 shows another example of the fluorescence emitted by secretory cells and secretions of *Abies sibirica* belonging to conifer plants from Gymnosperms. The fluorescence of terpenes from resin was observed on the surface of needles, cones and seeds. Seeds have no marked fluorescence. Resins and resin ducts of both cones and needles from *Abies sibirica* had,mainly, yellow fluorescence (maximum 550 nm), although resin ducts demonstrated the shifts - expressed maximum in green-yellow at 500 and 550 nm. There was a difference in the fluorescence spectra of the resin released on the surface and resin in the resin duct, as well as between resin ducts of needles and cone. Unlike resin fluoresced with maximum 530-550 nm, resin released from resin ducts of needles have two maxima at 475 and 575 nm. Moreover, the fluorescence spectra of the resin duct in a cone had maximum 465, 535 and shoulder 575 nm whereas free resin demonstrated weak-expressed two-component emission at 500-580 nm. The conifer species contain terpenes and sesquiterpenes. Among them, only sesquiterpene lactones (mainly proazulenes and. azulenes) have intensive fluorescence (Table 1.1), mainly, in the blue or yellow spectral regions. The chemical composition of resin is often very complex. Besides polysaccharides, it contains phenol (about 20 flavonoids) and protein components (Roshchina and Roshchina, 1989; 1993). Flavonoids and some terpenoids such as azulenes can fluoresce in the blue-green region of the spectra.

2.2.1.2 Reproductive organs of Angiosperms

Reproductive organs of Angiosperms are located in flowers, unlike Gymnosperms, and include anthers that are modified microsporangia with microspores (pollen grains) and pistils, which contain macrosporangia with

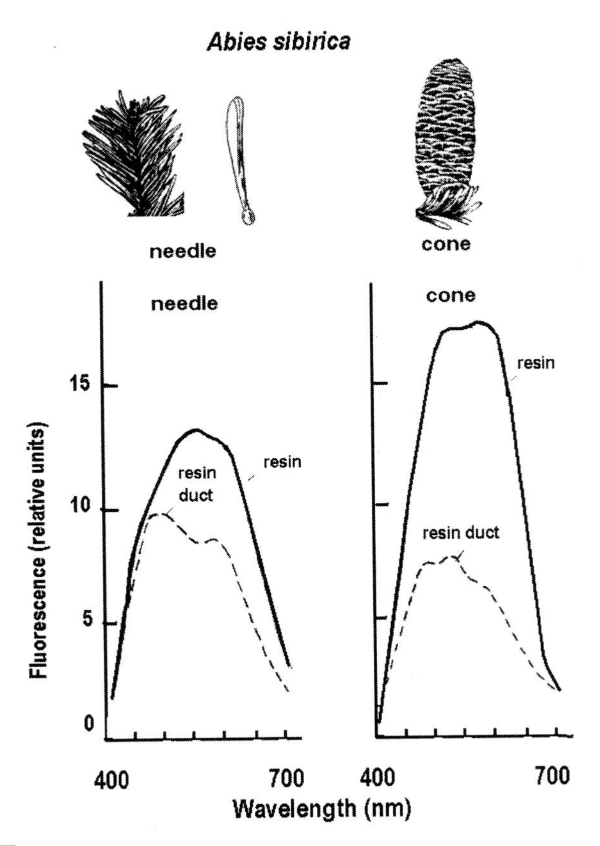

Fig. 2.6 The fluorescence spectra of the secretion and secretory cells of cones and needles of *Abies sibirica.* Adopted from Roshchina (2005b). Unbroken line–secretory cell with resin, broken line–secretion (resin). Non-secretory cell has only fluorescence of chlorophyll with max 680 nm. (non-illustrated).

macrospores (female gametes) within an embryonal sac. Pollens themselves are unicellular secretory structures, while multicellular anthers and pistils may have special secretory cells.

Autofluorescence of Angiosperms' microsporangia known as anthers with spores (pollen grains) was demonstrated since the 30's in the 20th century (Berger 1934; Asbeck 1955; van Gijzel 1961; 1967; 1971; Willemse 1971; Driessen et al., 1989) as well as pistil fluorescence (Kendrick and Knox, 1981). The phenomenon was never considered for understanding biological processes until recently. But rapid changes in the fluorescence may be an indicator of chemical reactions and relations.

Anthers and pollen. Male organs in flowers of Angiosperms are represented as anthers, where pollen is formed, matured and stored, and matured pollen, which is ready for fertilization. Anthers and pollen fluoresce at the excitation by UV-light (Fig. 2.7 Colour Fig. 9 in Appendix 2). Secreting cells of anthers contain slime which is released after the pollen has been liberated. Pollen is a male gametophyte, which is able to produce a pollen tube with male gametes called spermia. This microspore is also a unicellular secretory structure. As shown in Fig. 2.7 and Colour Fig. 9, the generative structures have fluorescence under ultra-violet or violet actinic light. Colour Fig. 9 shows the fluorescence of pollen grains seen as an LSCM image made by using laser excitation where the heterogenous parts of the pollen surface may be observed. The surface of

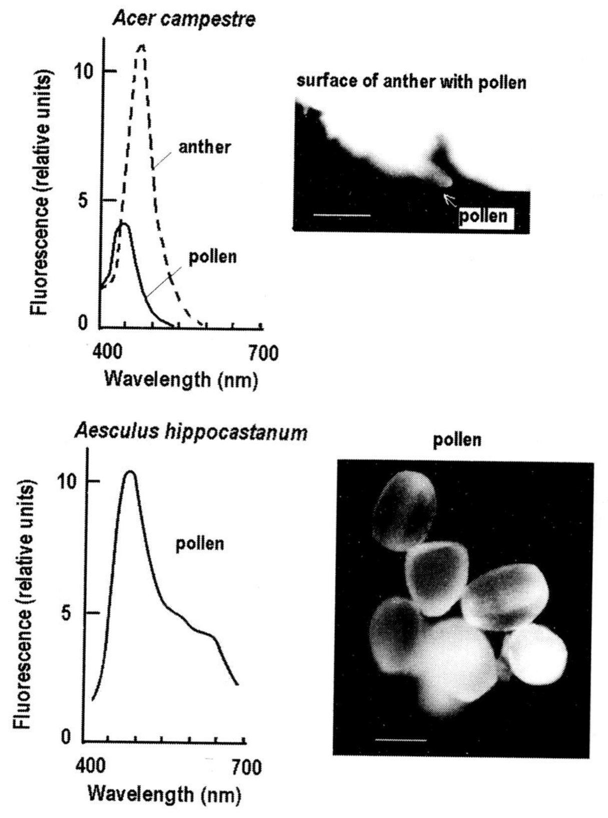

Fig. 2.7 The fluorescence spectra (left) and images under UV-light 360-380 nm in a luminescence microscope (right) of anther and pollen from *Acer campestre* and *Aesculus hippocastanum*. Adopted from Roshchina et al., (1997b; 1998a). Photograph has been made on film for air-shooting. Bar = 100 μm.

pollen varies in fluorescence intensity in different parts as seen under a usual luminescence microscope (Fig. 2.7) and under a laser scanning confocal microscope (Colour Fig. 9). Prickles fluoresce brighter, than other parts of the pollen exine which is seen clearly in colour Fig. 9 (Appendix 2). LSCM images also enables one to observe the pollen interior on the slice as seen for *Hippeastrum hybridum*.

The fluorescence spectra may demonstrate maxima in all parts of spectrum (Figs. 2.7 and 2.8). Many of the species have one maximum in blue as in *Artemisia vulgaris* (465-470 nm), in blue-green (480-500 nm) as in *Rosa canina* and green (500-530 nm) as in *Geum urbanum* and *Lycopersicon*

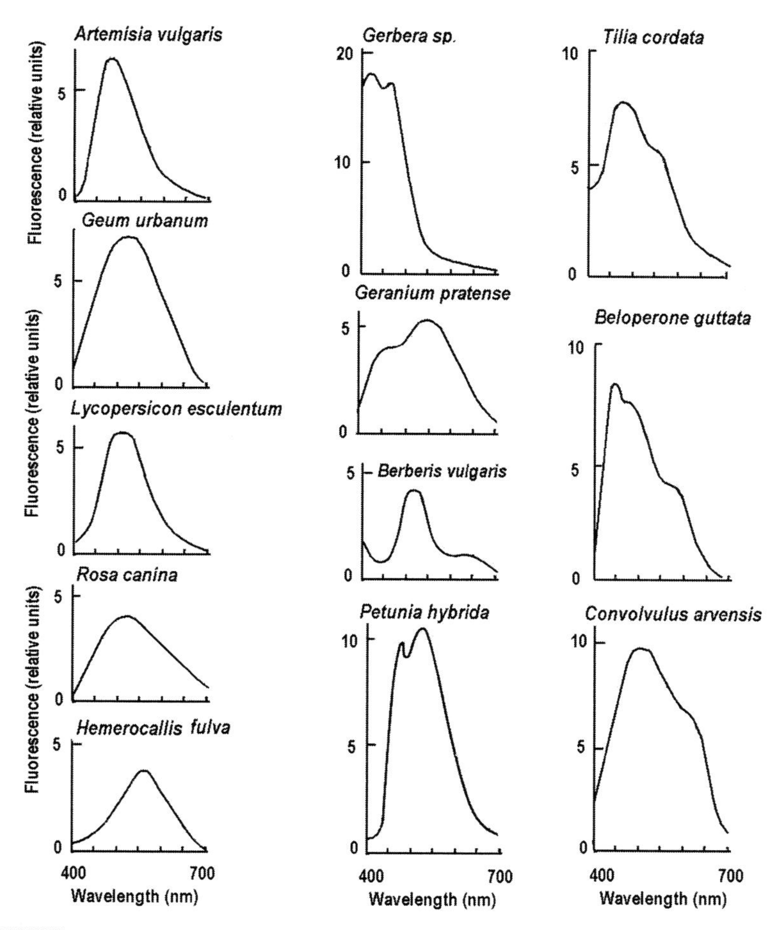

Fig. 2.8 The fluorescence spectra of pollen from various species

esculentum, or in yellow-orange as in *Hemerocallis fulva* (550-560 nm). Moreover, in Table 2.1, one can see unicomponent fluorescence spectra, peculiar to other species. Multicomponent fluorescence spectra, as shown in Fig. 2.8, also take place.

The decomposition of some spectra in Gauss curves demonstrates the presence of several fluorescing components. Pollen of *Gerbera* sp. has maxima 420 and 460 nm which is peculiar to terpenes and flavonoids, relatively (Roshchina et al., 1998a), while *Beloperone guttata* pollen demonstrates maximum 450 nm (terpenes), shoulder 500 nm (flavins) and maximum 570 nm (carotenoids and/or flavonoids). Other multicomponent fluorescence spectra show maxima 465-480 nm (flavonoids) and 535-560 nm (carotenoids) as for *Tilia cordata*, *Geranium pratense* and *Petunia hybrida* or maximum 530 nm (carotenoids, flavins) and 650 nm (azulenes) for pollen of *Berberis vulgaris*. Flavins and azulenes appear to contribute in the fluorescence of *Convolvulus arvensis*, because there are maximum 500-520 nm and shoulder 640 nm, relatively.

Fluorescenting pollen grains are clearly observed on the surface of anthers (Fig. 2.9) or the pistil stigma (Fig. 2.11), when they are excited by UV irradiation.

Anthers or pollen sacs, where pollen grains are forming, may have different maxima in the fluorescence spectra in comparison with those of pollen grains as can be seen for cacti examples (Fig. 2.9). Anthers of *Mammillaria dioica* K. Brand. or *Mammillaria sheldonii* Bod have maxima 530 or 480 and 545 nm, relatively, while both pollen peaks are 560 nm. Another picture is seen for cacti *Lobivia jajoiana* Backb. or *Echinocereus pentalophus* Rumpl., where anthers show maxima 485 nm with shoulder at 550 nm or 475 nm, but their pollen grains – 550 nm with small shoulder at 480 nm (Roshchina et al., 2000b). This may be due to a presence of different components in tectum (tissue, from which pollen is developing) of anthers. The maturing is often connected with a presence of carotenoids in anthers and pollen grains. It has been shown that the first two cacti from *Mammillaria* genus contain about 0.7-1 mg of carotenoid per 1 g of fresh weight whereas others – only traces (Roshchina et al., 2000b).

As seen from Table 2.2, pollen grains of various species have absorbance maxima in ultra- violet at (370-390 nm) as well as in violet (420-430 nm) and infrared (> 700 nm) spectral regions. Fluorescence excited by ultra-violet light 360-380 nm is in a visible spectrum depending on the plant species. The fluorescence maxima in the spectra are connected with

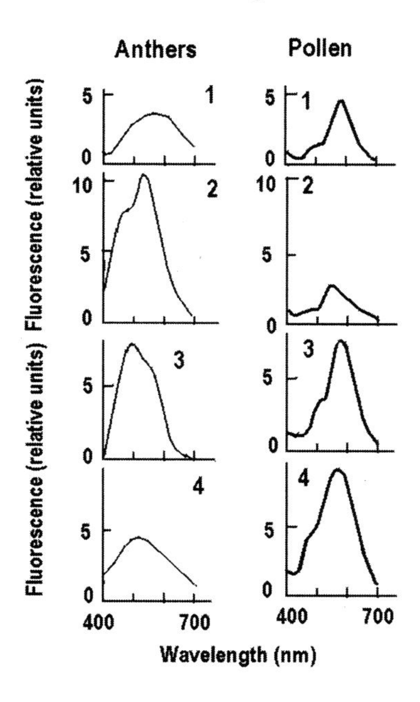

Fig. 2.9 The fluorescence spectra of anthers and pollen from various species belonging to the Cactaceae family. Adopted from Roshchina et al., (2000b). 1 - *Mammillaria dioica* K. Brand, 2 - *Mammillaria sheldonii* Bod., 3 - *Lobivia jajoiana* Backb., 4 - *Echinocereus pentalophus* Rumpl.

the presence of pigments in exine of the pollen grain – carotenoids, anthocyanins, flavonoids, azulenes and even chlorophyll, if the pollen is not matured.

The fluorescence of pollen is a sum of light emissions from rigid components of exine and liquid substances located in special channels of the microspore cover (see Figs. 1.2 and 1.5 in Chapter 1). Among them may be fluorescing componds such as phenols, carotenoids, azulenes, anthocyanes, etc (Roshchina et al., 1998a). The pigment composition and the position of the maxima in the fluorescence spectra of intact microspores may be changed during maturing of the pollen. Table 2.3 summarizes the pigment composition and the fluorescence maxima of studied pollen from various species and shows the variety of peaks, caused by different components of their surface and excreta. Their fluorescence spectra show maxima at 460-490, 510-550 and 620-680 nm. The first maximum could be related with phenolic compounds, the second one – with carotenoids (van Riel et al., 1983) and the third maximum – with

Table 2.2 The absorbance and fluorescence (λ excitation 360-380 nm) maxima in pollen of various plant species

Plant species (Latin names)	*Plant species (English names)*	*Maxima, nm*	
		Absorbance	*Fluorescence*
Amaryllidaceae *Hippeastrum hybridum*	Knight's star,	380, 434	480
Narcissus pseudonarcissus L.	Common daffodil	379, 430, 830	470, 675
Asteraceae Matricaria chamomilla *L.*	German camomile, wild camomile	380,450,855	435, 465, 570, 620
Betulaceae *Alnus* sp.	Alder	380, 434	520
Betula verrucosa Ehrh.	European white birch, weeping birch	379, 434, 841	440
Caprifoliaceae *Lonicera tatarica* L.	Honeysuckle	763	540
Caryophyllaceae Dianthus deltoides L	Maiden pink	379, 439	No
Geraniaceae *Geranium pratense* L.	Meadow geranium	380, 439, 855	430
Gramineae *Alopecurus pratensis* L.	Meadow foxtail	–	500, 680
Guttiferae *Hypericum perforatum* L.	Common St. John's-wort	385, 434, 846.	460, 550
Hippocastanaceae *Aesculus hippocastanum* L.	Common horse chestnut	379, 430, 870	470, 680
Hydrophyllaceae *Phacelia tanacetifolia* Benth.	Tansy phacelia	340, 442, 835	430
Iridaceae *Gladiolus sp.*	Gladiolus, cornflag	380, 439, 847	470
Crocus vernalis L.	Common crocus	–	550
Funkia sp. (Hosta).	Plantain lily	375, 440, 830	470
Liliaceae *Hemerocallis fulva* L.	Forrest's day lily, yellow day lily	379, 437,850	550, shoulder 620 470
Tulipa sp.	Tulip	380, 439, 849	470
Onagraceae *Chamerion angustifolium* (L.) Holub.	Fireweed	379, 434, 880	460
Papaveraceae *Papaver orientale* L.	Oriental poppy	379, 434, 847	470

Pinaceae *Pinus sylvestris* L.	Scotch pine	385, 454, 846	420
Plantaginaceae *Plantago major* L.	Ripple-seed plantain, great plantain	380, 843	540
Rosaceae *Malus domestica* Borkh.	Apple tree	380, 431	540
Rubus idaeus L.	Red raspberry, European raspberry	749	430

Sources: Roshchina et al., (1998a), Roshchina and Melnikova (1999)

Table 2.3 The pigment composition and maxima in the fluorescence spectra of intact pollen grains. Source: Roshchina et al., (1997b).

Plant species	*Total content of pigments, ng/mg of fresh mass*		*The fluorescence maxima of intact surface, nm*
	carotenoids	*azulenes*	
Aesculus hippo-castanum.	3.76 ± 0.02	6.0 ± 0.01	470, shoulder 580, 640
Hemerocallis fulva	11.1 ± 0.02	4.0 ± 0.03	550, shoulder 620
Matricaria chamomilla	6,3 ± 0,02	11.0 ± 0.04	465, 570, 620
Petunia hybrida yellow normal self-compatible clone	161,0 ± 0,06	5.8 ± 0.02	475, 550, 640
Petunia hybrida yellow self- incompatible clone	11.7 ± 0.03	traces	475, 550
Philadelphus grandiflorus Non-matured pollen	55.9 ± 0.05	traces	465-480, shoulder 665
Philadelphus grandiflorus Matured pollen	72.3 ± 0.06	6.0 ± 0.02	465-480, 510-520, 620

chlorophyll (Wolfbeis, 1985; Chappele et al., 1990) or azulenes (Murata et al., 1972; Roshchina et al., 1995). Azulenes are found in pollen of both Gymnosperms, in particular in *Picea excelsa*, and Angiosperms, for example, in *Aesculus hippocastanum, Matricaria chamomilla, Petunia hybrida, Philadelphus grandiflorus* and *Artemisia vulgaris* (Roshchina et al., 1995). Some examples are shown in Table 2.3.

There is a difference in the pigment composition between different clones of one and same species, for instance those belonging to *Petunia hybrida* (Table 2.3). Self-incompatible clone (in which the pollen is unable to participate in self-pollination) demonstrate lower concentrations of carotenoids and only traces of azulenes in comparison with a self-compatible clone. In this case, maximum 640 nm, peculiar to azulenes, in the fluorescence spectra of intact pollen grains is absent. This shows the necessity of the carotenoids and azulenes for pollen fertility.

The correlation between the amounts of carotenoids and azulenes and the arising of certain fluorescence maxima in pollen grains is also observed during the time that the pollen matures. In matured pollen, for example *Philadelphus grandiflorus*, the amount of carotenoids is higher, than in non-matured ones, while azulenes are present, mainly in matured microspores. Maxima 510-520 nm (it is likely to be linked with carotenoids) and 620 nm (peculiar to azulenes) arise only in matured pollen.

Pistils. A pistil is a structure, which consists of special pollen recognizing part known as a stigma, style and embryonal sac where the ovule is located. Abundant secretory ability is observed for a pistil stigma. The stigma of a pistil is a gland covered by specialized receptive cells able to recognize and to discriminate among pollen grains according to their genotype (Dumas et al., 1988). During interspecific matings it distinguishes "not self" i.e., pollen belonging to a species other than that of the pistil. This is generally rejected, for maintaining the stablility of the species. By contrast, in intraspecific mating, "non-self", which corresponds to allopollen, is accepted, while self-pollen is rejected. This latter process enforces outbreeding and characterizes "the self-incompatibility phenomenon". Photograph (Fig. 2.10) shows the fluorescence image of a pistil stigma and even of pollen grains lying on this stigma. After 2-5 min of the pollen addition (pollen from the same species) the fluorescence of the pistil surface increased significantly. It may be the response to the pollen from the same species. As will be shown in Chapter 5, the fluorescence intensity is also changed or not changed, if pollen grains from other species were applied to the pistil surface. Figures 2.11 and 2.12 demonstrate the use of microspectrofluorimetry which permits registering the fluorescence spectra of different parts of the pistil. Light emission along the pistil can be tested in detail, e.g. in tomato *Lycopersicon esculentum* (Fig. 2.11). The surface of the stigma has two maxima 460 and 525 or 595 nm (it depends on the degree of the opening of the pistil, the amount

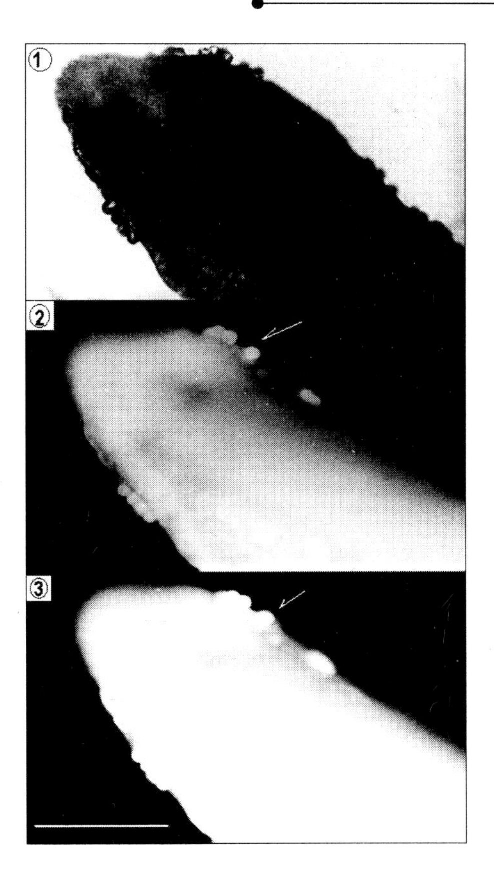

Fig. 2.10 The image in transparent light (1) and the fluorescent images of the pistil stigma of *Campanula grandiflora* L. and its pollen (shown with the arrows), lying on the stigma surface (2 and 3). Bar = 150 μm. 3 demonstrates the increased light emission from the pistil after 2-5 min of the addition of pollen grains from the same plant species.

of secretion released and is ready for fertilization). Pelliculae (1) on Fig. 2.11 release a secretion on the pistil stigma which leads to the appearance of the second maximum 590-595 nm in the fluorescence spectrum.The non-secreting part of the pistil stigma and parts of the style near the stigma (2-4) demonstrate similar spectra with the second maximum 530 nm. This shows that the secretions of pistil stigma contain compounds, which are not in non-secretory cells. Lower down to the style base, the maximum 525-530 nm becomes smoother. In basal one-seventh part of the style, the maximum gradually decreases and at last disappears. The

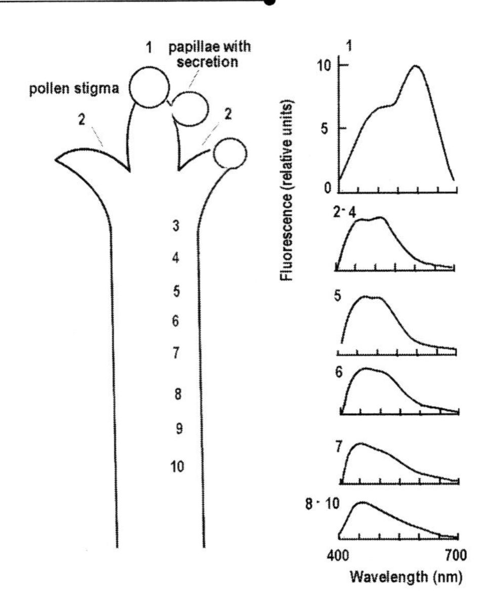

Fig. 2.11 The fluorescence spectra of the pistil of *Lycopersicon esculentum,* measured by moving the optical probe along the intact pistil from the stigma to its base. Adopted from Roshchina et al., (1998a). 1 — part of the pellicle of the stigma; 2 — the stigma surface; 3-10 — parts along the style.

microspectrofluorimetry registers the fluorescence spectra not only of a pistil, but even of pollen grains lying on its stigma (see Chapter 4).

Figure 2.12 demonstrates the fluorescence spectra of the pistil stigma from various plant species. Stigmas which were studied have one or two maxima of their spectra: one - in *Aesculus hippocastanum, Berberis vulgaris* or *Hemerocallis fulva* - (525, 560 or 475 nm, respectively), and two in *Petunia hybrida* (520 and 680 nm). It is clearly seen that the stigma pellicles may contain chlorophyll fluoresced with maximum 680 nm or not. While in pollen grain emission, numerous components are found. There are three maxima in both *Aesculus hippocastanum* and *Petunia hybrida* (500, 580, 620 and 450, 515, 650 nm, respectively). Pollen of *Berberis vulgaris* has two maxima at 525 and 645 nm, whereas pollen of *Hemerocallis fulva* has only one 565 nm maximum. Both the spectra forms and the maxima positions of the pistil fluorescence differ from those of the pollen grains. The difference depends on the nature of the species and the composition of the pollen grain surface. Blue fluorescence of pistils was also demonstrated earlier by Kendrick and Knox (1981). The observed changes in

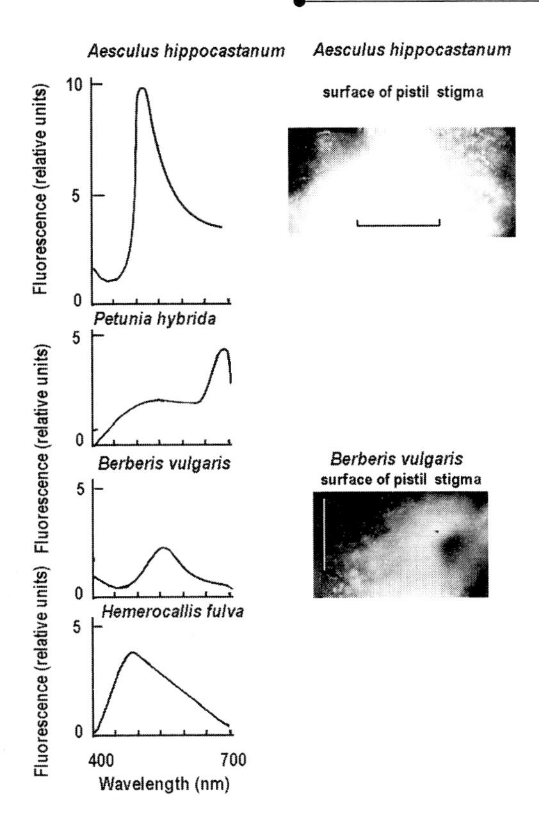

Fig. 2.12 The fluorescence spectra (left) and fluorescence images (right) of pistil stigma from various plants species. Adopted from Roshchina et al., (1998a). The photographs have been taken under a luminescence microscope (excitement by ultraviolet light 360-380 nm). Bars = 100 μm.

fluorescence in our work reflect the excretory and metabolic processes on the surface of the pistil and pollen. The comparison of the fluorescence maxima of pistils with others from different plant organs and cells is in Table 2.1.

In the secretion of the stigma, phenolic substances are present (Knox 1984) which can fluoresce in the same region of the spectra. Table 2.4 summarizes the pigment composition and the fluorescence maxima of studied pistils in different clones of *Petunia hybrida*, which differ in the content of carotenoids and azulenes. There is a difference of peaks, caused by different components of their surface and excreta. Their fluorescence

Table 2.4 The pigment composition and maxima in the fluorescence spectra of pistil stigmas in *Petunia hybrida*

Plant species	*Total content of pigments, ng/mg of fresh mass*		*The fluorescence maxima of intact surface, nm*
	carotenoids	*azulenes*	
yellow normal self-compatible clone	13.5±0.03	no	480, 520, 680
yellow self-incompatible clone	7.6±0.01	no	480, 520

spectra show maxima at 480, 520 and 680 nm. The first maximum could be related with phenolic compounds, the second one – with carotenoids and flavins (van Riel et al., 1983) and the third maximum – with chlorophyll (Chappele et al., 1990). The stigmas, depending on plant species, also contain phenols in the exudations (Martin, 1969; Knox, 1984), which fluoresce in the blue and yellow regions.

The approach to study the intact surface of plant generative cells could be the measurement of their fluorescence spectra at pollen germination and during pollen-pistil interaction (see Chapters 4 and 5).

Seeds. Some attention needs to be paid to the fluorescence of the formed embryo and seed as a whole, which arise after fertilization, although the emission of glandular cells will be specially considered in Section 2.2.2 of Chapter 2. The fluorescent pictures of fruits and seeds were studied in the laboratory of Zobel (Zobel and Brown, 1989; Zobel and March, 1993). Differences were shown in localization of fluorescent psoralens in fruits and seed coats of *Angelica archangelica, Daucus carota, a Sium suave* and *Psoralea bituminosa* (Zobel and March, 1993). All the fruits examined contained furanocoumarins on the surface. Fruit tissues contained very low *(Sium)*, medium *(Angelica)* or high *(Psoralea)* concentrations of these compounds. Seed covers showed low autofluorescence in *Angelica* but high in *Sium*. Not all the glands in the fruit or seed tissue exhibited autofluorescence. Embryos always contained furanocoumarins on their surface and in their tissues in varying proportions. Autofluorescence can be homogeneously distributed among all the cells of the embryo *(Sium)* or can be concentrated in certain cells *(Psoralea)*. The fluorescence in each case was compared with that of crystals and saturated solutions of psoralen, bergapten and xanthotoxin under similar conditions. Autofluorescence was observed after 10 min of irradiation. Embryos were removed from the covering tissue after 1 h imbibition as well as from dry unimbibed fruits, and autofluorescence on their surfaces was observed. Lignins can also give autofluorescence but it is of a blue colour, and these compounds would not be located on the surface of the embryo or seed, as they impregnate the secondary cell wall, mainly with tracheids and fibres. The fruit tissue showed very little lignification when a histochemical reaction characteristic of lignins (phloroglucinol-HCl) was applied. Reaction with chlor-zinc-iodine showed that the cell walls contained mostly cellulose. Although the authors (Zobel page 68) did not identify all the compounds deposited on the surface, it is most likely that the observed autofluorescence is due at least partially to furanocoumarins. These coumarins were removed in substantial concentrations by dipping fruits, seeds and leaves in almost-boiling water (Zobel and Brown, 1989). In *Daucus carota* yellow-green fluorescence was observed on the surface of trichomes and ground epidermis. The surface of the embryo showed a little stronger autofluorescence, than the rest of the cells of the embryo. A similar localization of fluorescence was observed in *Psoralea bituminosa Psozalea*, except for the embryo cells, which contained a distinct group of cells with brighter fluorescence than the others. Morphologically these cells do not differ from the neighbouring ones. In the latter species some glands were lined with yellow fluorescence along the walls. On the border between the embryo and the seed coat the fluorescing layer was more distinct.

2.2.2 Secretory Structures in Non-reproductive Organs

Secretory structures of non-reproductive organs are distinguished by external and internal tissues because secretions may be released from the organism or excreted within the body. Glandular cells, glands, glandular hairs and reservoirs of the secretions are present in external tissues, which are released by surrounded glandular cells, as well as nectaries and hydathodes (Vasilyev, 1977; Fahn, 1979; Roshchina and Roshchina, 1989; 1993). Secretory cells of glands and glandular hairs on flowers and leaves differ in the fluorescence spectra from the non-secretory cells of both the epidermis and leaf mesophyll. Sometimes idioblasts (single secretory cells) among non-secretory cells are also found in such tissues. Unlike external secretory organs, internal secretory structures include internal reservoirs and ducts, laticifers and idioblasts. The reservoirs filled with secretions are formed from intercellular spaces between living cells (schizogenous reservoirs) or arise instead of decomposed cells (lysogenous reservoirs).

Some of the reservoirs may be filled with mucilage, while lisogenous ones with of essential oils, especially in fruits such as those from the genus *Citrus*. Resin ducts and channels are characteristics of conifers, and sometimes they are found in the family Asteraceae. The structures are represented as long tubular intercellular space filled with resin.

We shall consider the fluorescence of some secretory structures of both types, based, mainly, on the chemical composition of the secretory cells.

2.2.2.1 Main features of glandular cells in various organs

Among glandular structures are glandular cells, glands, glandular hairs and reservoirs of the secretions, which are released by the surrounding glandular cells. Some schematic examples of the structures are given in Appendix 1. Glandular hairs differ in their anatomy and are distinguished as trichomes and emergencies. Trichomes are the derivatives of endoderm formed without participation of low-lying tissues, while in the structure of emergencies may be lower, deeper tissues. Glandular trichomes may have a head attached to a stalk (pedicel) , in this case called "capitate", or have no head and is called "non-capitated". Glands are usually sessile as seen in Appendix 1 or can also have a multicellular head on the pedicel (Vasilyev, 1977; Fahn, 1979; Roshchina and Roshchina, 1989; 1993). The examples of the fluorescing secretory structures seen in ultra-violet light (360-380 nm) of luminescence microscope are represented in some papers (Roshchina and Melnikova, 1995; 1999) for non-capitate trichomes of *Betula verrucosa*, emergencies *of Urtica dioica* and for the gland of *Mentha piperita*.

Comparing the fluorescence of the secretory cells in all organs in the same Angiosperm species is of special interest. Figure 2.13 demonstrates one of the examples (The fluorescence spectra in terpenoid-enriched species *Calendula officinalis*). In a blossoming plant there is emission maxima in blue at 440-450 nm for secretory cells from both ligulate (also has shoulder at 530-560 nm) and tubular flowers as well as in the husk of basket-like inflorescence and in the leaf glandular cells and hairs. Sometimes the yellow emission with maximum 530-550 nm occurs, especially expressed in the glandular hairs of the inflorescence husk. The presence of terpenoids such as sesquiterpene glycosides, in particular apicubebol (Ahmed et al., 1993), and various sesquiterpene lactones (Rodriguez et al., 1976; Kelsey and Shafizadeh, 1980; Fischer, 1991) as well

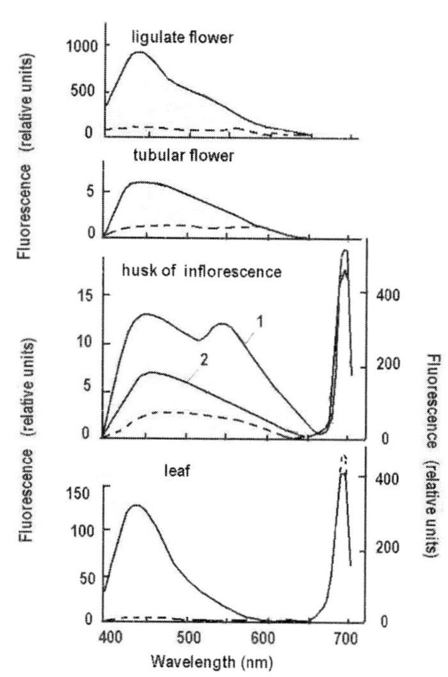

Fig. 2.13 The common view (left) and the fluorescence spectra (right) of secretory cells of *Calendula officinalis.* Unbroken line – secretory glandular cell of different structure; broken line – non-secretory cell. The separated scale was given for chlorophyll fluorescence at 680 nm. 1- capitate trichome; 2- non-capitate trichome.

as saponins (Plant Resources of Russia and Surrounded Countries, 1996) and salicylic acid (Golovkin et al., 2001) may be related to the emission. The contribution of chlorophyll is also seen in the inflorescence husk and leaf, although the parts of secretory trichomes, which lie at a margin, out of the red-fluorescing surface have no chlorophyll lightening.

There is a difference between the fluorescence spectra of various types of secretory glandular hairs of the same plant surface, in particular between capitate and non-capitate hairs of *Calendula officinalis*. As seen in Fig. 2.14, capitate and non-capitate trichomes have been distinguished not only in morphology and anatomy, but also in the chemical composition of the glandular secretions. Capitate hair demonstrates yellow fluorescence with maximum 565 nm, whereas non-capitate hair – blue-green fluorescence with smooth maximum 505 nm. There are two different types of secretory

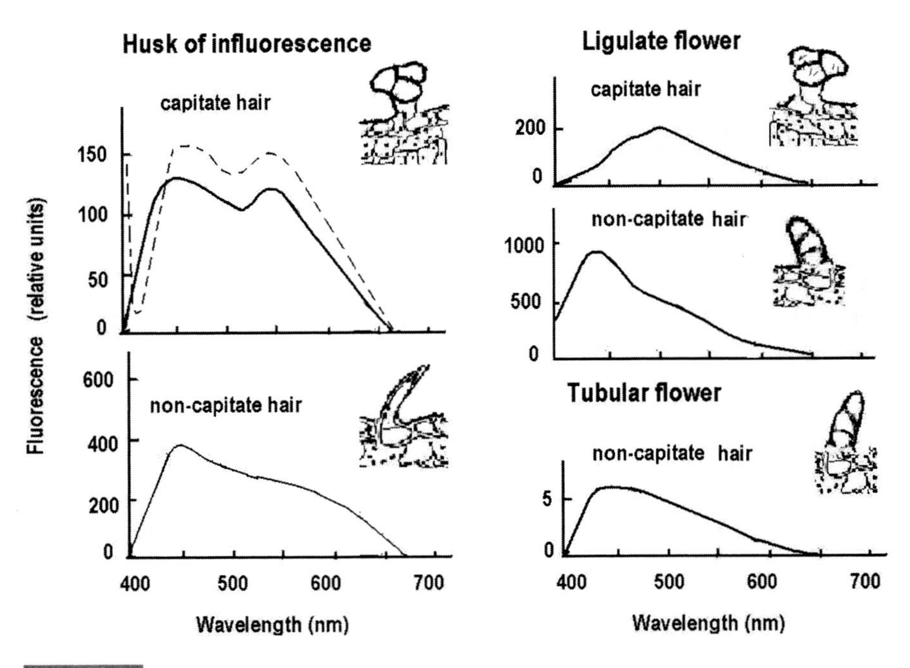

Fig. 2.14 The fluorescence spectra of glandular hairs on husk or petal of the ligulate flower from the basket inflorescence of *Calendula officinalis*. Left- capitate hair, Right – non-capitate hair. Unbroken line – glandular hair, broken line – non-secretory cell.

hairs on the surface of *Calendula officinalis* - capitate and non-capitate as well as non-capitate tubular (Fig. 2.14). Depending on the organ or tissue, they fluoresce in the blue or yellow spectral regions. Capitate hair on the husk of basket inflorescence has two maxima in the fluorescence spectra – 450 and 550 nm while non-capitate hair only one maximum 450 nm and shoulder at 550-600 nm. Non-capitate hairs of ligulate and tubular flowers fluoresce in blue and have only one maximum at 450 nm, and sometimes shoulder in the yellow region. But their capitate hairs emit, mainly in the yellow spectral part, with only one maximum 550 nm. Unlike the capitate hairs of the husk, one can see that yellow-fluorescing hairs of petals of both ligulate and tubular flowers have one maximum in the fluorescence spectra. This difference shows the various compositions of the secretions of the trichomes. Sometimes the fluorescing excretions are seen as crystals on the tissue surface. They are blue- or yellow fluorescing. The crystals are often observed on the surface of the head and on the pod (Base) of the capitate trichome (Fig. 2.14), but not on non-capitate trichomes. In secretory cells the 680 nm emission maximum typical of the chlorophyll is small or nearly absent.

The spectra of fluorescing glandular structures of leaves in *Mentha piperita*, *Lycopersicon esculentum*, and *Artemisia vulgaris* exhibit two or more maxima of the emission which show the different components in their secretions whereas non-secretory cells have only one maximum 680 nm which is typical of the chlorophyll (Table 2.1). The glands and glandular hairs of the species contain mainly monoterpenes (*Mentha*) or a mixture of terpenoids (*Lycopersicon*), or mixtures of sesquiterpenes (*Artemisia*) (Table 2.1). In the *Mentha* and *Artemisia*, spectra have three maxima: 475 nm (monoterpene alcohol menthol has a weak fluorescence at 440 nm), 525 nm (flavins) and 680 nm (chlorophyll), and 475 nm maximum (sesquiterpene lactones fluorescence in the blue region), 550 nm shoulder, and 680 nm maximum, respectively.

2.2.2.2 Oil cells, glands and reservoirs

Some glandular cells in the overground plant parts can accumulate oils, which contain, mainly terpenoids, and in a smaller degree phenolic compounds (Roshchina and Roshchina, 1989; 1993). Sometimes their secretions are so abundant that excreted in the spaces between cells forms oil reservoirs and even ducts.

Oil cells and glands of flowers. The fluorescence spectra of flower secretory cells of various species differ significantly (Fig. 2.15). The complex emission of the flower glandular cells may correlate with their contents. Contrary to the non-secretory cell, there is no 680 nm chlorophyll maximum or very small maximum. Terpenoids in essential oil-containing cells of *Rosa canina* and *Tagetes* spp., are perhaps, responsible for the 475 nm maximum of their fluorescence spectra. The glands on red petals of *Salvia splendens* flower shine mainly in the region 450-460 nm with maximum 465 nm and shoulder 550 nm whereas non-glandular cells have no fluorescence in this region. The emission is also due to monoterpenes of the essential oils Table 1.2 (Chapter 1). The position of the emission maximum varies in different *Tagetes* species. Secretory cells of ligulate flowers of *Tagetes patula* have fluorescence maximum at 600 nm, whereas those of *Tagetes erecta* – at 500 nm. There is also a difference between secretory cells of ligulate and tubular flowers of *Tagetes patula*. The former has maximum 600 nm, whereas the latter – 615 nm. Both maxima

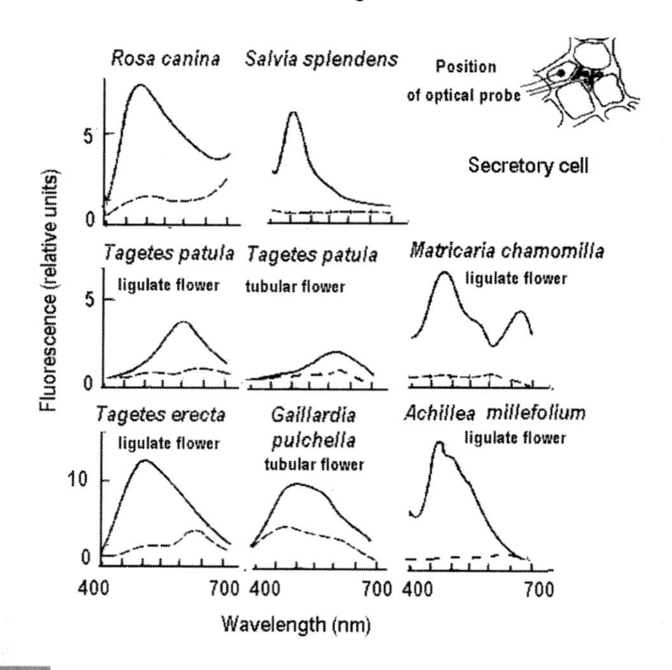

Fig. 2.15 The fluorescence spectra of petal glandular cells from flowers. Above - the position of optical probe on the non-secretory (light-drown cells) and secretory (dark-drown) cells. Below - the fluorescence spectra of the secretory cells (unbroken line) and non-secretory cells (broken line). Adopted from Roshchina et al., (1998a).

could belong to polyacetylenes, which occur in the species (Roshchina and Roshchina, 1993). Sesquiterpenes-containing glands, on a surface of ligulate flowers in *Achillea millefolium* and *Matricaria chamomilla* have fluorescence spectra (Table 2.1, Fig 2.15) with three significant maxima: 465, 500, and 535 nm and 460, 550 and 660-675 nm, respectively. The latter maximum in the red region may be due to the light emission of chamazulene (prochamazulene) or/and chlorophyll (Table 1.1, Chapter 1). Tubular flowers of *Tanacetum vulgare* and *Gaillardia pulchella* (Table 2.1) demonstrate the secretory structures, differing in the fluorescence spectra form. Secretory cells of first plant species show high maximum 475 nm and shoulder 600 nm (azulenes are found in the herb), whereas second species shows moderate 475 nm maximum (sesquiterpene lactones) and 580 nm shoulder (flavonoids).

Oil cells and reservoirs in leaves and sepals. Oil is found in secretory hairs such as in the secretory hair of *Artemisia* leaf, or in glands of the surface of *Pelargonium hybrida* or can be accumulated in oil reservoirs formed by surrounded glandular cells as in *Solidago virgaurea* which are seen from their fluorescence spectra (Fig. 2.16). Secretions of the cells and reservoirs include components, which emit in blue (at 420-450 nm as in *Solidago* or at 450-475 nm as in two other species) and in the green-yellow (the shoulder at 530-550 nm in *Artemisia* and the maximum 540-550 nm in *Pelargonium*) spectral regions. Sometimes maximum 680 nm, peculiar to chlorophyll is seen in the spectra. Non-secretory cells have mainly maximum 680 nm. Terpenoid-containing cells of many secretory structures contain terpenoid-enriched oils and are often found in the families of Asteraceae and Geraniaceae (the examples are seen in Table 2.1). Similar species such as *Artemisia vulagaris* and *Pelargonium hybrida* contain essential oils and oleoresins (phenols mixed with oil). Maxima in the blue region may be connected with terpenes, which fluoresce at 420-430 nm, whereas maximum 465-485 nm – to phenols. In *Artemisia* genus, among components of essential oils 1, 8-cineol (8.6 – 26.6%), carvacrol (2.2%), hexadecanol (~18.1%), hexadecanoic acid (11.2%), davanone (15%) prevail (Bicchi et al., 1985), but, according our data (Roshchina, 2003), the components have a weak fluorescence, if any. An emission of triterpenes and sesquiterpene lactones is more realistic (Table 1.1, Chapter 1). Genus *Pelargonium* contains terpenes and terpene alcohols such as α- and β-pinenes, linalool, geraniol, eugenol, borneol (Golovkin et al., 2001), which weakly fluoresce in UV-light. The maximum of fluorescence at 560-580 nm may belong to some phenols (Roshchina and Roshchina, 1989; 1993).

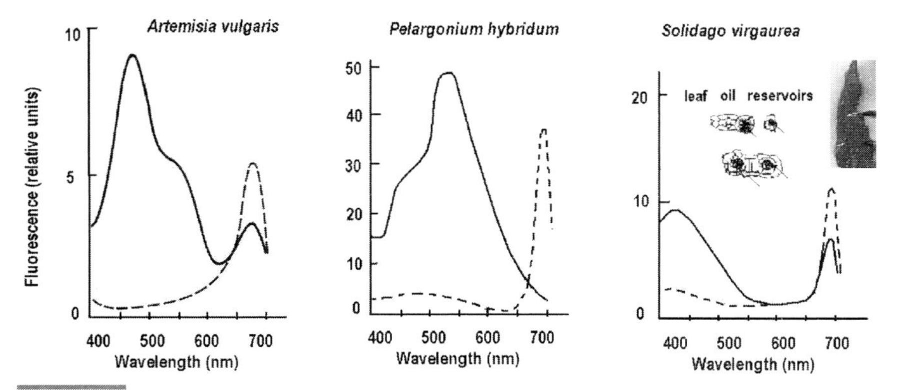

Fig. 2.16 The fluorescence spectra of oil–containing structures on leaf. Unbroken lines – secretory hair of leaf from *Artemisia vulgaris*, oil gland on leaf of *Pelargonium hybrida* and oil reservoirs on leaf of *Solidago virgaurea*, respectively. Broken line - non-secretory cells.

Internal secretory reservoirs storing oils and resin are widespread in the Asteraceae family, and one example is demonstrated for *Solidago virgaurea* L. and *Solidago canadensis* L. (Lersten and Curtis, 1989). As seen from Table 2.5, the fluorescence intensity in the glandular cells enriched in oils is higher, than in surrounded non-secretory cells. Main maxima related to terpenoid were in blue at 410-428 nm or in blue-green at 460-528 nm, if the emission, which was excited by ultra-violet light or violet light, respectively. There is a difference between species: the maxima of the oil reservoir emission for *Solidago canadensis* is in green, while for *Solidago virgaurea* – in blue. This is possible due to the different fluorescing components. In first species the effects may be related to flavonoids prevailing in oil mixtures of terpenoids with phenols. Actually, cell walls

Table 2.5 The maxima of fluorescence and its intensity in leaf oil reservoirs or in flower petal oil cells from *Solidago* genus measured by double beam microspectrofluorimeter (see Chapter 1).

Cell	*Maxima, nm*	$I_{520-540}$	$I_{640-680}$
	Solidago canadensis L		
Upper leaf side	410, 428, 680 (468, 680)	0.14 ± 0.027 (0.17 ± 0.05)	0.65 ± 0.12 (8.4 ± 0.50)
Oil reservoir of upper leaf side	515, 528	0.21 ± 0.03 (0.26 ± 0.02)	2.34 ± 0.7
Secretory hair of upper leaf side	528	0.27 ± 0.02	2.14 ± 0.06
Cell walls of stomata guard cells (formed the chink) on lower leaf side	470-480 (490-500)	(0.83 ± 0.02)	(9.85 ± 0.025)
Non-secretory part of upper leaf side	680	0.02 ± 0.007 (0.06 ± 0.004)	0.67 ± 0.13 (3.73 ± 0.90)
Non-secretory part of lower leaf side	410, 468, 680	(0.09 ± 0.020)	(3.12± 0.2)
	Solidago virgaurea L.		
Oil reservoir of leaf	410, 428, 680 (468, 680)	0.23 ± 0.06 (0.26 ± 0.06)	0.44 ± 0.06 (7.24 ± 0.9)
Secretory cells surrounded by the oil reservoir	410, 428, 680	0.20 ± 0.05 (0.25 ± 0.03)	0.8 ± 0.23 (7.28 ± 0.06)
Non-secretory cells of leaf	680	0.05 ± 0.009 (0.07± 0.037)	0.21 ± 0.04 (7.48 ± 1.0)
Cell walls of stomata guard cells (formed the chink)	470-480 (490-500)	0.09 ± 0.01	0.83 ± 0.2
Flower oil cell	410, 428, 680 (468, 680)	0.26 ± 0.01	0.44 ± 0.03

Excitement 360-380 nm (in brackets excit. 420-436 nm)

of stomata impregnated with phenols also emit in green. The non-secretory part of the leaf fluoresces in red which is related to chlorophyll. The fluorescence intensity of leaf oil-containing reservoirs was similar with oil cells of flower petals.

Oil cavities and ducts of *Hypericum perforatum* (Fig. 2.17) are seen on the leaf surface as dark spots. A cross-section of the leaf demonstrates the cavity filled with dark secretion, perhaps containing anthraquinone hypericin (Curtis and Lersten, 1990). The colour-less essential oil–enriched reservoirs in leaves and flower petals exhibited blue luminescence with maximum 460 nm. However, blue-coloured glands and reservoirs had only a weak fluorescence. A pure anthraquinone derivative, known as hypericin, was derived from the blue spot of the substance (Roshchina and Melnikova, 1995).

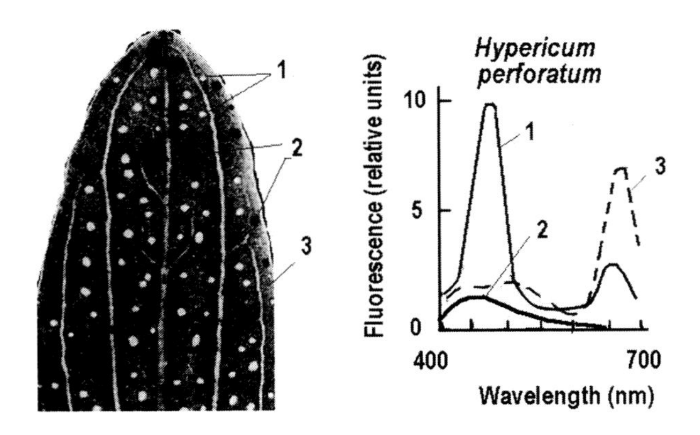

Fig. 2.17 The leaf image (left) and the fluorescence spectra (right) of the glandular cells of *Hypericum perforatum* containing phenols and quinones. 1 and 2 –secretory light and dark blue glands, respectively, 3 – non-secretory cells.

Oil cavities are peculiar to roots of some species belonging to the genera *Cicuta* (family Umbelliferae). In the secretory structures yellow oil contains polyacetylenes, in particular cicutotoxin and its derivatives (Anet et al., 1952; 1953). Cicutotoxin has an absorbance with maxima 230, 262, 275, 298, 310 and 355 nm and fluoresces with maximum 575-580 nm, which is peculiar to a conjugation of triplet (three -ene) -and double bonds (Roshchina et al., 1980; 1986).

Oil-containing cells in fruits. Various secretory structures, which include oils, are found in fruits of species belonging to the families

Solanaceae, Rutaceae, Berberidaceae, Palmae and others. They differ in the chemical composition. As seen in Fig. 2.18, the fluorescence of fruit secretory cell depends on the plant species and chemical composition.

Secretory cells of fruit from red pepper *Capsicum annuum* (Solanaceae) fluoresce with maximum 525 nm (due to the presence of lipophilic alkaloid capsaicin soluble in oil medium) whereas seeds and yellow-red non-secretory cells – with maximum 550 nm (fluorescence of carotenoids), although the latter also demonstrates a shoulder at 480 nm. Alkaloid capsaicin found in secretory cells of *Capsicum annuum* fruits has maxima in absorbance and fluorescence spectra, relatively 280 nm and 460 nm, like many phenols and phenol derivatives as well as alkaloids. The water and ethanol solutions of capsaicin fluoresce in blue – at 450-470 nm.

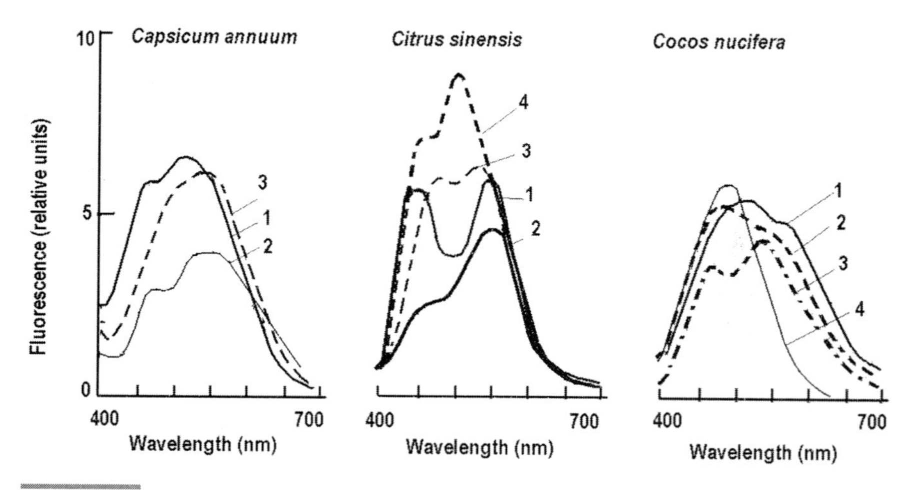

Fig. 2.18 The fluorescent spectra of secretory cells of fruits. Red pepper, Cayene pepper *Capsicum annuum* L.(Solanaceae) 1- secretory cell of internal surface of the fruits with seed-stalks; 2- non-secretory cells; 3- seeds; sweet orange *Citrus sinensis* (L.) Osbeck (Rutaceae). 1 and 2- glands on external and internal surface of fruit; 3-4 non-secretory cell of external surface of the fruit); cocos *Cocos nucifera* L. (Palmae). 1. The lignified cell of external fruit surface; 2- seeds; 3 and 4- secretory and non-secretory cells of the fruit pulp.

Unlike pepper, secretory (glandular) cells of external and internal surface of fruit from sweet orange *Citrus sinensis* have two different maxima in the fluorescence spectra at 440 (perhaps, fluorescence of terpenoids in oil gland) and 550 nm (the carotenoid contribution in the emission), while seeds – at 440 and 510 nm. In lignified fruit from *Cocos nucifera* we see bright fluorescence of fibrills on the surface of the fruit cover with two maxima 475 and 550 nm, while the lightening secretory cells and non-secretory cells of pulp have only one maxima at 475 nm and at 445 nm, relatively. Cells of mesocarp have two maxima in the fluorescence spectra at 450 nm and 540 nm.

Psoralens may be also present in oil-enriched fruits of *Angelica archangelica, Daucus carota, Sium suave* and *Psoralea bituminosa* (Zobel and March, 1993). *Sium* and *Angelica*, where they have bright blue fluorescence (Zobel and March, 1993). In fruits of *Berberis vulgaris* the characteristic fluorescence in green and yellow may be related to berberine (Roshchina et al., 1997a). Alkaloid of berberine from secreta of *Berberis* fruits (family Berberidaceae) emits with maximum at 545 nm in UV-light (540-580 nm depending on the wavelength of excitation) due to conjugation of the phenolic rings (Tables 1.1 and 1.2) Chapter 1.

The comparison of the fluorescence from different glandular oil-enriched structures. The oil-containing cells producing terpenoids are found in many genera and fluoresce, mainly, in the blue spectral region (Table 2.1). For example, in the genus *Tagetes* oil glands on leaves and husk of bracket blossom fluoresce with maxima 410 and 430 nm (characteristic emission region of the terpenoids, mainly monoterpenes), and chlorophyll at 680 nm also contributes. The oil glands of husk and leaves fluoresce in blue (maxima 408-410 and 430 nm) when excited with UV-light 360-380 nm or in blue-green (maxima when excited by violet light 420 nm. Yellow –fluorescing crystals of terpenoids are also found on the surface. Oil-enriched cells of leaves and crystals within the oil glands and reservoirs fluoresce in blue or blue-green 4-5 times more intensively, than surrounded non-secretory cells, which emit, mainly, in the red spectral region (Table 2.6). Sometimes, even yellow-fluorescing crystals are found in the oil structures, and their bright emission permits one to see them within the secretory cells. Fluorescent components appear to be flavonoids, which can emit in the blue and green-yellow spectral regions. Moreover, the phenolic compounds may be mixed with terpenoids in oil cells (Roshchina and Roshchina, 1989; 1993). Secretory oil glands of the *Tagetes* genus include terpenoids, polyacetylenes and phenols. *Tagetus erecta* contains sulphur-containing tiophenes terthienyl, flavonoid kaempferitrin, while *T. patula* - flavonoids patuletin and luteolin (Golovkin et al., 2001; Plant Resources of Russia and Surrounded Countries, 1996). Terpenoids and most polyacetylenes fluoresce in blue – at 420-450 nm whereas phenols, in particular some flavonoids (quercetin) may also have second maximum in the yellow-orange spectral region at 570-600 nm (Roshchina and Melnikova, 2001). In any case, we can distinguish bright fluorescence in secretory structures. The staining with $FeCl_3$ as reagent on the phenols showed the presence of phenolic compounds in the oil reservoirs and even crystals.

Another example of terpenoid location in oil-enriched secretory structures in the plants *Juglans mandshurica* and *Tanacetum vulgare* are demonstrated in Table 2.7, which shows the sharp difference between the fluorescence secretory structures in the green-yellow (or blue-green if the excitement 360-380 nm) and red spectral regions. We can see how

Table 2.6 The fluorescence intensity of oil-containing cells from plants of *Tagetes* genus

Cell	$I_{520\text{-}540}$	$I_{640\text{-}680}$
Tagetes erecta L.		
Oil gland of husk	0.24 ± 0.061 (0.73 ± 0.08)	0.71 ± 0.16 (2.7 ± 0.39)
Surrounded non-secretory cells	0.1 ± 0.008	2.9 ± 0.19
Oil gland of flower ligulate petal	0.06 ± 0.006 (0.05 ± 0.004)	0.05 ± 0.008 (0.07 ± 0.009)
Surrounded non-secretory cells	0.03 ± 0.006 (0.03 ± 0.006)	0.05 ± 0.006 (0.05 ± 0.01)
Crystals on the petal surface	0.24 ± 0.09	0.23 ± 0.05
Tagetes patula L.var.orange		
Oil gland of husk (upper side)	0.24 ± 0.061	0.71 ± 0.16
Oil gland of leaf	0.5 ± 0.02	7.12 ± 0.5
Surrounded by non-secretory cells	0.05 ± 0.012	2.18 ± 0.13
Oil gland of ligulate flower petal (upper side)	0.03 ± 0.005	0.08 ± 0.07
Surrounded non-secretory cells (upper side)	0.03 ± 0.002	0.05 ± 0.006
Crystals on petal upper surface	0.62 ± 0.04	0.61 ± 0.02
Non-secretory cells on lower part of petals (middle part)	0.02 ± 0.003	0.41 ± 0.04
Non-secretory cells on lower part of petals (red-fluorescing cell walls)	0.04 ± 0.005	0.34 ± 0.03

Excitement 360-380 nm. (in brackets-excitation by 420-436 nm)

secretory hairs and other secretory cells brightly fluoresce. Intensive emission is especially seen for the crystals within secretory cells. Non-secretory cells emitted in green with lower (several times) intensity than secretory ones, while the opposite picture is marked for the red spectral region. *Juglans mandshurica* contain terpenoids bornyl acetate, α- and β-pinenes, camphor, terpineol, as well as flavonoid quercetin, phenolcarbonic acids and their derivatives (Golovkin et al., 2001; Plant Resources of Russia and Sourrounded Countries, 1996). Phenols may be mixed with oil terpenoid components (Roshchina and Roshchina, 1989; 1993) and emit in the green spectral region. In pericarp of *Juglans mandshurica* chlorogenic, ferulic and other aromatic acids are found, which may fluoresce in this region. Secretory cells of *Tanacetum vulgare* contain monoterpenes borneol, geraniol, camphor, thymol, α- and β- pinenes, sesquiterpenoids (azulene, artemisiaketone, chrysantenone, partenine, tujone and others) and phenols elemicine, caffeic acid, flavonoids quercetin, luteolin, and jaceosidin in oil (Golovkin et al., 2001; Plant Resources of Russia and Surrounded

Table 2.7 The fluorescence intensity of oil-enriched secretory cells on leaves of *Juglans mandshurica* Maxim. and *Tanacetum vulgare* L

Cell	$I_{520-540}$	$I_{640-680}$
Juglans mandshurica		
Secretory green-fluorescing leaf hairs	0.18 ± 0.06 (0.68 ± 0.03)	4.08 ± 0.6 (9.39 ± 0.6)
Secretory green fluorescing leaf cell with crystal	0.91 ± 0.018	9.58 ± 0.75
Non-secretory red fluorscing leaf cells	0.07 ± 0.005 (0.28 ± 0.02)	5.16 ± 0.5 (2.99 ± 0.9)
Secretory cells of the fruit peel surface	0.15 ± 0.02 (0.33 ± 0.09)	0.74 ± 0.10 (3.3 ± 0.70)
Branched yellow-fluorescing gland from bundle secretory cells on the fruit peel surface	(0.54 ± 0.09)	(5.58 ± 0.90)
Non-branched yellow-fluorescing gland of the fruit peel surface	(0.72 ± 0.08)	(5.37 ± 0.90)
Non-secretory cells of the fruit peel surface	(0.55 ± 0.09)	(8.7 ± 0.80)
Tanacetum vulgare L.		
Green-fluorescing glandular hairs	0.58 ± 0.028	5.25 ± 0.6
Non-secretory red- fluorescing cells	0.17 ± 0.03	6.59 ± 0.9 (7.48 ± 1.0)

Excitement 360-380 nm. (in brackets excit. 420-436 nm)

Countries, 1996). Their intensive fluorescence is connected, mainly, with phenols.

As seen from Table 2.1, glandular trichomes of *Solanum tuberosum* contain green-yellow fluorescing secretions with maxima 465 and 540 nm or 525 nm. The common image of the lightening hairs is shown in earlier publication (Roshchina et al., 2000a) in Chapter 1. The trichomes protect the plant against aphids and other insects due to the inclusion of the phenols (Gibson, 1974; Wagner, 1990) and enzyme polyphenol oxidase (Ryan et al., 1982; Kowalski et al., 1992), which can oxidize the compounds, forming active oxygen species – free radicals and peroxides. Polyphenol oxidase, localized both intra- and extracellulary, oxidizes phenols to quinones at the expense of O_2 and utilizes H_2O_2 to oxidize a wide array of phenolic substrates. Among the substrates are chlorogenic, caffeic, hydrocaffeic and ferulic acids and, in a smaller degree, dioxyphenylalanine. The oxidation may influence the position of the maxima in the fluorescence spectra and colour of the emission.

2.2.2.3 Glandular structures enriched in special compounds

Glandular structures may be enriched in some special secondary metabolites, which prevail in the types of the secretory cells. For instance, among them are terpenoids as sesquiterpene lactones, phenols as flavonoids or alkaloids, etc. Some of the examples of their fluorescence will be represented below.

Terpenoid-enriched glandular structures. Table 2.1 demonstrated the maxima of the fluorescence from glands and glandular hairs of the species containing various terpenoids. For instance, there are, mainly, monoterpenes *(Mentha)* or a mixture of terpenoids *(Lycopersicon),* or mixtures of sesquiterpenes *(Artemisia).* In the *Mentha* and *Artemisia,* spectra have three maxima: 475 nm (monoterpene alcohol menthol has a weak fluorescence at 440 nm), 525 nm (flavins) and 680 nm (chlorophyll), and 475 nm maximum (sesquiterpene lactones fluorescence in the blue region), 550 nm shoulder, and 680 nm maximum, respectively. Triterpenoids are located in leaves and flowers from *Agrimonia pilosa,* mainly tetrahydroursenic acid and its derivatives (Koun et al., 1988), and in those from *Cucumis sativus,* respectively amyrins and cucurbitacines (Golovkin et al., 2001). The fluorescence (Fig. 2.19) is observed in their glandular cells – glands and hairs. In *Agrimonia,* one maximum at 455-460 nm may be peculiar to triterpenoids and flavonoids, such as apigenin

(Golovkin et al., 2001). Besides, vanillic acid, catechins and tannin ellagic acid may contribute to the fluorescence. Maximum 680 nm in the red spectral region shows the contribution of chlorophyll. In *Cucumis*, the fluorescence spectra differ between secretory hairs from the leaf and petal. Unlike secretory hairs of the flower, leaf glandular hair has no maximum 680 nm belonging to chlorophyll. Secretory hairs of the petal (and pistil as seen in Table 2.1) fluoresce, mainly in the yellow region (maximum 540-550 nm), but also have maximum 450-475 nm, whereas the secretion drop in leaf secretory hair demonstrates only one maximum at 470-475 nm, in the blue part that may be characteristic for triterpenoids.

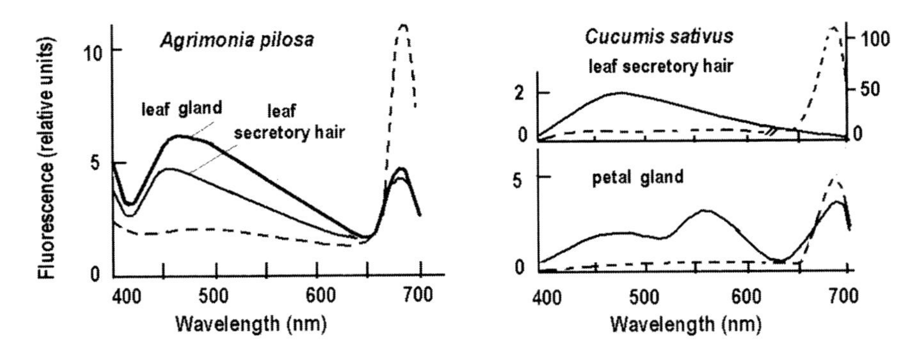

Fig. 2.19 The fluorescence spectra of glands and glandular secretory hairs in different parts of *Agrimonia pilosa* Ledeb. and *Cucumis sativus* L. Unbroken line-secretory cell; broken line- non-secretory cell. Y-scale is also done for chlorophyll maximum 680 nm in the variant with *Cucumis sativus.*

Sesquiterpene lactones are often concentrated in species of the family Asteraceae, in particular in the genera *Artemisia, Achillea, Gaillardia* and others (Rybalko, 1978; Fischer, 1991; Konovalov, 1995). As shown in Fig. 2.20, all parts of *Achillea millefolium* had fluorescent secretory cells. However, there was a difference between the emission of secretory cells from different organs that could be explained by different compositions of the secretions. Bright yellow- orange lightening with maximum 565 nm in the fluorescence spectra was observed in the flower, mainly in petals, and roots. On the contrary, surrounded non-secretory cells emitted in blue with maximum 465-475 nm. In leaves, secretory cells showed blue-green (maximum 475 nm and shoulder 535 nm) lightening whereas non-secretory

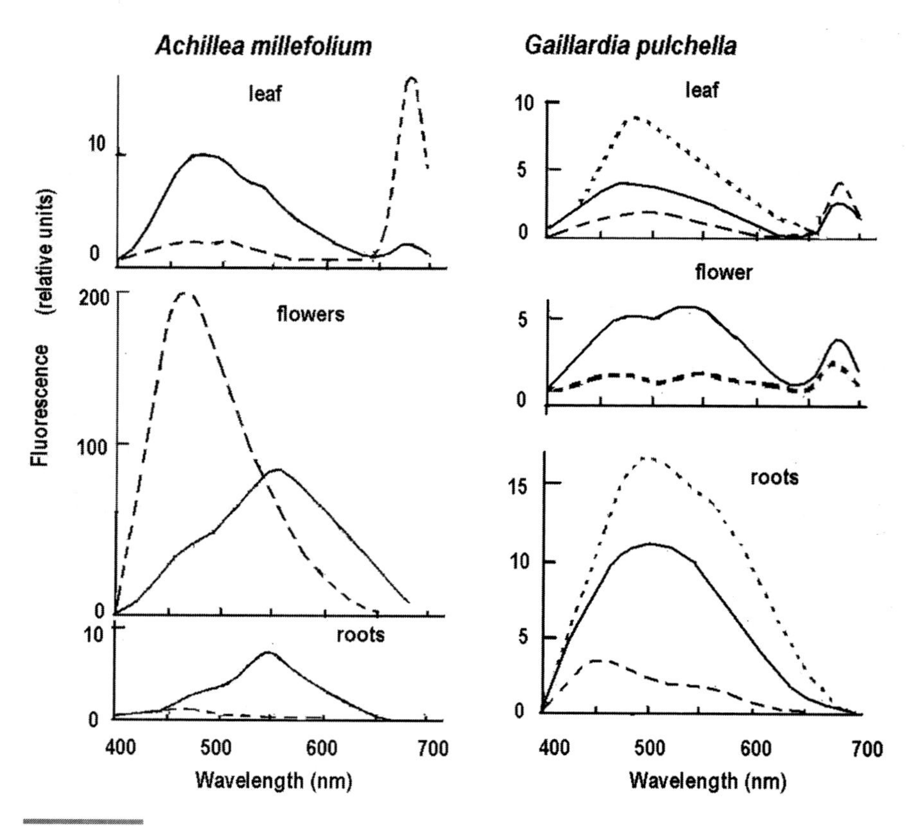

Fig. 2.20 The fluorescence spectra of secretory cells seen on the ligulate flower, leaf and root of *Achilea millefolium* and *Gaillardia pulchella*. Adopted from Roshchina (2005b). Unbroken fatty line – secretory hair; broken line – non-secretory cell; dotted line – crystal on the surface (secretion).

cells demonstrated red fluorescence with maximum 680 nm, peculiar to chlorophyll. In analysing the participation of concrete fluorescent compounds in secretory cells one can assume the contribution of azulenes and proazulenes. They fluoresce at 420-470 nm in a free state and – at > 565 nm if they are linked with cellulose (Roshchina et al., 1995; Roshchina, 1999a). Similar fluorescence was seen for the azulene-treated cells, while pure azulene fluoresced at 420-430 nm. The possible contribution of flavonoids in the secretory cells is minimal, if any, because the weak fluorescence of quercetin and rutin was with maximum 570-585 nm (600 nm for crystal form).

In the *Gaillardia* genus such as *Gaillardia pulchella* the fluorescence of components from secretory hairs also differed among organs (Fig. 2.20) which shows a different content of the secretions. In the leaf, blue emission with one maximum 475-480 nm was seen, but in the flower one could observe blue or green- yellow lightening with two maxima 475 and 550 nm, whereas in the root – mainly, green-yellow emission with maximum 500 nm. The secretions released on the leaf or root surface form crystals fluoresced in the blue-green spectral region. We can see their bright emission with maximum 500 nm, as in secretory hair, but more intensive, approximately 2-2.5 fold. Idioblast of the root has the fluorescence spectra similar with its secretory hair which means a possible similarity in the chemical composition of the secretions in both types of cells (Table 2.1). The main components of the plant secretions are sesquiterpene lactones, such as gaillardine, which fluoresces in the blue spectral region.

Secretory cells of *Gaillardia pulchella* contain sesquiterpene lactones, in particular gaillardine (see also Chapters 3 and 7). They demonstrate similar fluorescence with maximum 480 nm in leaf secretory hair and in root idioblast and secretory hair (also have the shoulder at 550 nm), whereas in petals of tubular and ligulate flowers more maxima, besides the above-mentioned one, at 550 nm. Secretory cells of the petals and leaf demonstrate the chlorophyll presence due to maximum 680 nm.

In Angiosperms there is the variability in the fluorescence spectra between different glandular organs. Moreover, the various maxima, belonging to different components of secretions, may be observed for one and the same organs. For instance, there is a difference between secretory cells of the upper and lower sides of the leaf of *Arctium tomentosum* (Colour Fig. 10). On the upper side yellow- and blue- fluorescing crystals are located on the leaf surface and in the secretory hairs (on the tip and below) whereas on the lower sides blue-fluorescing crystals are seen. Developing hairs have only blue fluorescence, and matured hairs also demonstrate yellowish 520 nm shoulder in the fluorescence spectrum. This shows the differences in the chemical composition of secretory hairs. Unlike the fluorescence of leaf secretory organs, flower secretory cells emit more intensively that those in leaves, approximately 100 fold. Different emission was observed even in the parts of the flower – generative (pollen and pistil) and non-generative (petals and sepals). Possible fluorescent components of the secretions may be in large degrees sesquiterpene lactones such as arctiopicrine and phenols (Golovkin et al., 2001).

Phenol– and quinone-containing glandular structures. Phenol-enriched secretory cells are often present in the scales of the tree buds (Roshchina and Roshchina, 1989; 1993). As seen in Fig. 2.21, the cells brightly fluoresce in the blue spectral region, sharply differing from the surrounding non-secretory cells. In the fluorescence spectra (Fig . 2.21), one can see only one maxima at 450-480 nm, belonging to phenols, mainly flavonoids. Bud exudations of the genus *Populus* contains flavonoids pinocembrin, pinobanksin and their derivatives (Greenaway et al., 1990) as well as caffeic acid derivatives (Wollenweber et al., 1987). The representatives of the family Betulaceae contain a lot of different phenols, mainly flavonoid aglycons (Wollenweber et al., 1991) as well as caffeic acid with derivatives (Wollenweber et al., 1987). All the compounds may contribute to the observed fluorescence at 470-490 nm. The buds of *Aesculus* contain flavonoids such as rutin and coumarins aesculin and aesculetin (Golovkin et al., 2001) which fluoresce in the blue and blue-green spectral regions (Table 1.1).

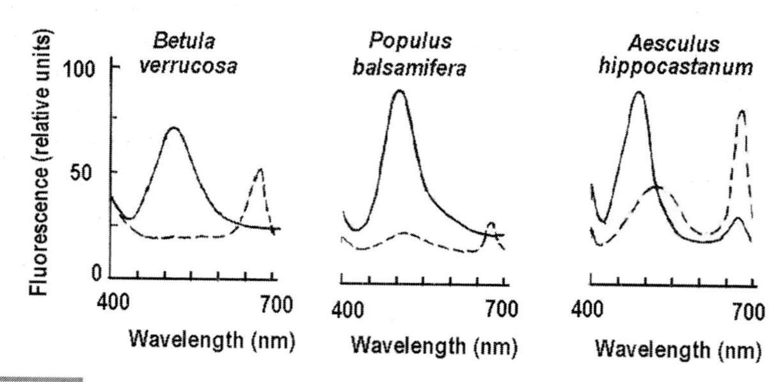

Fig. 2.21 The fluorescence spectra of the bud scales from tree species. Adopted from Roshchina and Melnikova (1995). Unbroken line – secretory cells, broken line – non-secretory cells. A. *Betula verrucosa*; 2. *Populus balsamifera*; 3. *Aesculus hippocastanum.*

Differences in autofluorescence of furanocoumarins (such as psoralen, bergapten and xanthotoxin) were found in fruits of *Angelica archangelica, Daucus carota. Sium suave* and *Psoralea bituminosa* (Zobel and March, 1993). All the fruits examined contained furanocoumarins on the cover surface. The emission was compared with that of crystals and saturated solutions of psoralen, bergapten and xanthotoxin under similar conditions. Autofluorescence can be also homogeneously distributed among all cells of the embryo *(Sium)* or can be concentrated in certain cells *(Psoralea)*. Embryos always contained furanocoumarins on their surface in varying proportions. Their concentrations in fruits were very low for *Sium*, medium – for *Angelica* or high – for *Psoralea*.

Among medicinal drug- containing plants, aromatic or phenolcarbonic acids, in particular salicylic acid, phenolic alcohols such as salidrosid (rhodiolosid), flavonoids such as quercetin, iridoid glycosides, tannin ellagic acid are concentrated in many species (Golovkin et al., 2001). These substances are found in secretions of *Swida* genus (S. *alba* and S. *sanguinea*), and it is likely that the fluorescence spectra of the plants may reflect it (Colour Fig. 11). Enalin (cornin), and tannins hydrolized, in particular gallic acid and ellagic acid are concentrated in many species (Golovkin et al., 2001). These substances are found in secretions of *Swida* genus (S. *alba* and S. *sanguinea*), and the fluorescence spectra of the plants may reflect it. Glandular hairs of Swida alba differ in their fluorescence as seen in Colour Fig. 11. Leaf trichome emits, mainly, in yellow at 550-560 nm, although the shoulder is present in green (510-550 nm).

The secretory cells, mainly glandular, of leaf and flower petals (see Table 2.1), contain phenolic and quinone compounds, which can be released into intratissuel secretory structures such as oil reservoirs and ducts (Curtis and Lersten, 1990). Oil- and resin- containing structures are also mentioned in Sections 2.1.2 and 2.2.5. The substances are located in different structures of genus *Hypericum*: first - in light glands, whereas the second (especially anthraquinone hypericin) in dark blue glands (see Fig. 2.17). Light glands fluoresce in blue with maximum 470-475 nm and the contribution of red fluorescence of chlorophyll is also seen (Fig. 2.17). Unlike light glands, dark blue glands demonstrated a weak emission at 450 nm (the intensity of the blue emission is similar with those emitted from non-secretory cells). Fluorescing constituents of the genus *Hypericum* are 3-epi-betulinic acid, caffeic acid, ferulic acid, docosanol, p-hydroxybenzoic acid, 3,4-dimethoxy benzoic acid, quercetin and its derivatives, and shikimic acid (El-Seedi et al., 2003). According to the above- mentioned characteristics the compounds may be first identified in plant excreta such as leachates and even on intact cells.

Alkaloid-containing glandular structures. Secretory hairs of leaves and flowers from *Nicotiana alata* contain alkaloids which determine the species resistance to insects (Fig. 2.22).

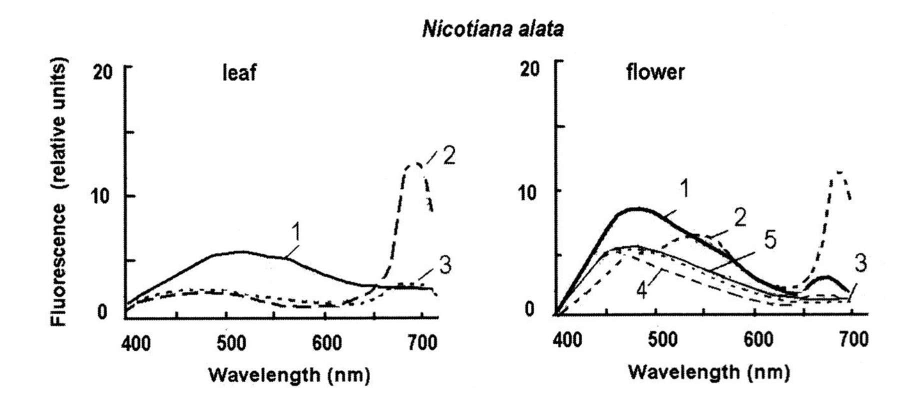

Fig. 2.22 The fluorescence spectra of leaf secretory and non-secretory cells of *Nicotiana alata*. Left- leaf. 1. Secretory hair on upper side; 2. Non-secretory cell on upper side; 3. Non-secretory cell of lower side; Right of flower. 1. Secretory cell of pistil stigma; 2. Secretory hair on upper side of petal; 3. Secretory cell on lower side; 4. Non-secretory cell of lower side, 5. Secretory cell of anther.

Among the compounds which are present in genus Nicotiana, are anabasine, nicotine and scopoletin (Golovkin et al., 2001). Secretory glandular cells demonstrated maxima in green (500-510 nm) as in the leaf or mainly in blue (470-480 nm) as in the flower. Only secretory hair of the petal has maximum at 550-560 nm.

The secretory hair of *Symphytum officinale*, is of interest due to its structure, as it is surrounded with red-fluorescing hill-like rosette cells (Colour Fig. 12). Under a luminescence microscope this complex structure is lightly observed. As seen in Colour Fig. 12, the secretory hairs of the flower petal fluoresce in green (maximum at 520 nm) and in red (maximum at 680 nm which belong to chlorophyll) whereas the structures in pistil stigma lack chlorophyll and lightening in green (maximum 520 nm). Pollen emission has maximum 470-480 nm and shoulder 580 nm. The secretory cells of *Symphytum officinale* contains alkaloids, among which cyanoglossin, that fluoresces at 460-480 nm (Roshchina and Melnikova, 1995), and flavonoid rutin (Golovkin et al., 2001) can also contribute to the emission. Besides the substances, slime secretions of the plant species contain tannins, which has green fluorescence with maximum at 520-530 nm.

2.2.2.4 Salt-containing glands

Salt-containing glands of fat hen *Atriplex patula* are concentrated on the lower side of the leaf (Fig. 2.23). The salt crystals located inside the gland also include fluorescent compounds.

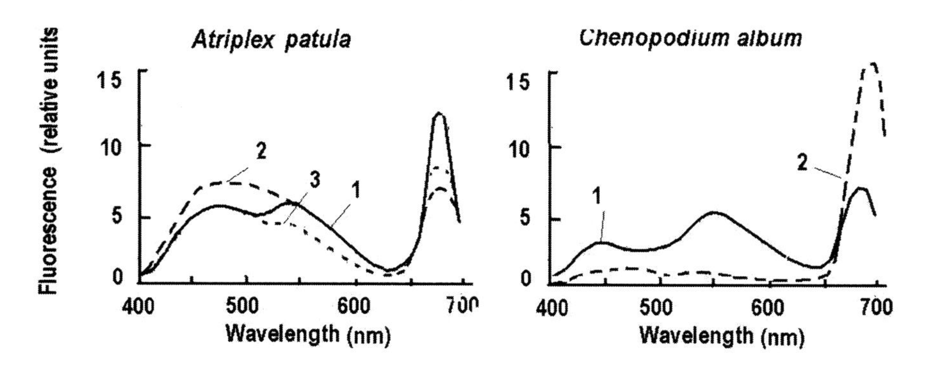

Fig. 2.23 The fluorescence spectra of salt glands on leaf *Atriplex patula* (Left) and *Chenopodium album* (Right). 1. Salt gland cell with crystal fluoresced in yellow; 2. Non-secretory cell. 3. Blue fluorescing gland cell with a crystal.

The yellow fluorescence of salt glands may be due to the presence of flavonoids, which impregnate the calcium crystals in the glandular cell. Moreover, sometimes blue fluorescence is seen.

The crystalline structure is clearly observed in the fluorescence image in a laser-scanning confocal microscope (Colour, Fig 13 Appendix 2). The LSCM image of the salt gland shows small bright crystals on the surface of the gland. The optical slices of the gland show big internal crystals within the glandular structure. The fluorescence colour (greenish-yellow) and intensity of the salt glands differs from surrounded non-secretory cells emitted in red (Table 2.8). As shown in Table 2.8, the green emission intensity in leaf salt glands is higher (7-10 times) whereas red fluorescence is lower (2-3 times) than in surrounded non-secretory cells.

The histochemical staining of the salt glands and their crystals with 2-5% $FeCl_3$ showed rose-violet colour, peculiar to phenols (Geissman, 1955). Perhaps, phenolic compounds contribute in the autofluorescence of the salt glands. Calcium sulphates and silicates usually have no fluorescence, but,

Table 2.8 The fluorescence intensity (I) of salt glands on the lower leaf side of *Atriplex patula* and *Chenopodium album*

Cell	*Green emission ($I_{510-530}$)*	*Red emission ($I_{640-680}$)*
Atriplex patula		
Salt gland with crystal	0.32 ± 0.024	2.56 ± 0.26
Salt gland (transparent part)	0.15 ± 0.02	1.31 ± 0.06
Crystal of the gland	0.32 ± 0.02	1.12 ± 0.06
Surrounded non-secretory cells	0.02 ± 0.003	6.18 ± 0.43
Chenopodium album		
Salt gland with crystal	0.19 ± 0.018	1.95 ± 0.10
Surrounded non-secretory cells	0.02 ± 0.003	3.02 ± 0.40

*$I_{510-530}$ and $I_{640-680}$ – the emission intensity at 510-530 and 640-680 nm, respectively. The excitation 420-436 nm.

when the crystalline salts are impregnated with phenols, the blue-green emission could appear.

The green-yellow fluorescing crystals of salts are also observed on the surface of seeds and cotyledons of the seedlings in the salt-enriched plant *Chenopodium album*. They are formed presumably on the medium containing sulphates (see Chapter 4).

2.2.2.5 Resin-containing structures

Resins are mixtures of substances (terpenoids, flavonoids and lipids), which are concentrated not only in specialized glandular trichomes, but also – on the inner surface of resinous ducts (mainly peculiar to conifer plants), and sometimes evacuated on the external surface of plants.

Resin in glandular hairs. Glandular hairs can contain resinous mixtures of substances (mainly hydrophobic such as terpenoids, lipids, but sometimes also phenols). Similar glands are found in various plants, and one example is the over-ground parts of *Rubus* genus, especially in *Rubus odoratus* L. (Table 2.1). Fluorescence of the head and multicellular stalk in glandular hair differs.The first has a weak blue fluorescence, if any, whereas the second brightly fluoresces in red due to the presence of chlorophyll. The glandular hairs on the stems, flower buds, flower calyx of purple-flowering raspberry, flowering raspberry, mulberry, rose-flowering raspberry *Ribes odoratus* L. have a powerful resinous scent somewhat like cedarwood (Genders, 1994). The composition of the head and of the cells of the stalk is not the same, as seen from different fluorescence of the cells.

The cell of a head fluoresce in blue, whereas the cells of the stalk – in red during abundant chloroplasts. In a composition of *Rubus* genus, there are fluorescent components such as phenolcarbonic acids (caffeic, n- and p-coumaric, ferulic, gallic, vanillic, protocatehuic, salicylic, etc), flavonoids (quercetin, isoquercetin, kaempferol, etc), methylsalycilate and derivatives of coumaric acid, terpenoids (piperitone, sabinol, ursolic acid, fragarine), tannin ellagic acid (Plant Resources of Russia, 1989; Golovkin et al., 2001). Essential oils of the resinous secretions (contain eugenol, geraniol, linalool, α - and β- pinenes, cinnamic aldehyde, but the fluorescence of the components themselves is small in blue, and is often not measurable.

Oleoresins are found in many plants, belonging to Angiosperms and contain glandular hairs as can be seen in the genera *Artemisia, Primula, Helianthus, Acacia* and others (Roshchina and Roshchina, 1989; 1993). Some sesquiterpene lactones found in the plants may fluoresce due to their chemical composition (see Chapter 3).

Resin cells and resin ducts. Resin is the main product of the secretory cells in conifer plants. For instance, resin of *Abies* genus is nearly 70-80% of the secretions (Roshchina and Roshchina, 1993). When the epidermal tissue of conifer species is ruptured, resin is released from the resin ducts, lying deeper in the needle mesophyll. This marks the resin ducts location. The resin-containing structures may fluoresce under ultra-violet or violet light of a luminescence microscope (Table 2.1). Excreted resin shows characteristic fluorescence maximum either in the blue region of the spectra (at 430-460 nm, typical for monoterpenes) or in the yellow region (at 500-525 nm). The lightening is peculiar to secretions consisting of resins and oils, in which terpenoids are the main components. The fluorescence of terpenes from resins was observed on the surface of needles and cones of conifer plants (Table 2.1). Resins and resin ducts of both cones and needles from *Abies sibirica* had, mainly, yellow fluorescence (maximum 550 nm) although resin ducts demonstrated the shifts – expressed maxima in green-yellow at 500 and 550 nm (Fig. 2.6). There was a difference in the fluorescence spectra of the resin released on the surface and resin in the resin duct, as well as between resin ducts of needles and cones. Unlike resin fluoresced with maxima 530-550 nm, resin released from resin ducts of needles have two maxima at 475 and 575 nm. Moreover, the fluorescence spectra of the resin duct in a cone had maxima 465, 535 and shoulder 575 nm whereas free resin demonstrated the weak-expressed two-component emission at 500-580 nm.

Transverse slices of needles show characteristic resin ducts (Fig. 2.24). The fluorescence of resin at the ends of resin ducts might also be seen through the transparent epidermal cells. It is similar to the fluorescence of the resin itself and to the brightly luminous parts of the fluorescing surface of the needle transverse slices (cuttings). In the slices, resinous cells, covering the resin ducts, also fluoresce in the blue region of the spectra. Terpenoids and phenolic compounds occurring in resin can contribute to

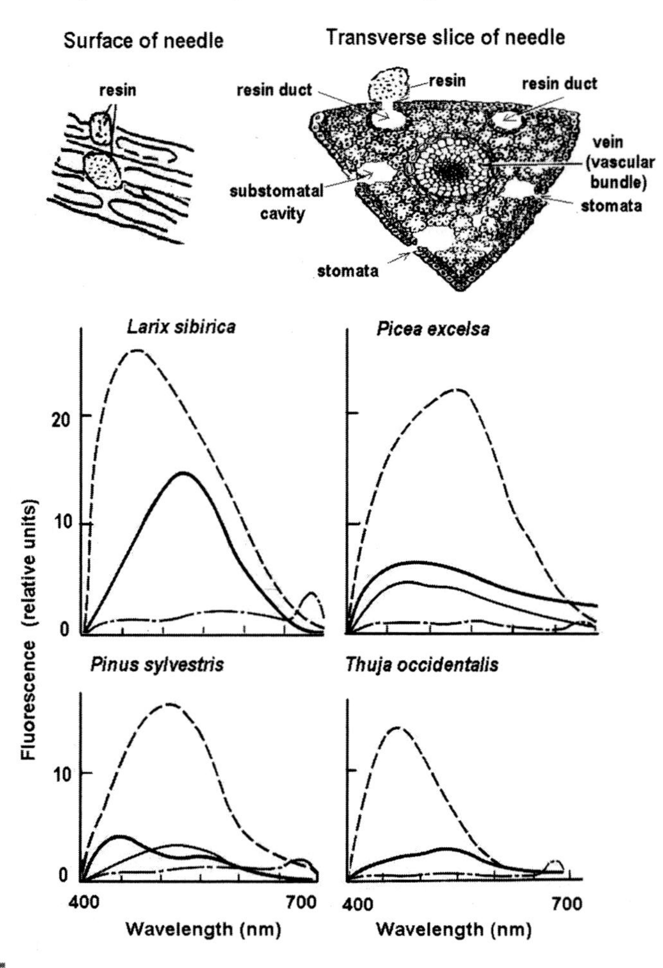

Fig. 2.24 The fluorescence spectra of conifer's resin released from resin ducts after rupture of the epidermis. Adopted from Roshchina et al., (1998a). Thick unbroken line - resin seen through the transparent epidermis, thin unbroken line — resinous cells on the slices, broken line - resin out tissues, broken-dotted line - non-secretory tissue.

blue fluorescence (Table 1. 1). In *Larix sibirica,* resin in resin ducts shows the emission maximum 525 nm whereas resin itself has maximum 465 nm. In *Picea excelsa,* maxima of resin in the ducts and evacuated resin are 460 and 550 nm, respectively (Table 2.1). Resin in resin ducts of blue needles of *Pinus pungens* also has additional shoulder 610 nm (Table 2.1). *Pinus sylvestris* contains the same components with a luminescence at 450, 560 or 535 nm, but the resin itself has only one component with fluorescence maximum 500 nm. As for for *Thuja occidentalis,* resin in the resin ducts fluoresce with 450 and 535 nm peaks, whereas exposed resin shows 460 nm peak.

The main emission in the blue region of the spectra might be a contribution of mixtures of terpenoids and phenols (Table 1.1) which also occur in resin. As one may learn from some books (Murav'eva, 1981; Golovkin et al., 2001), among possible fluorescent components of resin in conifer species studied are dihydroquercetin, menthone, bornyl acetate, α- and β- pinenes (in *Pinus sylvestris, Abies balsamea, Larix sibirica, Thuja occidentalis*), bornyl acetate and borneol (*Larix sibirica, Picea excelsa, Thuja occidentalis*), camphor and bicyclic terpenoid ketone thujone (in *Thuja occidentalis*), piceol or n-hydrooxyacetophenon, dihydroquercetin (*Larix sibirica*).

As a whole, terpenoid-containing secretory cells both in glands and in resin ducts fluoresce in the blue (450-460 nm) and green-yellow (500-550 nm) spectral regions (Table 2.1), The first is typical of the sesquiterpenes and monoterpenes of oils (Table 1.2). The conifer species contain terpenes and sesquiterpenes. Among components of essential oils found in *Pinus sylvestris* are α-pinene (22.2%), carene – (43, 4%) other monoterpenes (Chalchat et al., 1985), and also – sesquiterpenes such as azulenes (Kolesnikova et al., 1980). Monoterpenes have a weak fluorescence, while azulenes strongly emit in blue (420-430 nm). Shoulders or peaks in the red region (> 600 nm) may belong to azulenes. Azulenes are in essential oils of glands in some Asteraceae species (Konovalov, 1995), in needles of conifer plants (Kolesnikova et al., 1980), in pollen grains (Roshchina et al., 1995). Blue colour of the needles earlier was supposed as the surface waxes light reflectance (Reicosky and Hanover, 1978) which may be related to azulenes. Among them, only sesquiterpene lactones (proazulenes and. azulenes) have intensive fluorescence (Table 1.1), mainly, in the blue or yellow spectral regions. The chemical composition of conifer resin is often very complex. Besides terpenoids and polysaccharides, it contains phenols (about 20 flavonoids) and protein components (Roshchina and Roshchina, 1989; 1993). Flavonoids and some terpenoids such as azulenes can fluoresce in the blue-green region of the spectra.

2.2.2.6 Hydathodes and nectaries

Special secretory structures which release water or water-soluble substances are hydathodes and nectaries (Fahn, 1979; Roshchina and Roshchina, 1989; 1993). They differ in their structure and composition of the secretions – hydathodes, i.e. modified water-releasing stomata with underlying tissue epithem, and nectaries secreting mainly saccharides. They also show significant fluorescence (Table 2.1).

Hydathodes. Hydathodes of Angiosperms may be similar to the above-mentioned analogous structures of horsetail *Equisetum arvense* (Fig. 2.3). Some laminar hydathodes are found in the family Urticaceae (Lersten and Curtis, 1991). Colour Fig. 14 shows that the hydathodes of *Urtica dioica* L. contain not only water, but also blue-fluorescent components, which emit with maximum 465-470 nm. On the contrary, non-secretory cells show marked red luminescence only at 680 nm, typical for chlorophyll. The hydathodes of *Solidago virgaurea* are modified stomata, which cell walls emit in green at the excitation light 420 nm, but the interior of the hydathodes has no visible fluorescence (Colour Fig. 15).

The content of hydathodes is not well known, and the fluorescence test could give only preliminary information about it. Thus, phenolic substances of guttation water (Curtis, 1943) could contribute to blue luminescence of the secretory structures, although maximum 530 nm is also characteristic for some flavins and flavonoids (Table 1.1). In other cases, in particular as shown in Colour Fig. 15 for hydathodes, which exists as unclosed stomata, the guttation water has no fluorescence, but the cell walls of the structure lighten in green.

Nectaries. Nectaries are modified hydathodes, which release the sugar in the water droplets. However, nectar and nectaries may fluoresce, while sugar has no self-fluorescence (Roshchina et al., 1998a). Examples of the extrafloral nectaries and nectar fluorescence are shown in Figs. 2.25, 2.26. 2.27, 2.28, 2.29 and Colour Fig. 16. Nectary of *Impatiens balsamina* (Fig. 2.25) fluoresces in the blue-green region, whereas nectar itself shines only in the blue region with maximum 460 nm, demonstrating, perhaps, the presence of phenolic compounds in the water exudate revealed in some types of nectars (Baker et al., 1973; Baker and Baker 1975, 1983; Murphy, 1992). On the contrary, floral nectaries and nectar of *Convolvulus arvensis*

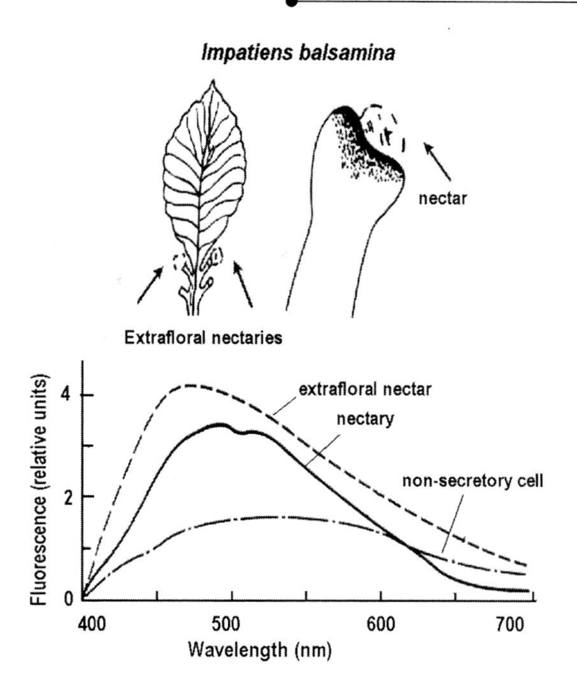

Fig. 2.25 The fluorescence spectra of extrafloral nectary and nectar located on the leaf petiole from *Impatiens balsamina.* Adopted from Roshchina et al., (1998a).

have emission in the yellow-orange region at 560-580 nm (Fig. 2.25). This may be related to the presence of polyacetylenes or some flavonoid fluorescing in the region (see Table 1.1 in Chapter 1).

Analysis of the emission of nectar from 102 species in 82 genera and 38 families showed most intensive fluorescence in UV-light for *Prunus persica*, *Allium porrum*, *Convolvulus arvensis*, *Daucus carota*, *Phacelia viscida*, and *Robinia pseudoacaci* (Thorp et al., 1975). The color of the emission varied from blue (nectar from mentioned species) to yellow (in *Muilla maritima*) or changed the color (in *Fagopyrum peploides*). The source of the blue-green fluorescence, besides phenols, may be terpenoids like in floral nectar droplets from *Maxillaria anceps* (Davies et al., 2005). This fluorescence is considered as likely useful for pollinators to locate and recognize rewards and to distinguish between reward-bearing and reward-less flowers (Cruden, 1972; Peumans *et al.*, 1997; Radice and Galati, 2003).

Maxima of the floral nectar fluorescence and the pollen fluorescence are also represented in Table 2.1 and Fig. 2.27. Floral nectar of *Bryophyllum daigremontianum* shows only one maximum 470 nm, which is due to the

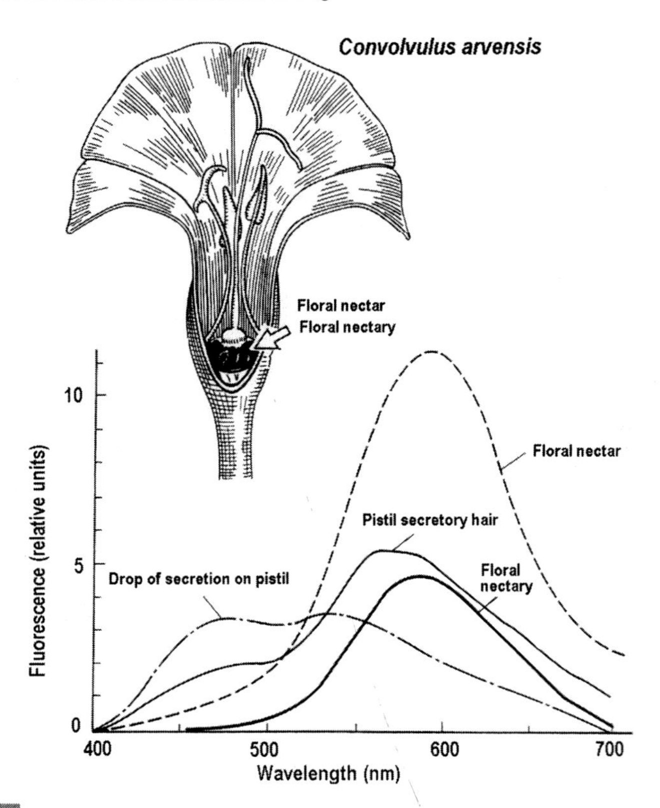

Fig. 2.26 The fluorescence spectra of floral nectary and nectar from *Convolvulus arvensis.* Adopted from Roshchina et al., (1998a).

phenol contribution whereas that of *Hoya carnosa* demonstrates two maxima, 495 and 610 nm, respectively. Floral nectar of *Epiphyllum hybridum* has no a fluorescence, perhaps, because it includes only non-fluorescent sugar.

The nectaries of *Passiflora* genus where both types of the nectaries are present–extrafloral and floral are of special interest (Fig. 2.28). Fluorescence spectra of the nectar and nectaries differ in the maxima position. Floral nectar fluoresces in the longer wavelength region, mainly in yellow whereas extrafloral nectar – in blue. This is evidence of the different chemical composition of the nectar and nectaries. Among fluorescent components of extrafloral nectar phenols-blue-emitted probably prevail.

Nectar-containing cells are clearly seen on the petals of *Alstroemeria* flowers (Colour Fig. 16) near red spots formed by anthocyanin-including

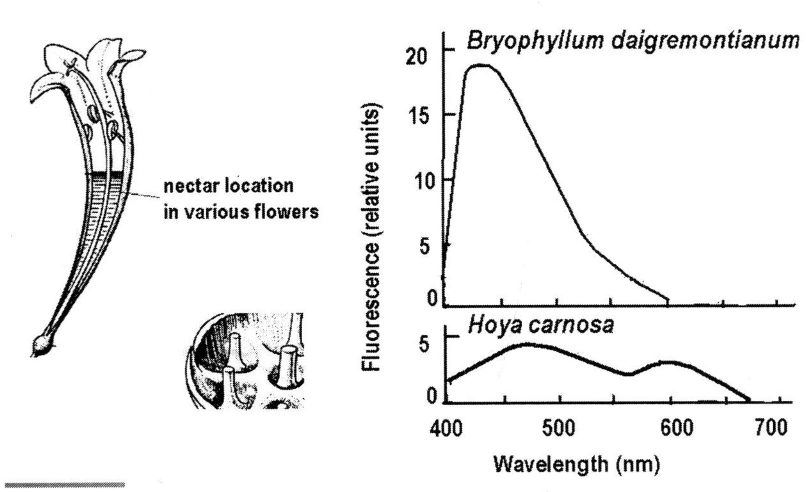

Fig. 2.27 The fluorescence spectra of isolated nectar from the flowers of *Bryophyllum daigremontianum* and *Hoya carnosa*

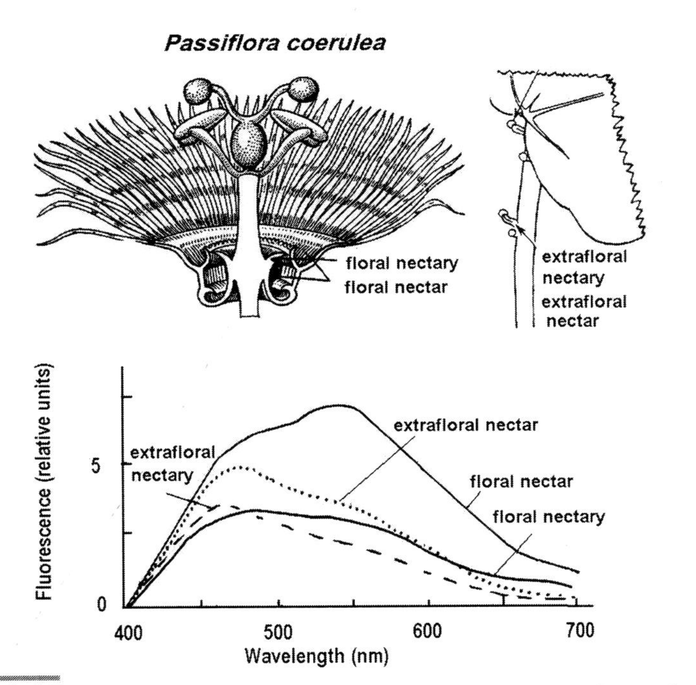

Fig. 2.28 The fluorescent spectra of nectar and nectaries of *Passiflora coerulea*. Adopted from Roshchina et al., (1998a). Nectar from extrafloral nectaries contains phenols, blue-fluorescing.

cells. The fluorescence spectra of secretory cells on the petals demonstrates bright-lightening secretory cells (one maximum at 470-475 or 500 nm) that differ from nectar-containing cells (two maxima at 520 and 620 nm) and anthocyanin-containing cells with weak fluorescence. None of these cells have the maximum 680 nm, peculiar to chlorophyll, unlike non-secretory cells located at the top of petals. In the fluorescence spectra of secretory cells of pistil there are three maxima (at 475, 535 and 680 nm).

Pollen also has maxima 465-470 and 535-540 nm, but lack of maximum 680 nm, peculiar to chlorophyll. Possible fluorescent compounds of nectar are flavonoids quercetin and kaempferol, which are present in the genus (Golovkin et al., 2001).

2.2.2.7 Mucilage (slime)-containing cells

Plant mucilages are represented by acidic or neutral polysaccharide polymers of higher molecular weight (Fahn, 1979; 1988). Secretion of mucilage may occur through various secretory structures – from single cells – slime idioblasts to specialized glands of carnivorous species.

Slime-secreting cells of non-carnivorous species. The cells of non-carnivorous plants contain slime (mucilage) which has a significant fluorescence unlike non-secretory ones. First of all secretory cells either have no the chlorophyll-related maximum at 680 nm, or if the cells of the underlying mesophyll are seen, contribute a small peak.

For example, the leaf slime cells of *Symphytum officinale* demonstrated the fluorescence maximum at 470-480 and shoulder at 530 nm (Fig. 2.29). The cells have a significant fluorescence with maximum in blue-green 490-

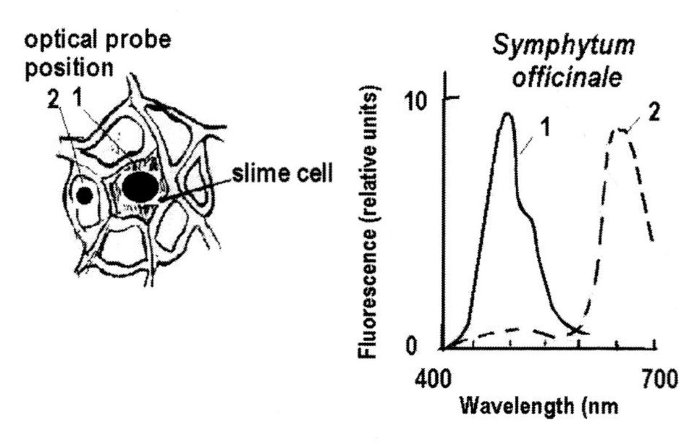

Fig. 2.29 The fluorescence spectra of leaf secretory cells from *Symphytum officinale*. Source: Roshchina and Melnikova (1995) and Roshchina et al., (1998a). 1. Slime cell; 2. Non-secretory cell.

500 nm with shoulder 530 nm, unlike non-secretory ones with one maximum 680 nm, peculiar to chlorophyll.

Slime-containing glands of carnivorous plants. Some plants, belonging to carnivorous species, can release slime by their special glands (Vasilyev, 1977; Fahn, 1979; Muravnik, 2000; Vasilyev and Muravnik, 1988). The glands of carnivorous plants, which release the trapping mucilage, are known to contain enzymes, polysaccharides, phenols, and other substances (Roshchina and Roshchina 1989, 1993). Perhaps, they contribute to the light emission. Colour Figs. 17 and 18 show fluoresced secretory cells of *Utricularia vulgaris* L. (family Lentibulariaceae) and *Drosera capensis* L. (family Droseraceae). Both the structures have a marked fluorescence with maximum at 560 nm. In *Utricularia,* the mucilage fluorescence is in a shorter wavelength region with maxima at 545 nm. Unlike this species, released slime of *Drosera* fluoresced the same as in the capitate trichomes. Tentacle hairs are lack of the fluorescent slime (Colour Fig. 18). Naphthoquinones of *Drosera* genus such as droserone and its derivatives or plumbagin (Finnie and van Staden, 1993) as well as 7-methyljuglone found in callus and organ culture of *Drosera spathulata* (Blehova et al., 1995) may contribute to light emission.

2.2.2.8 Laticifers

Secretory intratissular structures called laticifers are living cells with latex in vacuoles (Roshchina and Roshchina, 1989; 1993). Often the secretion – latex is seen after mechanical damage or cutting of leaves and stems of latex-containing plant species such as in the genera *Euphorbia*, *Taraxacum*, *Chelidonium* and others. Usually latex is milky white as in *Taraxacum officinale* or in species from the genus *Euphorbia*, but in plants from the genus *Chelidonium* – bright orange.

The secretory cells contain ing latex in *Chelidonium majus* L. may fluoresce. As shown in Fig. 2.30, the fluorescence of the laticifer of the species has a maximum at 590-595 nm in the flower and 550-575 nm in the stem, while surrounding non-secretory cells have peaks at 510 nm and 680 nm (chlorophyll). The orange fluorescence of latex and laticifer is observed both in the stem and the flower. In the flower, the maximum emission of latex and laticifer is shifted to longer wavelengths. The comparison of exuded latex and laticifer shows that the intensity of latex emission is higher, than that of a non-damaged laticifer. The latex

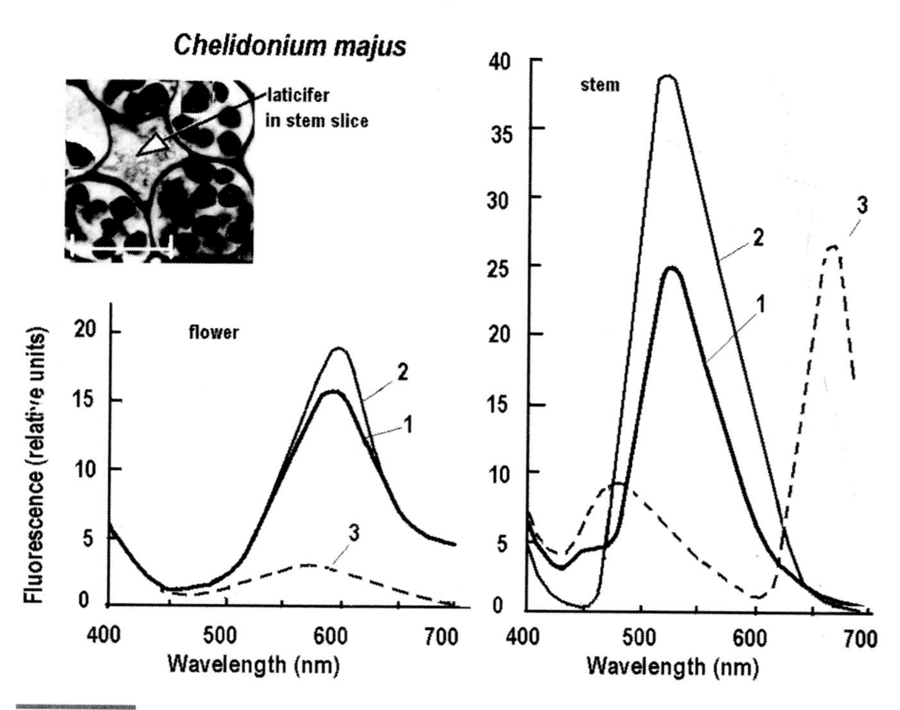

Fig. 2.30 The fluorescence spectra of latex and laticifer on transverse slices of the stem and the flower from *Chelidonium majus*. 1- laticifer; 2- latex, 3- non-secretory cells. Upper side (left) - common view of laticifer (shown with the arrow) in transparent light of the microscope. Bar = 100 μm

fluorescence grows during the first 30-40 min after the beginning of exudation and achieves a plateau, when the orange latex coagulates in the air, changing to a brown viscous mass (Roshchina and Melnikova, 1999). It may depend on the redox state of the secretion released. When it is oxidized by air, the colour of the latex becomes browner, and its fluorescence changes too. Latex of this species includes phenols, dopamine, and various alkaloids derived from the amine such as berberine and chelerythrine (Roshchina and Roshchina, 1993). Besides alkaloids, mainly isoquinolinic (berberine, chelerythrine, sanguinarine, allocryptopine, chelirubine, dihydrosanguinarine, canadine, coptisine, protopine, etc), tannins (gallic acid), phenolcarbonic acids (caffeic, ferulic) o-coumaric acid, flavonoids (quercetin, rutin, kaempferol) and others are found in the species (Golovkin et al., 2001). As a whole, yellow-orange luminescence corresponds to berberine or chelerythrine fluorescence with maximum at 520-550 nm (excitation by UV-light 360-380 nm) or 550-580 nm (excitation by light 420 nm).

Phenol- and quinone-containing laticifers may fluoresce presumably in the blue or blue-green or even green-yellow region, as well as amine-containing cells (Table 1.2). This is demonstrated in some species belonging to Chenopodiaceae, Urticaceae and Solanaceae. The laticifers of milky species *Euphorbia viminalis* and *Taraxacum officinale* often have green-yellow or blue fluorescence. The comparison of the fluorescence spectra of latex-containing plants such as *Euphorbia viminalis* and *Taraxacum officinale* (Fig. 2.31) shows that latex fluoresces in a shorter region, than the laticifer, perhaps, due to the contribution of the cell wall or other compounds. These species differs in some components, but the fluorescence maximum is the same for released secretion – 450-460 nm and for the laticifer – 530 -550 nm. The oxidation on the air that appears to take place also explains the difference between released secretion and internal secretion, which is contained within a laticifer. In *Euphorbia viminalis*, there are fluorescing components such as tannins (gallic acid), flavonoids and their glycosides (quercetin, quercetin-3-galactoside, kaempferol), while in *Taraxacum officinale* – flavins (riboflavin), coumarins (scopoletin, coumaric acid), flavonoids and their glycosides (luteolin, quercetin, kaempferol), phenolcarbonic acids (caffeic) (Golovkin et al., 2001).

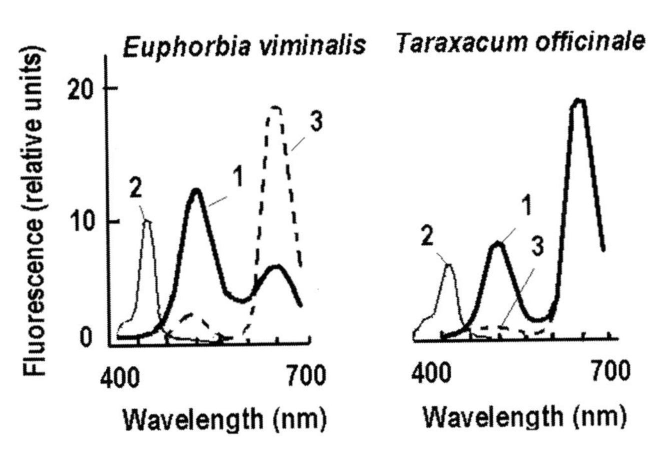

Fig. 2.31 The fluorescence spectra of latex-containing plants. Adopted from Roshchina and Melnikova, 1995). 1– laticifer; 2– drop of latex; 3– non-secretory cell.

2.2.2.9 *Root secretory cells*

Among root secretory cells, there are idioblasts (single cells surrounded by non-secretory cells), glandular cells in complexes and secretory hairs (trichomes). Often the tip of the root intensively releases fluorescing substances, especially from seedlings at seed germination, and therefore may be considered as secretory tissue. Figure 2.32 shows the example and demonstrates the difference in the fluorescence spectra of two species of *Chlorophytum commosum* (emitted with two maxima at 465-470 and 545 nm) and *Tagetes erecta* (fluoresce with one maximum at 550-570 nm). This picture is due to the different composition of the secretions. The first species contains alkaloids whereas the second – thiophenes terthienyl and polythienyls (Golovkin et al., 2001). Although flavonoids such as kaempferitrin, luteolin, patuletin may also contribute in *Tagetes* genus (Golovkin et al., 2001). The surrounded non-secretory cells fluoresce in the shorter wavelength region, rather than in secretory cells. In roots of *Tagetes patula,* cells fluoresce in green-yellow and red. They also differ in the emission intensity measured by a double-beam microspectrofluorimeter (Table 2.9). The most intensive green fluorescence is seen for green-fluorescing cells, unlike red- or brownish- fluorescing surrounded cells. The difference appears to be due to the different composition of the emitted cells. Green-yellow emission may be related with polyacetylenes or/and flavonoids solved in oils.

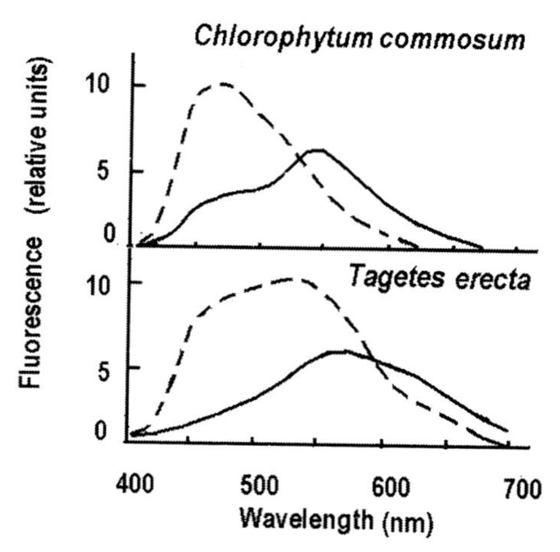

Fig. 2.32 The fluorescence spectra of the root secretory cells of *Chlorophytum commosum* and *Tagetes erecta.* Unbroken line- secretory cell; broken lines – non-secretory cell.

Table 2.9 The fluorescence intensity of root secretory cells from *Tagetes patula* L. (excitation by the light 420 nm)

Cell	$I_{520\text{-}540}$	$I_{640\text{-}680}$
Green-yellow-fluorescing cells	0.47 ± 0.04	0.73 ± 0.06
Red-fluorescing cells	0.24 ± 0.05	0.42 ± 0.09
Surrounded by brownish-fluorescing non-secretory cells	0.17 ± 0.04	0.31 ± 0.05

Among alkaloid-enriched cells, the fluorescence from the root of *Ruta graveolens* L. is very bright (Roshchina, 2005b) which is demonstrated in Colour Fig. 19 Appendix 2. The root tip shows the yellow-orange fluorescing excretions with high maximum 595-600 nm, while other smaller maximum 465-470 nm is in blue. Unlike secreting tip meristem, root hairs demonstrate small fluorescent components with maximum 590-600 nm, and higher peak at 465-470 nm. Single secretory cells, called idioblasts, are also seen in the root of the species and fluoresce as crystals, mainly, in the yellow-orange spectral region with the highest maximum 595-600 and smaller – at 465-470 nm. Non-secretory cells emit in blue-green with only one maximum 475 nm. The fluorescence of secreting root cells depends on the chemical composition of their secretions and may be changed among various rue cultures. For instance, cells of pRi-transformed roots of *Ruta graveolens* L. cultivated *in vitro* contain acridone alkaloids, whose location is seen due to their orange fluorescence (Kuzovkina et al., 1999). In non-differentiated cells of callus and roots, acridone alkaloids are accumulated in idioblasts or excreted on the cellular surface and into the nutritional medium, and fluoresce as seen in Fig. 2.33. The fluorescence spectra are similar with those of the root secretory cells: high maximum 595-600 nm in the orange spectral region and smaller (if any) maximum in blue at 460-465 nm. A similar composition of the fluorescent secretions is proposed. Non-secretory cells fluoresce mainly in blue with peak 460-465 nm, and smaller maximum is seen in orange 595-600 nm.

2.2.2.10 Comparison of the fluorescence between secretory cells in various organs

The fluorescence of secretory cells may differ in various organs. Taxon-specific differences are also present. Some examples are given in Fig. 2.34. If the root and leaf secretory cells are compared, particularly of *Sedum* sp. or the seed and leaf of *Thuja occidentalis*, the fluorescence spectra demonstrates the difference in the emission, whereas between the secretory

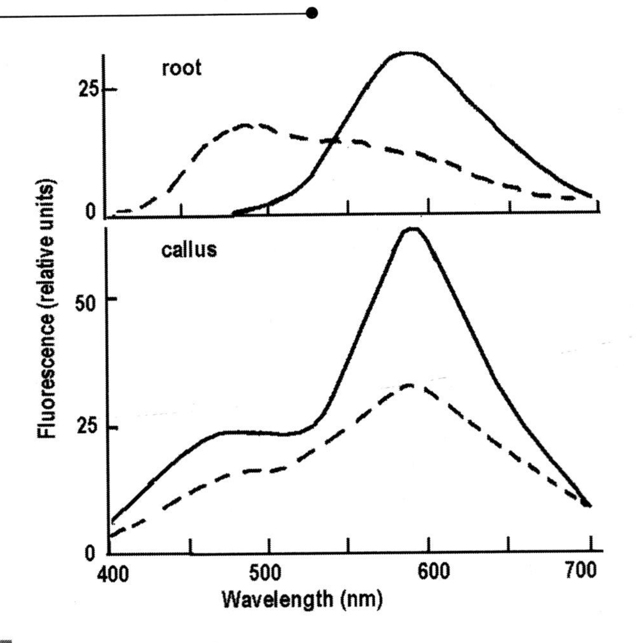

Fig. 2.33 The fluorescence spectra of root and root callus from *Ruta graveolens* L. Unbroken line – secretory cells; broken line – non-secretory cells

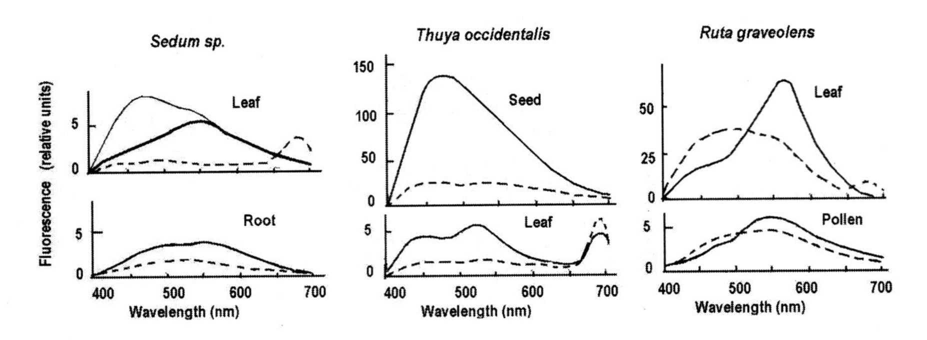

Fig. 2.34 The fluorescence spectra of secretory cells of different organs. Unbroken thin and thick lines – secretory cells, broken line – non-secretory cell or anther cell, unlike pollen.

cells of the leaf and pollen of *Ruta graveolens* a similarity is seen. Unlike secretory cells of sedum leaves, root secretory cells have two smooth maxima at 480 and 550 nm, perhaps, due to the different fluorescent compounds. Secretory cells of the leaf may have maximum (at 465 nm) in blue and maximum or shoulder (at 550 nm) in the yellow region of the spectrum as seen for the first cell whereas the second cell demonstrates only one peak 550 nm in yellow. In *Thuja* secretory cells of seeds, there is only one emission maximum 450-460 nm, unlike the leaf, where second maximum is seen in green-yellow (520-535 nm) belonging to another component. On the contrary, secretory components of both pollen (as well as anther) and leaf that fluoresce with peaks 550-575 nm of *Ruta graveolens* may be similar.

Another example is the difference between secretory cells of leaves and flowers (Fig. 2.35). In the species *Malva* and *Lavatera* the fluorescence maxima of flower secretory cells in blue are shifted to longer wavelengths, and the chlorophyll maximum 680 nm is absent in comparison with secretory cells of leaves. The position of the emission maximum in *Epilobium* flower (in blue) is approximately the same as in leaves, but there

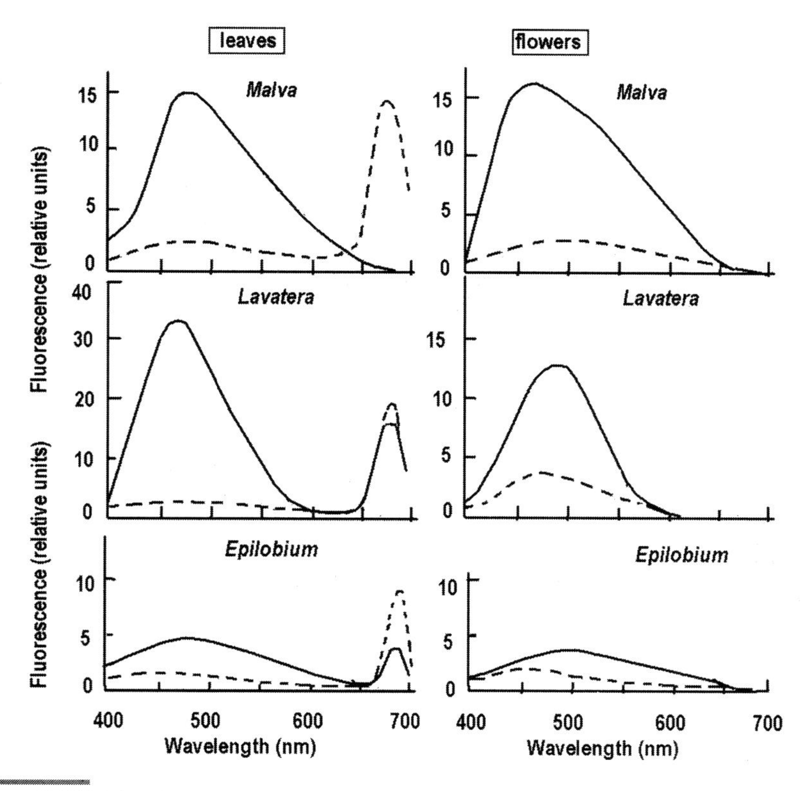

Fig. 2.35 The fluorescence spectra of secretory cells on leaf (hairs) and flowers belonging to *Malva* sp., *Lavatera trimestris* (fam Malvaceae) and *Epilobium hirsutum* (fam Onagraceae). Unbroken line – secretory cell, broken line – non-secretory cell.

is no maximum 680 nm too. In plant secretory cells, chlorophyll often contributes if the cell is not completely full with a secretion.

There is also taxon-specific differences between secretory cells of various species. As shown in Fig. 2.36, an example is seen for *Lamium* genus. The fluorescence maxima in *Lamium album* are 460 and 560 nm, while in *Lamium maculatum* second maximum is small and smooth. Secretion of the *Lamium maculatum* has only maximum in blue as well as the secretory cell of the anther.

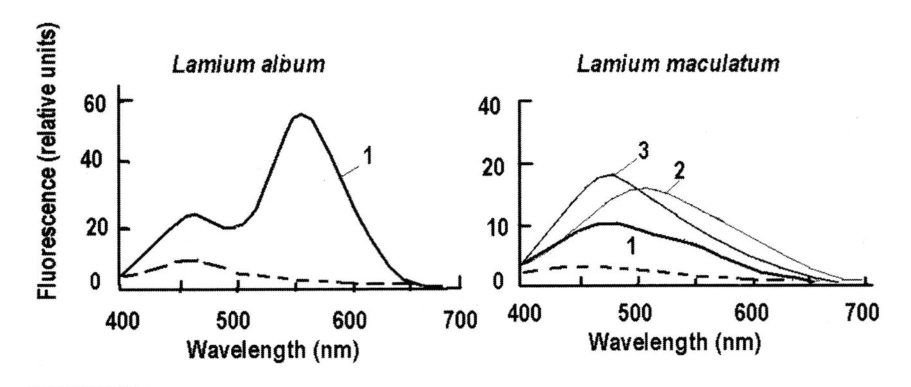

Fig. 2.36 The fluorescence spectra of secretory cells on flowers belonging to *Lamium* genus (fam Lamiaceae). 1 – secretory cell on petal; 2 – secretion; 3 – secretory cell of anther. Broken line- non-secretory cell.

Conclusion

Autofluorescence excited by ultraviolet, violet or blue light is peculiar to many plant secretions that shows the phenomenon in intact secretory cells. The fluorescence of secretory cells is well distinguished among non-secretory cells so there is no necessity to stain the cells with artificial fluorescent dyes. It is possible by non-invasive observation of the secretory cells, based on their autofluorescence to analyze the structures in the processes of development and under the influence of various factors. Fluorescent secretory cells differ in the colour of the fluorescence, position of the maxima in the fluorescence spectra and intensity of the emission. It depends on the chemical composition of the secretions and anatomical features of the secretory structures as well as taxa of the plant.

CHAPTER 3

Fluorescence of Secretions and their Individual Components

Analysis of the fluorescence emitted from secretory cells and secretions of the cells is based on the comparison of the emission spectra with those of the individual compounds of the plants. The fluorescence of secretions released on the plant surface from the secretory structures has been discussed in Chapter 2 with some examples (Table 2.1). Among the substances, which are found in secretions, and are known as growth regulators, attractants or repellents of insects, allelochemicals, and others (Roshchina and Roshchina, 1989; 1993). Spectral characteristics of some classes of the compounds found in plant secretions are given in Table 1.1 (Chapter 1). In this chapter, common aspects concerning fluorescence of the secretions or extracts from the secreting surface as well as individual compounds contained in the secretory cells will be considered as well as factors influencing the emission.

3.1 SPECTRAL ANALYSIS OF EXCRETIONS AND EXTRACTS FROM SECRETORY STRUCTURES

Secretory cells include fluorescent secretions in crystalline or hardened and liquid forms. In some cases, the light emission of excretions from the secretory cells may be analyzed on object glasses under a luminescence microscope and/or their fluorescence spectra could be registered. The analyzed examples of natural secretions released may be divided into several groups: 1. crystals on the surface or within transparent secretory structure; 2. Sporopollenin of microspores; 3. liquid excretions such as slime, nectar, guttation water, resin seen on the plant surface or released on the object glass (slice). The main information dealing with natural excretions has been given in Chapter 2, where intact secretory cells were analyzed and compared with the released secretions. In this section, some new aspects will be included.

In many cases, internal secretions are rarely excreted out of the secretory structures, and it requires special methods of analysis. Liquid components of various secretions may be extracted by the solvents and compared with the fluorescent components in intact secretory cells *in situ*. The duration of extractions is an informative indicator, as to where the secretions are located: in external structures on the tissuel surface or within the tissue. Unlike the above-mentioned natural excretions, water extracts may additionally include water-soluble cellular components, which are not contained in secretory cells. In other words, the fluorescence of these extracts could serve only as a model system for some secretions. The examples are as follows: 1. washings or leachates from the surface (leaching no more than 1 h) or the infusions from the tissue during a longer time; and 2. extracts by organic solvents, both found in plants (natural ethanol, in particular) and artificial.

3.1.1 Crystals

Crystals are usual forms of the excretions of many plants. Salts as crystals of calcium oxalate is abundant as crystalline depositions in legumes (the family of Fabaceae) and *Begonia* genus (Horner and Zindler-Frank, 1982), the crystalline resin is also seen on the surface of conifer plants (see Fig. 2.24, Chapter 2) as well as crystals of acridone alkaloids on the surface of roots (Colour Fig. 19, Chapter 2) and root cell cultures from *Ruta graveolens* (Kuzovkina et al., 1999). Crystalline excretions are often seen within transparent cells (see Colour Fig. 20, Appendix 2). As a whole, the fluorescence of many crystals *in vivo* has not been studied enough.

Crystalline excretions are seen, for instance, on the surface of motherwort *Leonurus cardiaca* L. (Fig. 3.1). The fluorescence spectra of intact cells, secretory cells and crystal have been compared. In the crystal spectra there are maxima 450, 530 and 605-610 nm. Thus it can be seen that the crystals of secretions may be multicomponent (in this case there are approximately three components, and it needs special decomposition in Gauss curves as is demonstrated in Section 3.2.1.5) system, which may include various compounds. The highest maximum in the orange-red spectral region is correlated with common fluorescence colour of the crystals. Only yellow fluorescing secretory hair of petals has maximum at 600-610 nm, but its other maximum 470 nm is higher, and the fluorescing picture is seen as yellow. Another flower secretory hair fluoresces in blue

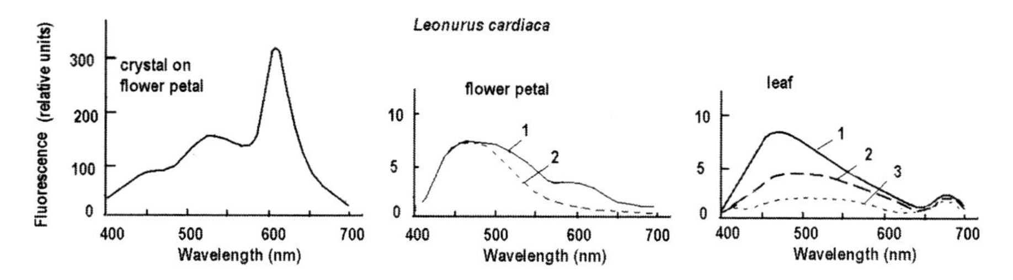

Fig. 3.1 The fluorescence spectra of crystals, leaf and flower secretory and non-secretory cells of *Leonurus cardiaca L.* Left – crystal on flower petal; Middle – flower petal. 1. Yellow fluorescing hair; 2. Blue fluorescing hair. Non-secretory cells have no light emission. Right - leaf. 1. Secretory cell on lower side; 2. Secretory cell on upper side; 3.Non-secretory cell.

(maximum 460 nm) as well as secretory cells of the leaf. Therefore, crystals within the secretory structure or are excreted on the plant surface may show a prevailing (whose concentration is highest in the secretion) component of the secretions with maximum 600-610 nm. Alkaloids leonurine, leonuridin, leoneorinin, stachydrine, betonycine and turicin are found in the medicinal plant motherwort *Leonurus cardiaca* (Duke, 2002; Dictionary of Natural Products on CD-ROM, 2004) and may contribute in the fluorescence. Moreover, it contains the glycosides of leonurin, stachidrine (β-betain tetra ammonium base of gigrinic acid), cardiac glycosides, saponins, and tannins are present as well as flavonoids apigenin, kaempferol, quercetin and rutin (Golovkin et al., 2001; Medical and Aromatic Plants-Industrial Profiles, Series, 2000-2004). The fluorescence of the crystals at 520-550 nm may be due to fluorescence of tannins and partly, to flavonoids, while saponins fluoresce weakly in the blue spectral region, if it could be registered. The weak fluorescence of the derivative of proline stachidrine often is impossible to measure.

Crystals on the plant surface often consist of different substances as shown in Fig. 3.2 for leaf secretory hairs of *Arctium tomentosum* and *Lycopersicon esculentum* as well as crystals on the intact root surface of *Ruta graveolens*. Blue-fluorescing crystals are seen in secretory cells of multicellular hairs of the leaf. In other parts of secretory hairs the fluorescence spectra have maxima in blue (450-460 nm) and in the green-yellow (500-530 nm) spectral regions whereas crystals fluoresce with only one maximum 450-460 nm. Sometimes crystals of the upper side of the leaf lighten in yellow (maximum 530 nm). Therefore, we can see possibilities of individual compounds from multicomponent secretion to crystallize in

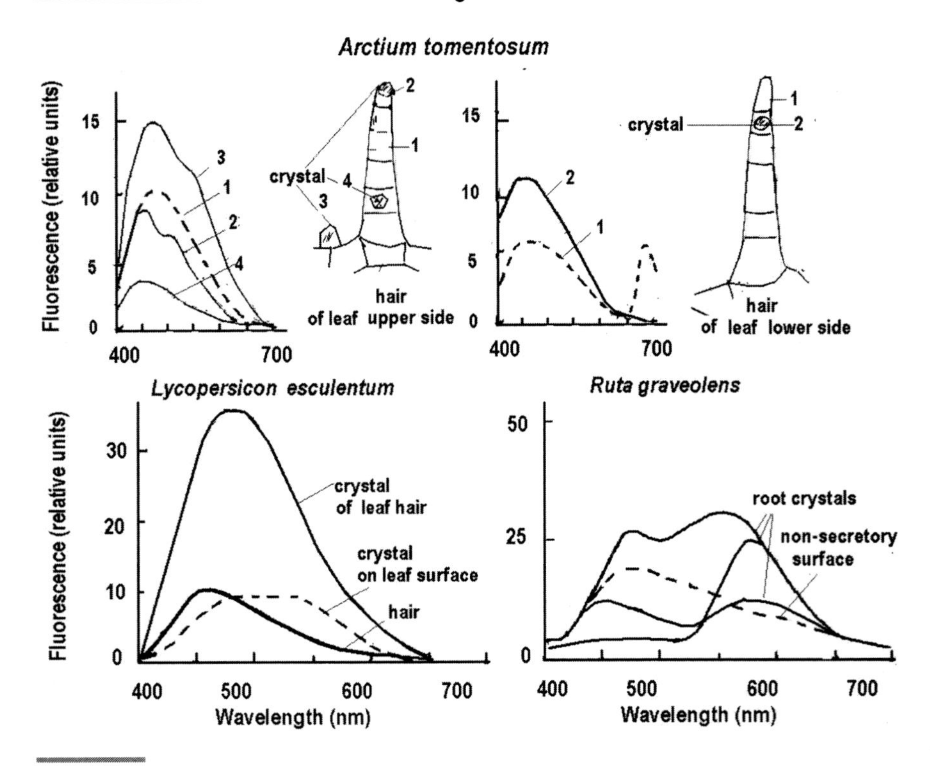

Fig. 3.2 The fluorescence spectra of crystals in leaf secretory hairs of *Arctium tomentosum* and *Lycopersicon esculentum* as well as those on the intact root surface of *Ruta graveolens. Arctium tomentosum.* Left- upper side of leaf. Right- lower side of leaf. Position of crystals are shown on the surface.

different parts of the secretory hair. Among components of the secretions from *Arctium tomentosum* may be tannins, alkaloids, sesquiterpenes (Golovkin et al., 2001). Crystals may be released on the leaf surface or accumulated within secretory hair as in *Lycopersicon esculentum* (Fig. 3.2). Moreover, the crystals contain different chemical products that fluoresce, in particular in blue with one maximum at 485-490 nm (flavonoids) in the hair and in greenish-yellow with two smoothed maxima 485 and 540 nm, when the secretions are evacuated on the leaf surface. Alkaloids such as tomatine (Roshchina and Roshchina, 1989; 1993) can also contribute to the fluorescence. Many compounds found in rue *Ruta greveolens* (Aliotta and Cafiero, 1999) may fluoresce too. Emitted crystals of acridone alkaloids (Eilert et al., 1986) are also seen on the root surface of *Ruta greveolens* (Fig. 3.2, and Colour Fig. 19 in Appendix 2) among different secretory (hairs and idioblasts) and non-secretory cells. In the fluorescence spectra they show two maxima at 465 and 550 nm. In some cases, crystals emit with maximum 590-595 nm which appears to be associated with alkaloids such as rutacridone (Colour Fig. 19, Chapter 2). Three types of the rue root crystals with maxima: 1.610 nm; 2.480 and 530 nm; 3.490 and 590-595 nm can be seen. The maxima were compared with the fluorescence of crystals, peculiar to compounds found in rue. Flavonoids rutin and quercetin also have maximum at 610 nm (sometimes additionally maximum at 470-480 nm was seen), furanocoumarin psoralen – at 530 nm, whereas alkaloid rutacridone emitted with peak at 590-595 nm. The substances in crystal forms are also clearly seen on the root surface (Kuzovkina et al., 1975) and in the root tissue culture, for instance in transformed callus (Baumert et al., 1982; Kuzovkina et al., 1979). It could be applied for the rue investigation, especially on well-luminescent cells of rutacridone-enriched roots of *Ruta graveolens*, originating from callus R20 which produced this alkaloid (fluorescence with maximum 590-595 nm) predominantly (Kuzovkina et al., 1999). Depending on the clone of the plant species, there are also various fluorescent crystals in the root tissue, filled with furocoumarins (Zobel and March, 1993).

Blue autofluorescence (excited by UV-light 365 nm) was found in a highly crystalline array of cellulose microfibrills arranged parallel to the cell surface in root epidermal mucilage released from the surface of *Zea mays* and *Triticum aestivum* (Schildknecht et al., 2004). Cellulose is the base of the crystals impregnated with fluorescent phenols. Besides, green-fluorescing crystals are also found in salt glands on the leaves of some species (Chapter 2, Section 2.2). Table 2.8 shows how intensively the crystals fluoresce on their surface. The potential of crystallization for amassing high levels of exudate is emphasized because this aspect may be characterized by the rate of the secondary metabolites' formation. In many cases, secretory hairs may form fluorescent crystals within their cells. The capacity of secretion accumulation in the crystalline form may be related to the structures of secreting cells.

3.1.2 Sporopollenin of Microspores

Plant micropores, both of vegetative from Cryptogams and generative (pollen) from Phanerogams, are covered by sporopollenin, a hardened secretion of sporangium or anther (see Chapter 2, Section 2.2.1.2.). The example of the fluorescence from pollens and anthers of *Hibiscus rosa-*

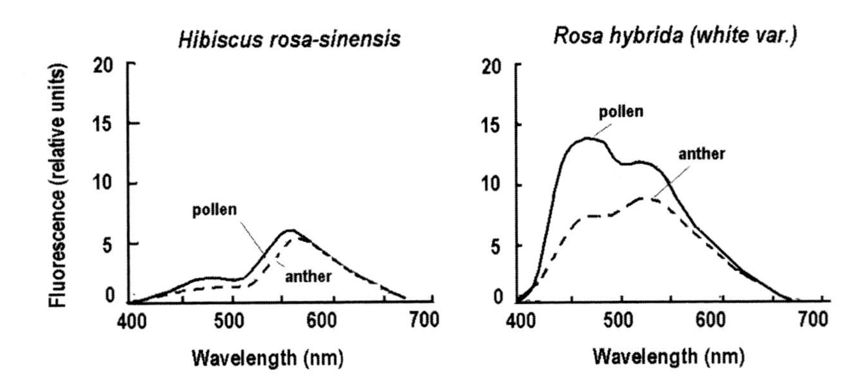

Fig. 3.3 The fluorescence spectra of pollen and anthers from *Hibiscus rosa-sinensis* and *Rosa hybrida (white var.)*

sinensis and *Rosa hybrida* (Fig. 3.3) demonstrates the similarity between the emission maxima in both structures.

Pollen from the second species has more intensive maximum in blue at 455-460 nm and in yellow at 530 nm, than in the anther, while in the first species there is no difference between pollen grain and anther. Light emission excited by UV-light represents the overall luminescence of a multicomponent system (van Gijzel, 1971; Roshchina et al., 1996; 1997b). It includes: (1) the surface luminescence of sporopollenin, exine, including such fluorescing components as phenols, carotenoids, azulenes, pyridine nucleotides, flavins and others; (2) the luminescence of the components of the intine – the membrane present under sporopollenin (they may be represented, in the main, by pectin compounds); and (3) the luminescence of the components of the cytoplasm and organelles. The main role in the global fluorescence spectrum is probably played by the sporopollenin components, since their removal leads to a considerable reduction in fluorescence in the blue region (see Chapter 1, Section 1.3, Fig. 1.7). Willemse (1971) considers that the contribution to the total luminescence of the components of the intine, the cell wall and cytoplasm with organelles is low compared with the potent fluorescence of sporopollenin itself. In addition, it should be borne in mind that ultraviolet light exciting fluorescence penetrates with difficulty into the pollen grain through sporopollenin. Sporopollenin consists of polymers of terpenoids (carotenoids) or/and phenols. The rapid changes occurring in sporopollenin lead to appreciable shifts in the fluorescence spectra during

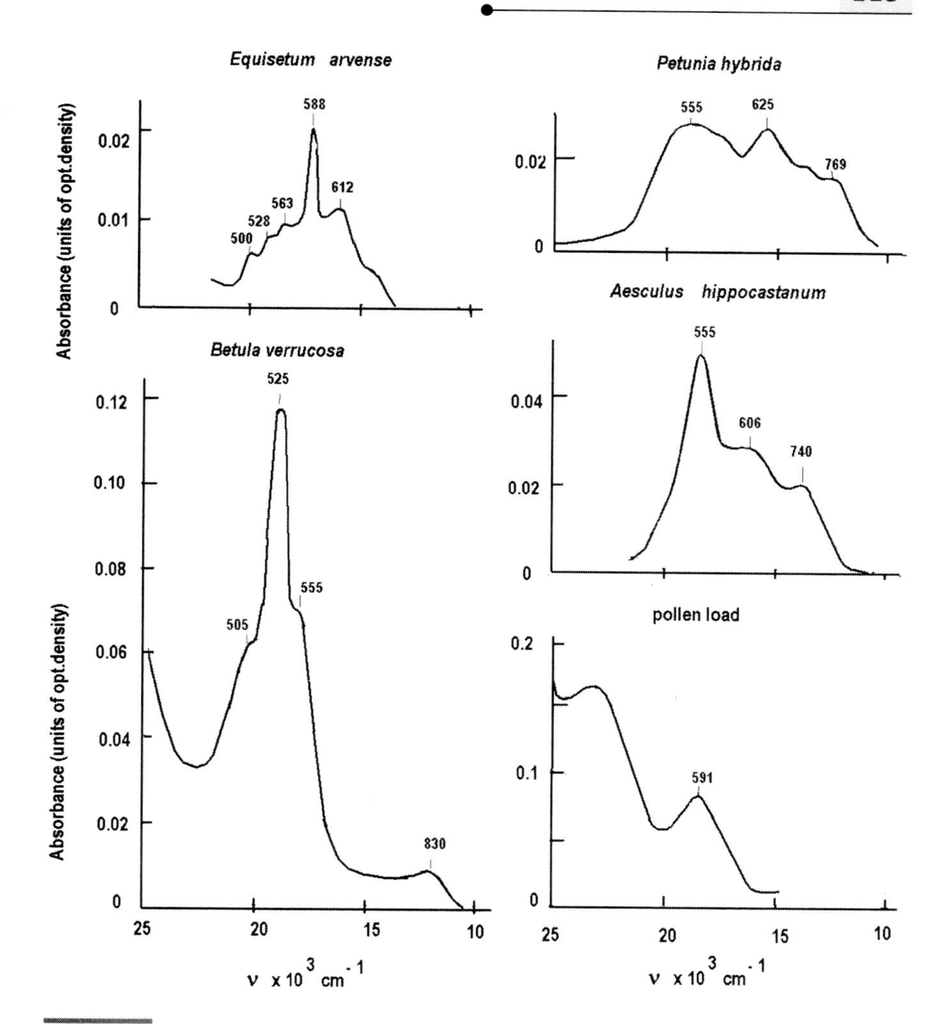

Fig. 3.4 The absorbance spectra of the ethanol solutions of azulenes purified from vegetative microspores of *Equisetum arvense*, native pollen grains from *Betula verrucosa*, *Petunia hybrida* and *Aesculus hippocastanum* as well as from pollen load (bee-collected). Sources: Roshchina et al., 1995; 1997d; 2002.

contacts of the pollen grains of different species (Roshchina et al., 1995; 1997b; 1998a; Roshchina and Melnikova, 1996; 1999) or pollen grain with the pistil stigma (Roshchina et al., 1995; 1997b) or under oxidative stress (Roshchina and Roshchina, 2003). The multicomponent composition of sporopollenin, depending on the ratio of the components, is also represented by different fluorescence spectra. Moreover, there are many inclusions in pores of the pollen exine, in particular carotenoid bodies, azulenes and anthocyanins (Roshchina et al., 1995; 1998a; 2002; Roshchina, 2003). The content of the carotenoids and azulenes in some pollens can be seen in Table 2.3 (Chapter 2). This enables us to judge the contribution predominantly of any particular component to the overall fluorescence. The remaining luminescence in the blue region after removal of the water-insoluble pigments indicates the predominant participation of these substances in the fluorescence of pollen in the yellow and red regions of the spectrum. Blue pigments azulenes, found in pollens (Roshchina et al., 1995; Roshchina, 1999a) and vegetative microspores of Cryptogams (Roshchina et al., 2002) are of special interest. Their absorbance spectra are given for some objects in Fig. 3.4 and show characteristic maxima at 550-610 nm. Although there are also maxima in ultra-violet spectral region at 379-385 nm, in blue – at 430-460 nm or in infra-red with $\lambda > 700$ nm (740-765 and 841-880 nm). They fluoresce predominantly in the blue region of the spectrum with maximum at 420-440 nm, but, if are bound with cellulose of the cell wall, red fluorescence with maximum 620-640 nm occurs too. Fluorescence of the compounds will be discussed in Section 3.2.1.2. Pollen sporopollenin of fluorescence in blue with maxima 430-440 nm is related to terpenoids (in particular azulenes), whereas with maxima at 460-480 nm – to flavonoids (445-460 nm) or folic (450 nm) or arachidonic (470 nm) acids. Green and green-yellow fluorescence may be due to the presence carotenoids (500-525 nm), flavins (riboflavin 526 nm) and flavoproteins (520-540 nm). If azulenes are bound with cell wall cellulose they may emit in red with maxima 620-640 nm (Roshchina et al., 1995; Roshchina, 1999a).

Unlike pollen grains, the composition of sporopollenin of vegetative spores of Cryptogams is not known yet. There are only the above-mentioned data related to azulenes of *Equisetum arvense* (Roshchina et al., 2002). The main fluorescence of microspores is in blue with maximum 440-460 nm, in yellow – 540-550 nm and in red – 680 nm (chlorophyll) as seen in Chapter 2, Section 2.1.1.)

3.1.3 Liquid Excretions

Liquid secretions may include secretory drops natively excreted from secretory structures on the cellular surface or products contained in the guttation water or leached by water or more lipophilic components such as ethanol, methanol, essential oils and other substances which serve as natural solvents.

3.1.3.1 Natural secretions

There are several types of the liquid secretions which fluoresce on the surface of secretory cells such as resins, slime, latex and nectar (see Chapter 2). Resin fluoresces, mainly, in the blue region of the spectra due to the presence of terpenes and phenols (Roshchina and Roshçhina, 1993), for example in *Abies sibirica*, but resin ducts and resinous cells of needle and cone, relatively, have fluorescence in the yellow part of the spectra (see Chapter 2, Section 2.2.2.5). Sometimes, resin also contains yellow-fluoresñing components, probably, proazulenes or other terpenoids, or phenolic components emitted in orange or yellow-orange (Roshchina et al., 1998a). The chemical composition of resin is often very complex. Besides polysaccharides, it contains phenol and protein components. About 20 flavonoids (5,7,4'-trihydroxy-3, 6,8,3'-tetramethoxyflavone; 5,7,3'-trihydroxy-3, 6,8,4' - tetra-methoxyflavone, etc) have been isolated from resins of various plant species (Hradetzky and Wollenweber, 1987).

Drops of slime also fluoresce on the surface of slime cells in comfrey *Symphytum officinale* (Roshchina and Melnikova, 1995). Alkaloids (such as cynoglossin, lasiocarpin) as well as tannins, digallic acids and resins may contribute in the fluorescence of the slime cells in the blue-green region.

Latex may be released and fluoresces after the mechanical damage (cutting) of laticifers (Roshchina et al., 1998a). While latex of *Chelidonium majus*, which includes alkaloids chelerythrine and sanguinarine, fluoresces with maximum 540-550 nm. The fluorescence of their laticifer coincides with the maximum. Unlike cases, in which liquid excretions have no light emission, the latex from *Taraxacum officinale*, *Euphorbia virgata*, *Euphorbia viminalis* and those from *Chelidonium majus* fluoresce in blue and orange as described by Roshchina et al., (1998a).

Excretion of nectar by pistils of some species was also observed in connection with light emission (Roshchina et al., 1996; 1998a). All types of nectar plants such as *Hippeastrum hybridum* and *Epiphyllum hybridum* have no fluorescence, perhaps, due to the presence of sugars, which do not fluoresce in the visible region of the spectrum. The fluorescence of pistil nectar was found in *Hoya carnosa* and *Bryophyllium daigremontianum* (see Chapter 2, Section 2.2.2.6). It is presumed that this occurs due to the presence of phenols or alkaloids in the nectar composition.

Another example is excretion from moistened pollen (Fig. 3.5). The first drop of the secretion contains a lot of compounds, including fluorescent with maxima at 420-470 nm. Kinetics of the fast process shows that after 10 sec of moistening the increase in the fluorescence at 450-470 nm arises which lasts maximally 10 sec and then decreases up to the original level in many species. Figure 3.5 shows it for the *Cyperus papyrus* pollen, but the same picture was demonstrated for the *Hippeastrum hybridum*, *Petunia hybrida*, *Acacia decurrens* and *Tulipa hybrida*. The composition of the first drop, which can recognize foreign or self pistil or foreign pollen, is not known yet. There is a proposition that there is an emission of phenolic compounds, for instance in *Cyperus* – vanillic, coumaric, ferulic and chlorogenic acids as well as catechins and flavonoid luteolin (Golovkin et al., 2001).

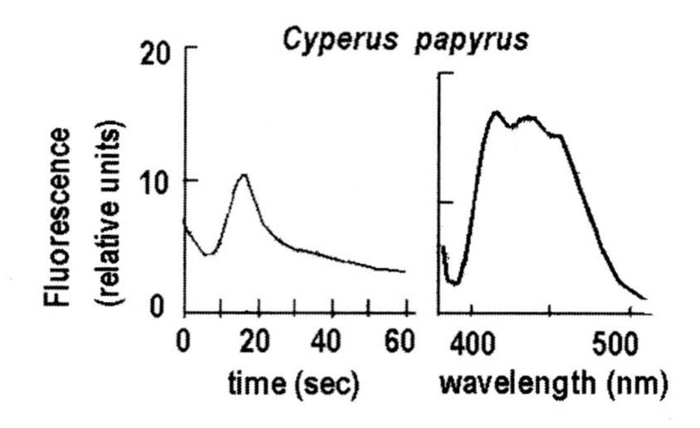

Fig. 3.5 The kinetic of the fluorescence at 450 nm (left) of the first excretion (by 0.3 ml of water) from *Cyperus papyrus* pollen (2 mg) and its fluorescence spectrum (right) measured by Perkin Elmer fluorimeter 150. Excitement by light 365 nm.

3.1.3.2 Water extracts

Some secretions may be evacuated by water. The extracts (washings or leachates from the surface, leaching no more than 1 h, or the infusions from the tissue during a longer time) serve as a model system. Washings (lechates) and infusions from flowers and leaves of various species contain flavonoids, which fluoresce with maximum 450 nm, peculiar to the bright fluorescing in the blue spectral region secretory cells of the leaf, but the fluorescence differs from the blue-emitting glands (Roshchina and Melnikova, 1995). Water extracts from bud scales of *Populus balsamifera*, *Betula verrucosa* and *Aesculus hippocastanum* contains phenolic components fluorescing in blue at 450-470 nm (Roshchiha and Melnikova, 1995). Some examples of fluorescence from water extracts of herb plants' leaves are represented in Fig. 3.6. The washings or leachates from the leaf epidermal tissues with secretory cells show the fluorescence with one maximum at

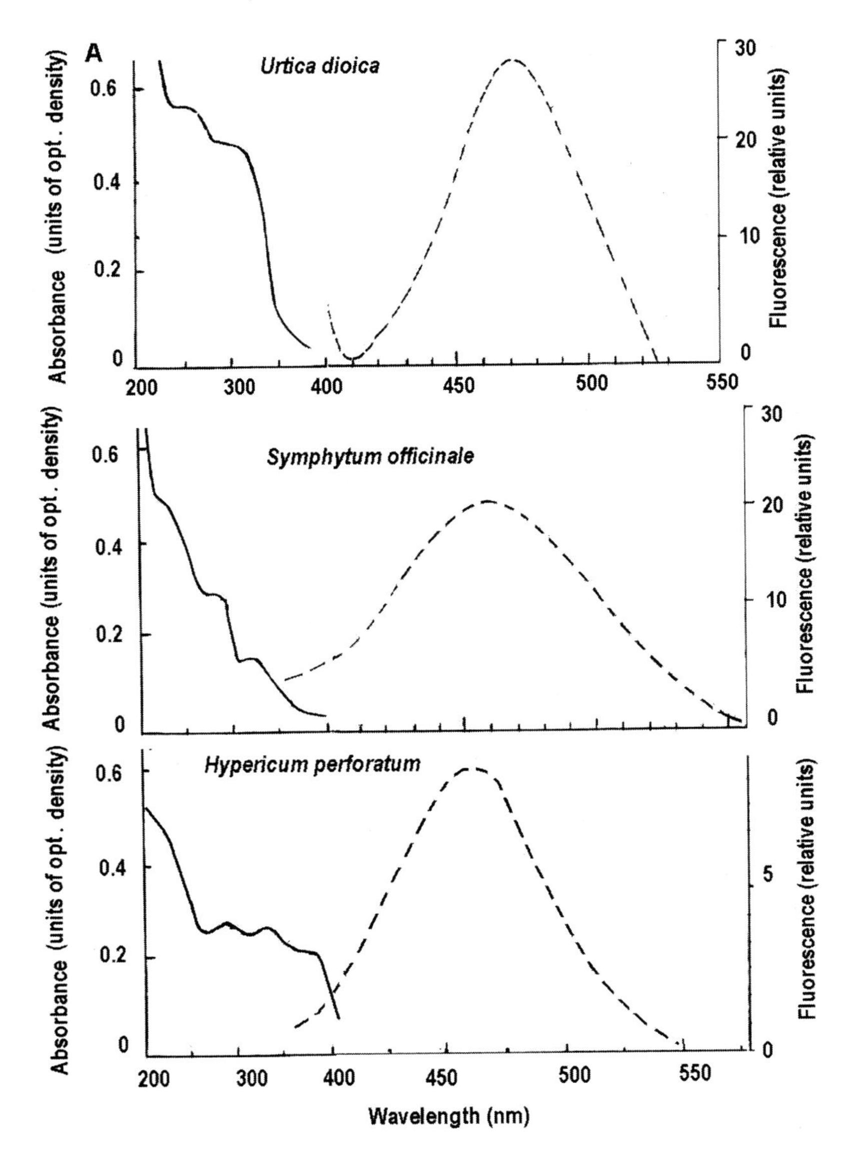

Fig. 3.6 The absorbance and fluorescence spectra of water extracts from leaves of various plant species. Washings (2 ml per 100 mg of tissue, 3 times during 1 h) from leaf epidermal surface. 1. *Urtica dioica*; 2. *Symphytum officinale*; 3 *Hypericum perforatum*.

450-460 nm for three plant species – *Urtica dioica, Symphytum officinale* and *Hypericum perforatum,* although there is a difference in the chemical composition. It may be due to the presence phenolic compounds which fluoresce in blue at 450-460 nm in stinging emergencies of *Urtica dioica* and in slime of *Symphytum officinale* as well as in oil cavities and ducts of *Hypericum perforatum* (Roshchina et al., 1997a, b, 1998a; Roshchina, 2001a, b). An analysis of the fluorescence of intact pollen and water extracts from it enabled us to evaluate the contribution to luminescence of various phenolic components too, in particular anthocyanins, which can emit not only in blue with maximum 450-470 nm, but in orange-red with maximum 585-605 nm at excitation by light 360-380 nm (Roshchina and Melnikova, 1996; Melnikova et al., 1997). See also Section 3.2.1.1 given below.

3.1.3.3 Extracts by organic solvents

Secretory components related to water-insoluble substances are extracted with more hydrophobic or more lipophilic organic solvents. Acetone and ethanol may solve both hydrophilic and hydrophobic compounds whereas chloroform, benzene and oils – only hydrophobic.

Buds scales of woody plants such as of *Populus balsamifera, Betula verrucosa* and *Aesculus hippocastanum* are enriched in the lipophilic compounds (Roshchina, V.D. and Roshchina, V.V., 1989; Roshchina, V.V. and Roshchina, V.D., 1993). In particular, their ethanol and acetone extracts fluoresce in blue at 450- 470 nm (Roshchina and Melnikova, 1995). The acetone and ethanol extracts of lipophilic pollen compounds of the some species contain flavonoids and azulenes, which can fluoresce in the blue (maxima at 450- 470 nm) spectral region, whereas carotenoids emit in green-yellow with maxima 520-560 nm (Roshchina et al., 1995; 1997a, b; 1998a).

Many lipophilic products of secretions are in oil ducts, resin ducts and other terpenoid–containing secretory structures. They may be extracted from the structures by non-polar solvents such as chloroform. The extraction from the secretory cells, which are located on the surface, depends on the time of exposure with these solvents and the rate of the solvent penetration into deeper tissues. Figure 3.7 shows that at exposures, which lasted more than 10 min, the extract may include more chlorophyll, than is contained in secretory structures located on the surface. At exposures of 60 min chloroform extracts the pigment from mesophyll. 10 min –extracts from oil glands of the *Tagetes patula* and

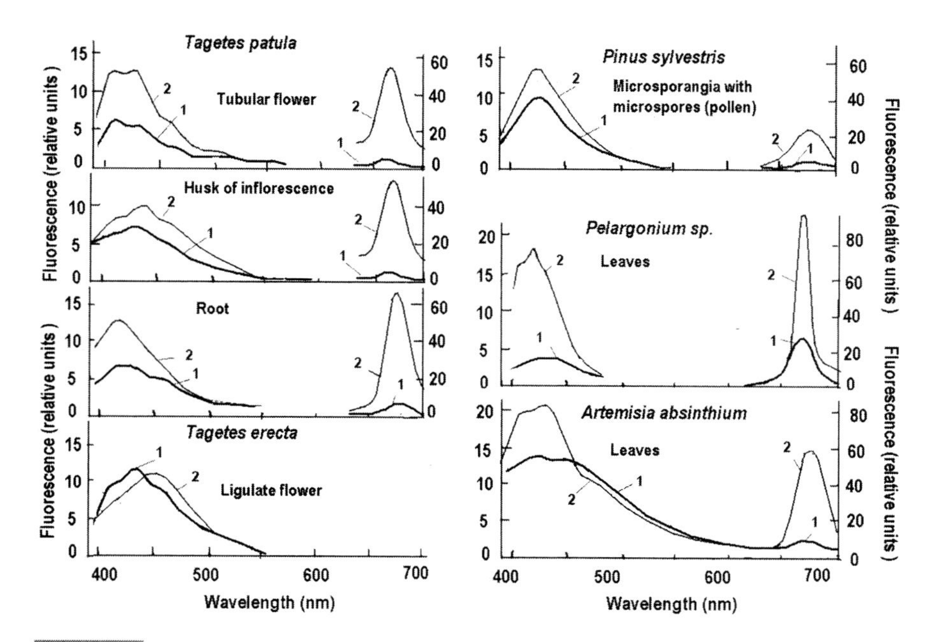

Fig. 3.7 The fluorescence spectra of the chloroform extracts (1:10 w/v) from the various plant parts enriched in the terpenoid-containing secretory structures. The fluorescence excitation – 380 nm. 1- and 2- extraction during 10 and 60 min, relatively. Y-scale for blue-yellow spectral region (left), for red spectral part (right).

Tagetes erecta, contain the chlorophyll peak 675-680 nm, excepts ligulate flower petals, in the fluorescence spectra (Fig. 3.7). The terpenoid-related emission in the spectra are seen in violet-blue (410-450 nm) and, in a lesser degree in the blue-green regions. Maxima of the emission for tubular flower of *Tagetes patula* in the 10 min-extracts were 408, 435 and shoulder 465 nm whereas in the 60 min extracts, the same extracts of husk of the inflorescence showed maxima 410, 432 and 450 nm. Unlike the flower examples, root extracts from the plant species demonstrate maxima 406, 425 and 465 nm. Extracts from the ligulate flower in other species, *Tagetes erecta*, had the peaks of the emission 410, 432 and 450 nm, and no maximum of chlorophyll. The extractions from microsporangia with microspores (pollen) of *Pinus sylvestris* also contain blue-fluorescing compounds with maxima 430-435 nm while the extractions from *Pelargonium* leaves (enriched in oil glands) – with maximum 430 nm (10 min extract) and maxima 410, 425, 430 nm (60 min extract) with the additional pick of chlorophyll. The extracts of leaves from *Artemisia absinthium* demonstrated weak maxima 430 and 460 nm after 10 min of exposure in the solvent whereas after 60 min - maxima 414 and 430 nm with the shoulder 465 nm. When the excitation was 420-430 nm, all maxima, except of chlorophyll, shifted to the longer wavelength region at 460-490 nm. The chromatography in the thin layer of silicagel or Whatman paper 1 separated the green band emitted in red (summed chlorophylls and azulenes) and weak yellowish band (terpenoids), which fluoresced in blue with maxima at 430-435 and 460-465 nm.

Internal secretory reservoirs storing oils and resin are widespread in the Asteraceae family, and one example is demonstrated for *Solidago virgaurea* L. and *Solidago canadensis* L. (Lersten and Curtis, 1989). Intact oil reservoirs fluoresce in blue (maxima 410 and 428 nm), when excited by ultra-violet 360-380 nm or blue-green (maximum at 450 and 470 nm when excited by violet light 420 nm) as seen in Table 2.5 (Chapter 2) and in Fig. 3.8. Extraction during 10 min shows similar emission, but after a longer extraction some components with emission maxima at 485 and 528 nm appeared (Fig. 3.8). Chlorophyll maximum 680 nm arose more, than 100

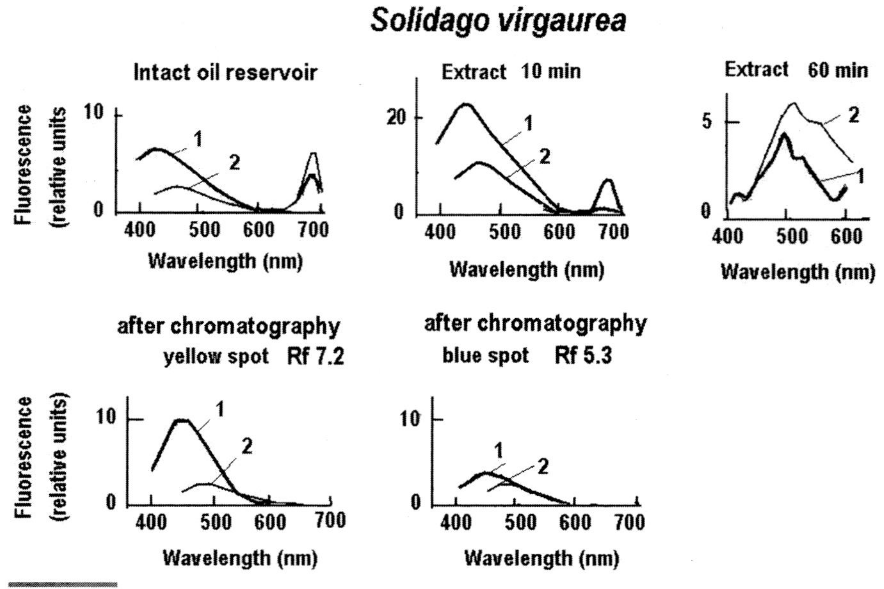

Fig. 3.8 Analysis of the fluorescence spectra of intact leaf oil reservoirs in *Solidago virgaurea* and fluorescence spectra of extracts from the leaves made during 10 and 60 min by chloroform. 1 and 2 excitation by light 360-380 nm and 420 nm, relatively.

times which was difficult to include in one and the same spectrum of the extract. Therefore, practically 10 min-extraction is desirable for releasing oil secretory products from epidermal secretory structures without chlorophyll of mesophyll cells lying below. In absorbance spectrum of 10 min-extract, there are maxima, peculiar to carotenoids (430 and 450 nm) or to azulenes (610 and 615 nm) and to chlorophyll (650-660 nm), whereas chromatographically separated colour spots have maximum at 357 nm (yellow spot with Rf 7.2, possible terpenoid-phenol oil) or 611 and 614 nm (blue spot with Rf 5.3, possible azulenes). As seen in Fig. 3.8, only the yellow spot intensively fluoresces in blue-green with maximum at 450 or 470 nm, depending on the wavelength of excitation. The blue spot (perhaps, azulenes) fluoresces in blue with maximum 450 nm. Azulenes often are mixed with oils in reservoirs (Roshchina and Roshchina, 1993).

3.1.3.4 Fractionation of liquid extractions

Water, acetone, ethanol, chloroform and benzene extracts from vegetative microspores of *Equisetum arvense* and pollen grains of *Hippeastrum hybridum* were also analyzed, according to the following scheme (Roshchina et al., 1996):

Material + distillate (1:10) 30min → **Water fraction**
(supernatant after centrifugation)
↓
Pellet (drying on the air) + acetone or ethanol (1:10) 30min → **Acetone or ethanol fraction**
(supernatant after centrifugation)
↓
Pellet (drying on the air)+ Benzene or chloroform(1:10) 30min→ **Benzene or chloroform**

Water fraction, acetonic fraction and benzene fractions were collected as well as diffusates from the surfaces, and their fluorescence spectra were analyzed (Fig. 3.9).

Water fraction from vegetative microspores has no marked maxima in the fluorescence spectra, while the same fraction from pollen and pistils fluoresced in the blue spectral region with maxima at 460-480 nm which is peculiar to water-soluble compounds such as NAD(P)H and flavonoids (Roshchina et al., 1998a; 2003a). All these components were found in pollen (Stanley and Linskens, 1974) and in the pistil stigma of various plant species (Knox, 1984; Vogt et al., 1994). Similar maximum in blue (460-480

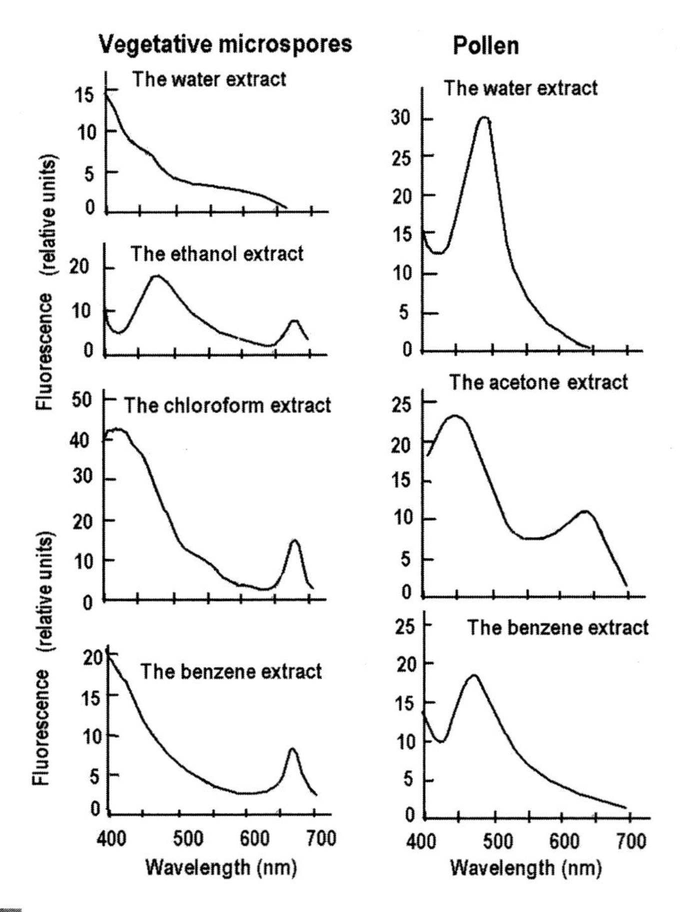

Fig. 3.9 The fluorescence spectra of the extracts by water and organic solvents from vegetative microspores of *Equisetum arvense* and from pollen *Hippeastrum hybridum.*

nm) is peculiar to the ethanol and acetone extractions of both vegetative microspores and pollen grains, and, additionally to maximum in blue, have maximum of the emission at 620-650 or 680 nm. The contribution of chlorophyll is possible in this case. Unlike vegetative microspores, usually matured pollen grain is lack of chlorophyll, which fluoresce at 675-680 nm. Therefore, acetone-extracted component belongs to other chemical groups of compounds, perhaps, to azulenes. Azulenes are found in pollen (Roshchina et al., 1995; Roshchina, 1999a) and vegetative microspores of horsetail (Roshchina et al., 2002). The chloroform fraction from vegetative microspores, besides maximum at 675 nm, has maxima at 412 and 425 nm which is characteristic for terpenoids. The benzene fraction of pollen contains strong lipophilic compounds, which emit with maximum 470-480 nm. The contribution in the blue fluorescence in water, ethanol and acetone fractions may be due to phenolic residues of components of cellular walls and exine of pollen (Hartley and Tones, 1976). As most lipophilic fraction was analyzed, one can see that all benzene extracts from pollens fluoresce with maxima in blue at 460-470 nm (peculiar to terpenoids) and in green-yellow at 500-540 nm (carotenoids) which is peculiar to terpenoids including carotenoids, relatively (Chappele et al., 1990; Roshchina et al., 1998a; Roshchina and Melnikova, 1999; 2001; Roshchina and Roshchina, 2003). Moreover, the components of lipid peroxidation and free radical reactions may contribute in the light emission of all the extracts (Merzlyak, 1989). Free radicals were found on the pollen surface (Dodd and Ebert, 1971). This possibility is shown in model experiments (see Chapter 5, Section 5.1.2.).

Water excretions from pistils of *Hippeastrum hybridum* have no fluorescence or, have weak ones with maximum at 440-460 nm (Roshchina et al., 1996; Roshchina, 1999a). Perhaps, this is due to the presence, mainly of sugars, which do not fluoresce in the visible region of the spectrum. Unlike acetonic extracts from pollens, analogous fraction from pistils had not only weak emission in the blue region (460-470 nm) of the spectrum, but also – intensive emission in red at 680 nm. Significant maximum 680 nm is related to chlorophyll and is also seen in the fluorescent spectra of intact pistil. Benzene fraction of pistils fluoresces with two maxima in blue at 460-470 nm and in green-yellow at 500-540 nm which is peculiar to terpenoids and carotenoids, relatively (Chappele et al., 1990; Roshchina et al., 1998a; Roshchina and Melnikova, 1999; 2001; Roshchina and Roshchina, 2003).

3.2 FLUORESCENT SUBSTANCES OF SECRETIONS

Observed fluorescence of intact cells is a sum of emissions of several different groups of substances, both excreted out or accumulated within the cell and linked on the cellular surface. The visible emission from secreting cells may occur from the secretory products themselves located on the cell surface, in extracellular space or within secretory vesicles and vacuoles. Some contribution in the process is also due to structural components of the cells – the cell wall, plasmalemma and cytoplasm with some organelles (see Chapter 1, Section, 1.3).

3.2.1 Fluorescence of Secondary Metabolites Found in Secretions

The fluorescence spectra of the secretory compounds can differ strongly, depending on the emitting fluorophore. The absence of fluorescence is either due to the lack of electron systems as in the case of fatty alcohols, some steroids, and pheromones or due to the presence of rather short and non-fluorescent π-electron systems, as in the case of fatty acids; aldoses and others; the mono-, di-, and some trienes, vitamin C; and most of the amino acids (Wolfbeis, 1985). But most of the plant secondary substances can fluoresce in the visible region of the spectrum. If such a compound has no fluorescence, the cause perhaps, also is either due to the lack of electron systems as in the case of fatty alcohols, some steroids, and pheromones or due to the presence of rather short and non-fluorescent π-electron systems. In addition, the presence of a transition metal ion in a fluorescent system may render it non-fluorescent (e.g., heme and related transition metal complexes). The rules for the fluorescence are as follows. Substances can emit in all states if have discrete energetic spectra, otherwise the energy transforms to warm energy instead of fluorescence. According to Pheophilov (1944), many substances, in particular dyes, fluoresce in rigid solutions (sugar, gelatine, plastic, etc) and in an absorbed state. The possibility of the dyes to fluoresce in such a state is observed if their structure allows rotation of two big large parts of the molecule surrounding the chain of conjugated double bonds, which condition the fluorescence colour of the compounds. The light emission depends on the chemical groups of the compound (Table 1.1, Colour Fig. 21). Aromatic hydrocarbons fluoresce if they include amino-, alkyl-, and oxy- groups (Goryunova, 1952). But the inclusion of nitro-, nitrozo-, halogen- groups may quench of the fluorescence (Zelinskii, 1947). The ageing pigments, which are formed in the reaction of malondialdehyde with the free amino acids, in particular in plants tissues fumigated by ozone, fluoresce in blue (Roshchina and Roshchina, 2003). In many plant excretions, there are non-fluorescent polysaccharides, which are usual in slime or mucilages. The slime also include proteins, in many cases are non-fluorescent too.

3.2.1.1 Phenols

Phenols are compounds, which have one or more phenolic groups (benzene ring with the hydroxyl groups). The class of the substances is subdivided in several groups: aromatic acids, flavonoids, coumarins, furocoumarins or furanocoumarins, tannins and their derivatives The most important single group of phenolics in plants is the flavonoids, which consist mainly of catechins, proanthocyanidins, anthocyanidins and flavons, flavonols and their glycosides, for instance glycosides of anthocyanidines are known as anthocyanins. Among the variety of naturally occurring oxygen ring compounds coumarins, furocoumarins and flavonoids fluoresce intensively (Wolfbeis, 1985). Chromenes and chromones also have native fluorescence. 2,2-Dimethyl-2H-chromene is non-fluorescent, but the introduction of an 8-methoxy group gives rise to fluorescence (Wolfbeis, 1985). Flavonoids fluoresce in blue, but some of them also have maxima in green and yellow-orange (see below). Examples of the fluorescence spectra for main groups of phenols found in plant secretions are represented in Fig. 3.10.

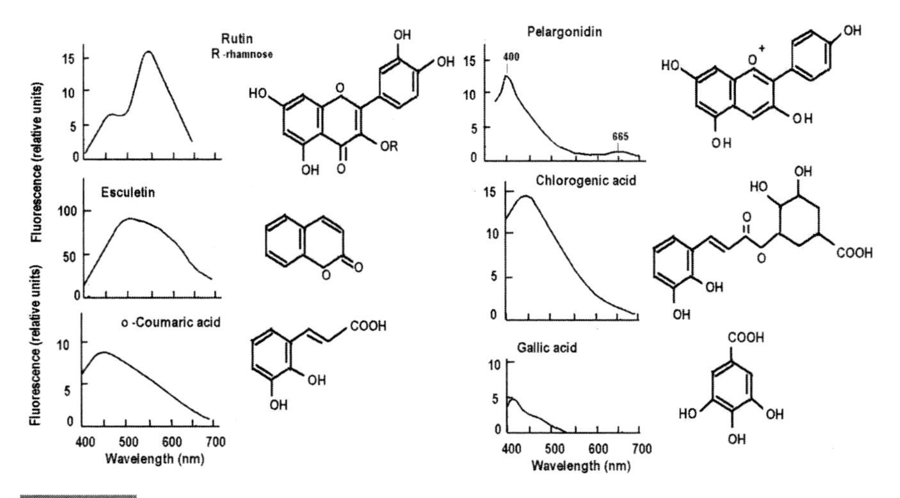

Fig. 3.10 The fluorescence spectra of some secreting phenols (2 mg per ml) solved in water (pelargonidin), ethanol (esculetin, o-coumaric, chlorogenic and gallic acids) or dimethylsulphoxide (rutin).

All groups of phenols are found in glandular hairs and epithelium of many plant species (Roshchina and Roshchina, 1989; 1993). They may be concentrated in the vacuolar space of the trichomes as water-soluble glycosides of flavonols and anthocyanins, but their lipophilic aglycons usually are excreted out of the cell in oil and resinous secretions. A lot of phenols, mainly, flavonols are found in buds of woody plants (Wollenweber, 1984; Wollenweber et al., 1987; 1991). Kaempferol is secreted by pistil stigma, when the flower is ready for fertilization (Vogt et al., 1994). The roots of many plants secrete and excrete various phenolic compounds (Rao, 1990). Some phenols when released in the cell wall form polymers, as it is seen for lignins.

Aromatic acids. Aromatic acids are mainly derivatives of amino acids – phenylalanine and tryptophan, and usually have one benzene ring. Components of the cell wall include ferulic acid and carbohydrate esters in many plants (Hartley, 1973) which may fluoresce. Blue-green fluorescence is related to ferulic acid as a main fluorophore of epidermis (Morales et al., 1996; 1998). The phenomenon in particular was studied well in the family Gramineae by the use of ultra-violet fluorescence microscopy (Harris and Hartley, 1976). Chlorogenic, ferulic and caffeic acids solved in ethanol may fluoresce in blue with maxima at 450-460 nm. Aromatic acids and their derivatives may be included in many mixtures of plant secretory products (Roshchina and Roshchina, 1989; 1993).

Polymeric phenols. Polymeric phenols – lignins of cell walls (formed by earlier secreted monomers) often demonstrate blue lightening in ultra-violet light (Zobel and March, 1993). Among products secreted through the cell wall is hydrogen peroxide which induces the accumulation of insoluble fluorescent material on the surfaces of cell walls of living coleoptyl segments (Schopfer, 1996). When polymeric material containing phenolic components (cinnamic acid derivatives) accumulates in the cell walls, H_2O_2 takes part in formation of phenolic cross-links, the fluorescence of the surface increases and becomes bright greenish-yellow. The peroxidase-catalyzed polymerization of the pollen sporopollenin also needs H_2O_2 (Scott, 1994), and the pattern of light emission during the process of pollen development (Roshchina et al., 1997a) may be related with the peroxidation reactions.

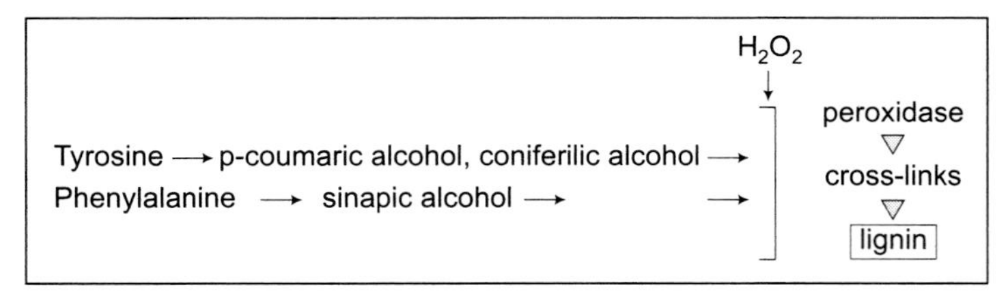

Flavonoids. The group of flavonoids includes compounds having double conjugated benzene rings with an attached third ring: flavons, flavans, flavonols and anthocyanins, etc. Flavonoids are found in cell walls, vacuoles, chloroplasts (Roshchina and Roshchina, 1993). Among them, coloured pigments anthocyanins may be found in cytoplasm and vacuolar sap in the form of crystals or bodies as well as stored in cell walls (Yasuda and Shinoda, 1985).

Many flavonoids fluoresce in the green spectral region if pH is higher than 8, but the emission disappears or decreases at lower pH (Wolfbeis, 1985). Addition BrO_3 makes the 5-hydroxyflavones become a greenish-yellow fluorescent. The fluorescence spectra of flavonol (Tyukavkina et al., 1975) and other flavones (Frolov et al., 1975) were analysed (Frolov et al., 1978). Kopach et al. (1980) recorded the fluorescence properties of galangin, kaempferol, morin, quercetin, and rutin in methanol and in a solid state. The emissions at room temperature are sometimes weak, but increase markedly at 77°K. The colour is yellow, as is typical for flavonol derivatives. Unlike other reports, which state that many flavonols show two fluorescence bands (violet/blue and green/yellow), this work gives only one fluorescence maximum for methanol solutions at room temperature, but two for temperatures of 77°K. According to Wolfbeis (1985), the following generalizations on the fluorescence properties of flavonoids can be made: 1) 3-, 6-, 7-, and 4'-Hydroxyflavones fluoresce strongly, 2'-, 3'-, and 8-hydroxyflavones fluoresce weakly, 5-hydroxyflavones are practically non-fluorescent; 2) Flavones exhibit exceptionally large Stokes' shifts; 3) The fluorescence of flavonols (3-hydroxyflavones) is green whereas others fluoresce blue. 4) 5-Hydroxyflavones become increasingly fluorescent with an increasing number of oxygen functions being in the molecule. 5-Hydroxyflavones begin to fluoresce in green when the thin-layer chromatogram spot is sprayed by methanol solution of diphenylboric acid p-aminoethylester. Complexes of flavonoid aglycons with metal salts gave various colours of the emission from blue (hesperidin, flavones) to blue-green (quercetin) or even yellow (rutin) observed by Inglett with co-workers (1959). These effect of the emission enhancement results from interactions of the excited-state of fluorophore with free (plasmon) electrons in the metal that can increase the extent of resonance energy transfer (Aslan et al., 2005).

In rigid, crystalline state flavonoids, in particular flavone quercetin has maximum 580 nm, although most of them also fluoresce in blue 450-470 nm, but in some cases the shifts observed from the blue to orange spectral region, in particular after the fumigation with ozone (Roshchina and Melnikova, 2001). The presence of two maxima in the flavonoid spectra of the solutions is not clear and explained by the interactions with solvent and formation of complex associates (Kopach et al., 1980). According to Kopach et al. (1980) quercetin has maximum 581 in a rigid state and 543 nm in methanol. Quercetin complexes with proteins, in particular with albumins, emit with maxima from 450 to 600 nm if excited by light 290 nm (Mishra et al., 2005). Moreover, the proteins stimulated the fluorescence intensity of quercetin. This fluorescent flavonoid and its derivatives as well as apigenin, naringenin and others are found in pollen of many plant species (Stanley and Linskens, 1974; Zerback et al., 1989). Some flavonoids do not demonstrate any fluorescence at room temperature. It is non-emitted transition, which in particular is characteristic for galangin and kaempferol (Kopach et al., 1980; Wolfbeis, 1985), but with the excitation by light 438 nm, kaempferol fluoresce with maximum 478 nm and kaempferol-3-rhamnoside excited by light 403 nm – with maximum 525 nm (Ribachenko and Georgievskii, 1975). Rutin may also fluoresce with two maxima at 460 and 600-610 nm (Fig. 3.10). Fluorescence of flavonols in hydrocarbon solutions is controlled with water traces. In the absence of water, ethanol and other proton-accepting molecules, only green or violet fluorescence is observed.

Among flavonoids, flavanons such as flacosid (7-β-D-glycopyranoside-8 (3-methylbut-2-ehyl)-4',5,7-trioxiflavanon) from leaves of *Phellodendron amurensis* Rupr. (Family Rutaceae) have weak bluish emission, while its derivative alpisarine (2–C-β-D- (glycopyranoside)-1,3,6,7, -tetraoxixanthone) from the herb of *Hedysarum alpinum* L. and *H. flavescens* L. (family Fabaceae) demonstrate more significant fluorescence with maximum at 430 nm. As mentioned above, flavonols rutin and quercetin fluoresce in bright blue and may also have maximum in orange as well as flavons (Bondar et al., 1998).

Anthocyanins may fluoresce not only in blue, but with maximum in the longer spectral region (Santhanam et al., 1983) as well as in green (at 510 and 515 nm), yellow- orange and even red (555, 570-585, 610 nm). For example, under ultra violet light of the microscope the yellow fluorescence of anthocyanin pelargonin (aglycon) and its glycoside pelargonidin is seen. Pelargonidin in 96% ethanol (pH < 7) has the emission maxima 402 and 430 nm as well as small maximum 585 or 665 nm (Fig. 3.10). In water-ethanol mixture (50:50) maxima 410, 468 and 590 nm arose. The anthocyanin fluorescence depends on the formation of complexes or conjugates with metals or other cellular components.

The fluorescence of various flavonoids in their mixtures differs from the emission from the individual compound. Flavons stimulate the emission of anthocyanins whereas guercetin quenches it. On the contrary, other phenols can stimulate the luminescence of the anthocyanin complexes or conjugates with metals. The anthocyanins, which have no free hydroxyl groups in C 15 position can fluoresce in the phase water-citric acid with maxima 510, 515, 555, 570-585, shoulder 610 nm. When the compounds applied to Silicagel plates or Avicell ST microcrystallic cellulose, delphinidin, petunidin and malvidin fluoresce in blue, green and red, while cyanidin and peonidin – in red, and pelargonidin – in orange – red spectral region (Wolfbeis, 1985). The exposure of petunidin in NH_3 leads to shifts in their fluorescence from the green to yellow spectral region. According to Lynn and Luh (1964), the fluorescence spectra of the anthocyanin pigments (excitation by light 300 nm) extracted from Bing cherries (*Prunus avium* L. var. Bing) with 1% methanolic HCl and separated by paper chromatography showed characteristic maxima that can be used to diffrentiate the anthocyanins. The cyanidin-containing pigments (cyanidin 3-rutinoside and cyanidin 3-glucoside) have a characteristic fluorescence peak at 520 mμ, whereas the peonidin ones exhibited a peak at 610 mμ.

Coumarins. The smell of herbs in many meadow plants, for example clover and alfalfa, is due to the presence of coumarins. The fluorescence of coumarins is described in the review of Wolbeis (1985). Coumarin itself is said to be non-fluorescent in a fluid solution, but emits weakly (max 348 nm) in an ethanol glass at 77 K. o-Coumaric acid emits in blue with maxima at 440-450 nm depending on the solvent (Fig. 3.10). Coumarin derivative esculetin has the characteristic absorbance maxima at 258, 300 (shoulder) and 350 nm, while fluoresces at 460 nm (Roshchina and Melnikova, 2001). 7-Alkoxy coumarins generally have a purple fluorescence, whereas 7-hydroxy and 5.7-dihydroxycoumarins tend to fluoresce in blue.The addition of monovalent cations, which form complex with the carbonyl group, results in a dramatic increase in intensity. Introduction of a hydroxyl group makes all coumarins fluorescent at room temperature, except 8-hydroxycoumarin. Several oxycoumarins fluoresce depending on the solvent and position of the substituent, which determines their intensity and the location of the emission band (Fedorin and Georgievskii, 1974a). A comparison of fluorescence maxima in dry methanol solution shows that both 3- and hydroxycoumarin fluoresce in UV-light, while the 7-, 5-, and 6-isomers exhibit visible fluorescence. In aprotic solvents the emission maxima of alkoxy derivatives are similar to those of the hydroxy derivatives, but 4-methoxycoumarin- are non-fluorescent. In water the spectra of 7-hydroxy- and 7-alkoxycoumarins differ largely. The fluorescence intensity of umbelliferones is greatly enhanced by induction of electron-delocalizing substituents such as phenyl in position 13 (Wolfbeis, 1985). The native fluorescence of natural coumarins occurs in the fruit secretory cells of the genus *Citrus*, in particular in the peel of grapefruits, mandarins, and lemons. The oils of lime, mandarin, lemon, and bergamot contain compounds mixed with terpenoids.

Furo(no)coumarins. Furocoumarins are derivatives of coumarin base with an attached group of furan. The chemical nature of the compounds is described in the monograph of Kuznetsova (1967). Among the two main groups of furocoumarins the aflatoxins are highly fluorescent, while psoralens show more intense phosphorescence than fluorescence. Furanocoumarins are concentrated in various secretory structures of *Ruta graveolens* L. – schizogenous receptacles of sepal, petal and ovary of the flower, fruit, leaf plate, and ferns, in glandular hairs of flower sepal, petal and style, and in root idioblasts (Andon and Denisova, 1974). Differences were shown in histological localization of psoralens in fruits and seed coats of *Angelica archangelica, Daucus carota. Sium suave* and *Psoralea bituminosa* (Zobel and March, 1993). Autofluorescence related to furocoumarins can be homogeneously distributed among all cells of the embryo *(Sium)* or can be concentrated in certain cells *(Psoralea)*. The fluorescence in each case was compared with that of crystals and saturated solutions of psoralen, bergapten and xanthotoxin under similar conditions. Furanocoumarins can play several roles in seeds and fruits. Friedman et al. (1982) found xanthotoxin produced by *Ammi majus*. The compound appears to be a strong germination inhibitor, acting more strongly on seeds of other species or act as autoinhibitors in the embryo, as they were found to inhibit mitosis (Zobel and March, 1993). The second role would be a defence role when these compounds are extruded to the surface (Zobel and Brown, 1990) and have been known for many years as antimicrobial agents (Murray et al., 1982; Towers, 1987a, b). The third role is to prevent other seeds from germinating in the vicinity of seeds containing coumarins.

The tissue of rue *Ruta graveolens* contains furanocoumarins such as psoralen, xanthotoxin and bergapten, which are located, mainly, in the leaf epidermis and in small amounts – in glands of the stem (Zobel and Brown, 1989). The slices of fruit skin, leaf, and stems demonstrated the fluorescence in epidermal cells. Psoralen fluoresces in blue, xanthotoxin looks dull orange, whereas bergapten - bright blue. Cuticle layer lightens greenish, epidermal glands–light greenish and mesophyll cell walls–green. The experiments showed that blue fluorescence of epidermis is the main origin of furanocoumarins.

Stilbenes, indoles, naphthoquinones and anthraquinones. Phenolic compounds such as stilbenes, indoles, naphthoquinones and anthraquinones, have aromatic rings. The stilbenes and anthraquinones fluoresce in blue from 419 to 440 nm, whereas indoles are lack of light emission in the visible spectral region (Wolbeis, 1985). The indole derivative biogenic amine serotonin shows (Table 1.1 in Chapter 1) two characteristic maxima in absorbance spectra at 285 nm and shoulder at 310 nm and weak maximum 430 nm in fluorescence spectra. While anthraquinones, hypericin, which is found in glands of flowers and leaves of *Hypericum* genus, fluoresce with maxima 608 and 650 nm (Fig. 3.11). Its fluorescence was analyzed by Andreoni et al. (1994) and Benes et al. (2001).

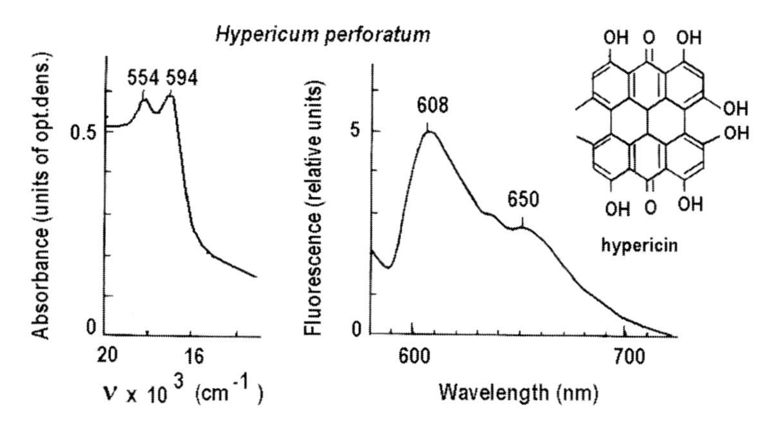

Fig. 3.11 The absorbance and fluorescence spectra of acetonic extracts from *Hypericum perforatum*. (Fluorescence excited by light 380 nm).

Tannins. Tannins are polycondensed phenols, which are known as constituents of cell walls and also play roles as effective allelochemicals and medicinal drugs. Among tannins one can distinguish the hydrolyzable and nonhydrolyzable ones (condensed). They are distinguished depending on the type of their phenolic nuclei and modes of the nucleus linkage. The first class of tannins is constructed on the base of multiatomic alcohol (for instance glucose), of which hydroxyl groups are esterified by gallic acid or relative compounds such as hexaoxydiphenic acid (Swain, 1965). Gallotannins are first derived from galls formed on the leaves and branches of *Rhus semialata* and *Quercus infectoria*. Among hydrolyzable tannins, many contain benzoid acids, such as syringic, synapic, protocatechuic, vanillic, ferulic and their derivatives. Dilactones such as valoneaic acid in valonia *Quercus aegilops* is also present. Second class, nonhydrolyzable tannins, are formed by condensation of gallocatechins and similar compounds (Swain, 1965).

The example is gallic acid included in hydrolysable tannins. Gallic and valoneaic acids emit with maximum 420-425 nm under UV-light 360 nm. Water solutions of some tannins such as chamaeriol from *Chamaerion* (*Chamerion*) *angustifolium* Holib. also fluoresce in blue at 420-450 nm if the excitation is 360-380 nm and in blue-green at 480-500 nm under violet light 420 nm (Roshchina, 2005c).

Tannins are abundant in seed pods, fruits, bark and wood as well as in heartwood and sometimes in leaves, twigs and galls on leaves (Swain, 1965). They are present in idioblasts (Buvat, 1989), and various cells of roots, tubers and rhizomes (Swain, 1965).

3.2.1.2 Terpenoids and their derivatives

Among the main groups of substances, which have conjugated double bonds and belong to terpenoids, are various compounds: from mono-, di-, triterpenes, sesquiterpenes and sesquiterpene lactones with derivatives such as alcohols and aldehydes to carotenoids and polymers of terpenoids or hydrogenased sesquiterpene lactones. Secretory cells often include terpenoids in a composition of mixtures. For instance, oil cells contain mono-, di-, triterpenes as well as sesquiterpene lactones (Roshchina and Roshchina, 1993). Exine of cover in pollen and vegetative microspores of *Equisetum arvense* includes earlier secreted carotenoids and azulenes (Roshchina et al., 1995; 2002). Terpenoids have various functions in plants – from pigments and toxins (Halligan, 1975; Goodwin and Mercer, 1983) to messengers (Harrawiji et al., 2001). Moreover, sesquiterpene lactones show high biological activity (Rodriguez et al., 1976; Rüngeler et al., 1999) inhibiting development and growth of plants (Fischer et al., 1989; Fischer, 1991; Schmidt, 1999).

Figure 3.12 shows the fluorescence spectra of some representatives of the compounds – monoterpenes, carotenoids, azulenes and proazulenes (sesquiterpene lactones). The emission also depends on the solvent. Monoterpenes and its hydrogenated derivatives usually have a weak fluorescence as a whole, and predominantly in the blue spectral region, in particular as can be seen for menthol (the plate-like absorbance in region 250-280 nm, and fluorescence weakly – at 435-460 nm). Carotenoid β-carotene, besides peak in blue, has maximum in the yellow orange region – at 590-600 nm. Sesquiterpene lactones such as azulenes and proazulenes, in particular austricine, inulicine and hydragenated sesquiterpene lactone ledol, fluoresce with maxima in blue at 408-450 nm.

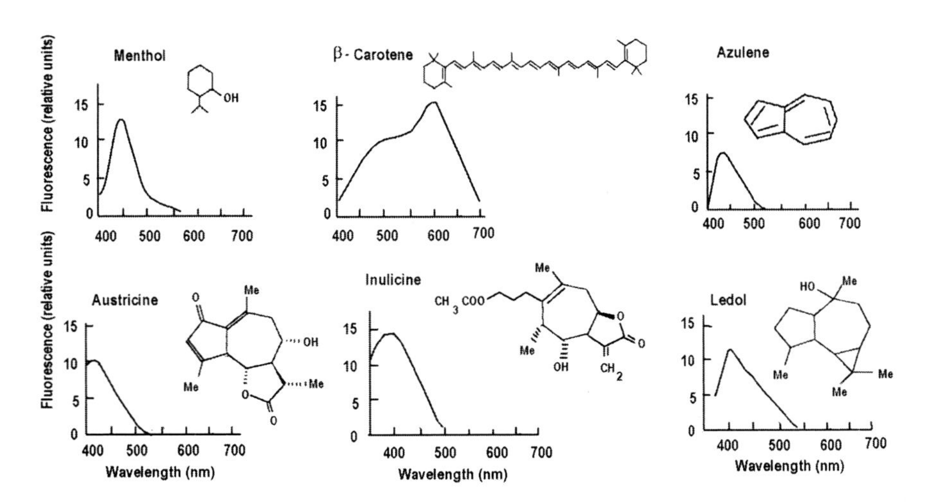

Fig. 3.12 The fluorescence spectra of representatives of main terpenoids (1 mg/ml) solved in 100% ethanol

Terpenoid-containing secretory cells both in glands and in ducts fluoresce in the blue (450-460 nm) and green-yellow (500-550 nm) spectral regions (Roshchina et al., 1998a, d). The first is typical of sesquiterpenes and monoterpenes of oils. Shoulder in the red region (>600 nm) may belong to derivatives of sesquiterpene lactones such as azulenes bound with cell wall. Blue pigments azulenes are found in essential oils of glands in some Asteraceae species (Konovalov, 1995), in needles of conifer plants (Kolesnikova et al., 1980), in pollen grains (Roshchina et al., 1995) and in other secretory tissues and cells, for example in clovers leaves (Roshchina, 1999a) and in oil glands from *Solidago* leaves (unpublished data of Roshchina). The blue-green colour of conifer needles appears to related to blue azulenes. The compounds are found in the fractions of isolated chloroplasts (Roshchina, 1999a, b).

Some medical drugs, such as terpenoids are valuable for the fluorescent analysis in pharmacology. Diterpenes such as hydrogenized diterpene forscolin from *Coleus forscohlii* Briq. (Family Lamiaceae), which is known as a stimulator of the adenylate cyclase activity and accumulation of intracellular cyclic AMP that serves as hypotensive cardiac-vessel drug (Golovkin et al., 2001), also fluoresces in blue maximum 455 nm (Roshchina et al., 2007b). Below we shall look at carotenoids and sesquiterpene lactones, which may show significant fluorescence in secretory cells.

Carotenoids. Carotenoids, yellow and orange pigments found in chloroplasts from all photosynthesizing cells, are also found in many plant secretory cells, especially in the form of carotenoid bodies in the exine of pollen grains (Stanley and Linskens, 1974) and in the glandular oil-containing cells of fruits (Roshchina and Roshchina, 1993). As a whole, carotenoids of plants and animals have weak fluorescence (Karnaukhov, 1988). However, their crystals emit significantly, mainly, β-carotene in green (maximum at 520 nm) or in greenish-yellow (Roshchina and Melnikova, 2001; Roshchina, 2003). Their lightening depends on the media. β-Carotene in ethanol emits with maximum 527 nm (Fig. 3.12), in chloroform – with maximum 546 nm and in solutions of the phosphatydylcholine derivatives – with maximum 534 nm. β-Carotene has a weak fluorescence as a whole, but, if it is included in lipid-water film (thick 175 mm), the fluorescence is visible during 17 h (Riel et al., 1983). In chloroplasts of the main secretory cells the fluorescence of carotenoids are masked by the emission of chlorophyll and phenolic compounds such as quercetin.

Sesquiterpene lactones. Sesquiterpene lactones such as proazulenes and azulenes (Rybalko, 1978; Konovalov, 1995) are present in many secretory structures – glandular trichomes, oil cells and ducts, and resinous cells (Roshchina and Roshchina, 1993). They are recognized as medicine and biologically valuable compounds of herbs contained, mainly in the plants of the Asteraceae family. Azulenes, blue pigments, are usually components of essential oils from secretory cells (Wagner and Wolff (eds), 1977; Reihling and Beiderbeck, 1991). Among the sesquiterpene lactones and their derivatives of special interest is is fluorescence of azulenes which are often present in plant oils in various species of Asteraceae and are known as inflammatory medical drugs and antioxidants and have a wound-healing activity (Rekka et al., 1996; 2002). Azulenes are also found in blue bodies of Bryophytes, in particular in liverwort (Nakagawara et al., 1992; Siegel et al., 1992). The pigments demonstrate protective effect in antiulcer medicine (Racz-Kotilla et al., 1968).

The azulene fluorescence in various media has been studied in earlier papers (Viswanath and Kasha, 1956; Griesser and Wild, 1980). The emission of proazulenes gaillardine, grosshemine and austricine under UV-light 360 nm from the species *Gaillardia pulchella* and *Achillea millefolium* were also examined (Roshchina et al., 1998b; Roshchina, 2004a, 2005b). The examples of the absorbance and fluorescence spectra for some of them are shown in Figs. 3.4, 3.12 and 3.13. The maxima are given in Table 3.1.

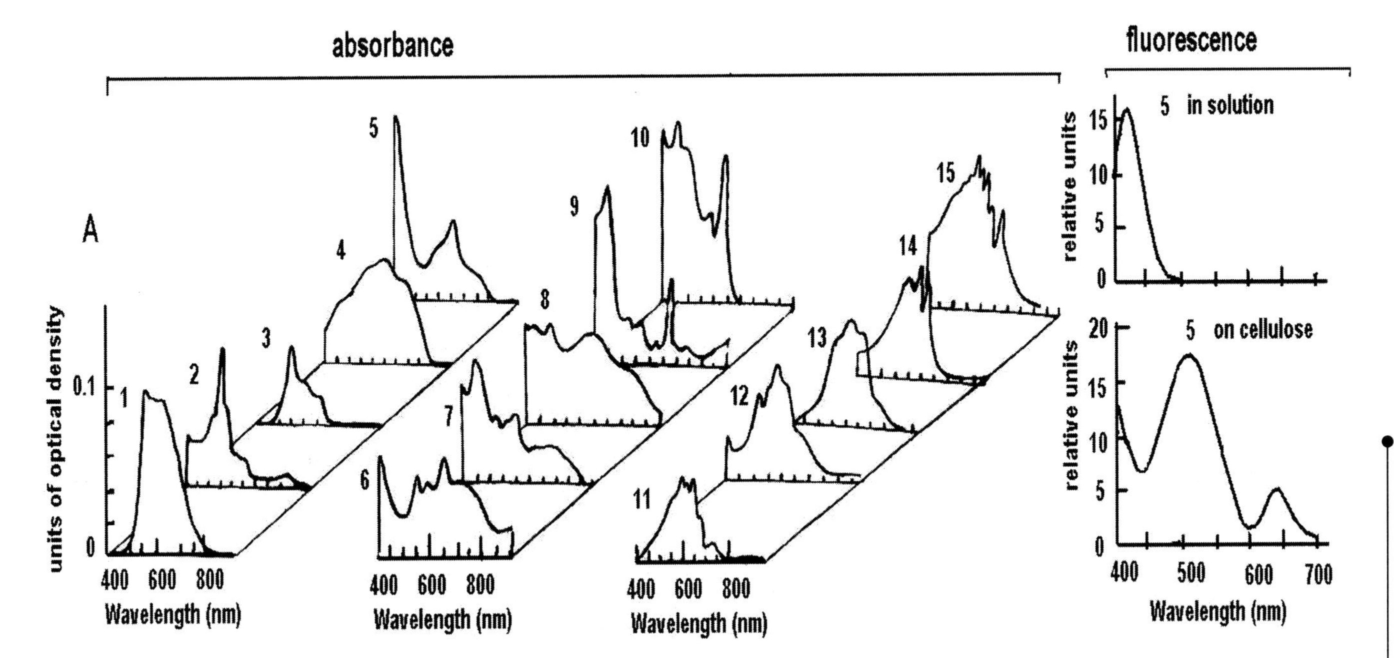

Fig. 3.13 The absorbance and fluorescence spectra of azulenes from secreting plant tissues. Azulenes in ethanol, from pollen (1-5) or in ethanol isolated from secretory cells of leaves and needles (6-10) or in hexane, individual azulenes (11-15) [Source: Roshchina, 1999a). 1. *P hiladelphus grandiflorus.* 2. *Betula verrucosa.* 3. *Aesculus hippocastanum.* 4. Pollen load, bee- collected. 5. pollen load, bee-collected. 6. Leaf of *Hypericum perforatum.* 7. Needles of *Picea excelsa* var blue. 8. Needles of *Picea excelsa* var. blue. 9. Chloroplasts of *Trifolium repens.* 10. Chloroplasts of *Pisum sativum.* 11. Azulene. 12. Se-chamazulene. 13. Se-guajazulene. 14. 2-Mefhylazulene. 15. 6-Methylazulene. Rf for paper chromatogram: 1-10: 0.9; 0.95; 0.95; 0.79; 0.1; 0.25; 0.16; 0.32; 0.2; 0.01. Sample 5 fluoresces in the solution and after the impregnation into cellulose filter.

Table 3.1 The absorbance and fluorescence maxima (nm) of sesquiterpene lactones and its derivatives (excitation by light 360 nm) solved in 96% ethanol

Terpenoid	*Absorbance*	*Fluorescence*
Artemisinine	298	395, shoulder 430
Austricine	256, 350	406
Azulene commercial	580	420
Deacetylinulicine	208-212, 225-230	446
Gaillardine	208	415-420
Gelenine	220, shoulder 285	420
Grosshemine	256, 338	400
Inulicine	208(220), shoulder 270	400, shoulder 440
Ledol	200-210	400-408, shoulder 460
Santonine	240-250, shoulder 270	408, 450
Taurine	200	440
Tauremisine	200, 295	406, shoulder430

As seen from the illustrations, many sesquiterpene lactones fluoresce in the blue spectral region which is characteristic for terpenoids as a whole. Main maxima were at 406-440 nm depending on the chemical nature of sesquiterpene lactone. Substances related to the guajacol group such as gaillardine has only one maximum in blue, whereas inulicine from psilostachanas – also the shoulder. The maximum and shoulder are characteristic for artemisinine and tauremisine. Azulene fractions extracted from pollen loads (Roshchina et al., 1995) or from vegetative microspores of *Equisetum arvense* (Roshchina et al., 2002) fluoresce with maxima 438-456 nm and 420 nm, relatively, whereas commercial azulene – at 420 nm. The fluorescence could be also associated with the presence of γ-lactone as fluorophore in the molecules. Metals stimulate it (Zhang et al., 2006).

Sesquitepene lactones are found in idioblasts of leaves from plants, belonging to the families Magnoliaceae, Lauraceae, and Asteraceae. Especially rich is of genus *Artemisia*, as well even in mosses (Rodriguez et al., 1976; Roshchina and Roshchina, 1989; 1993; Konovalov, 1995). The compounds are biologically active and act on the DNA, RNA and protein synthesis. As a result, the germination of seeds and spores of root parasites is inhibited or stimulated (Fisher et al., 1989). Due to the presence of unsaturated γ-lactone, the compound show cytotoxic, antimicrobial and phytotoxic activities as well as deterrents of the insect feeding. The plant growth can be inhibited by the lactones. Mechanism of the action are associated with the presence of exocyclic methylene conjugated to a γ-lactone or to functional groups such as epoxide, hydroxy or o-acyl adjacent to the α–CH_2 of γ-lactone.

Azulenes are derivatives of sesquiterpene lactones, which are solvent in organic solutions and have been known from the classical studies of Ruzicka and Rudolf since the 1920s as blue pigments produced by distillation of plant-derived essential oils. Although the chemical composition, paths of the biosynthesis, and physical properties of azulenes have been studied in sufficient detail (Gordon, 1952; Hellbronner, 1959; Mochalin and Porshnev, 1977), their biological role and function in plants are still unknown. Azulenes may protect microspores against unfavourable external factors as antioxidants. Rekka et al. (1996) reported the significant antioxidant activity of the natural azulene derivatives chamazulene and guaiazulene. All the tested molecules were found to inhibit lipid peroxidation by 100% at 1 mM. They were also found to considerably inhibit lipoxygenase activity.

Recently azulenes have been found in pollen grains (Roshchina et al., 1995; Roshchina, 1999a). Earlier, pollen of lupine, narcissus and crocus exhibited red fluorescence, originally attributed to chlorophyll (Ruhland and Wetzel, 1924), but the absence of the latter in matured pollen was demonstrated later (Stanley and Linskens, 1974). Moreover, it was also impossible to explain this phenomenon merely by the fluorescence of water-soluble anthocyanins having blue colour at pH >7, because the yellow and orange pollens studied contained negligible amounts of these pigments. Further studies of the spectra of intact pollen of coniferous and herbaceous plants (van Gijzel, 1971; Satterwhite, 1990) revealed red florescence at 570-685 nm, but no relationship was found between this fact and the defined sporopollenin complex. The study of the lipophilic extracts from vegetative microspores of *Equisetum arvense* and generative microspores (pollen) of various species as well as pollen loads, bee-collected show the presence of azulenes in exine (Roshchina et al., 1995; 1997c; 2002). Moreover, azulenes have been found in isolated chloroplasts (Roshchina, 1999a) that are shown in Fig. 3.13. All blue pigments have the absorbance maxima at 570-620 nm. The components were found and purified chromatographically (Roshchina et al., 1995; Roshchina, 1999a; Roshchina et al., 2002a). Most of the azulenes which were found fluoresce in blue at 420-430 nm, but after the impregnation into cellulose filter as a model of the cell wall (Fig. 3.12) the shift in the blue fluorescence to longer wavelengths was observed, and the new maximum 620-640 nm in the red spectral region appeared (Roshchina et al., 1995).

3.2.1.3 Polyacetylenes

Polyacetylenes are aliphatic or heterocyclic components with triple bonds. They are often present in the genera *Tagetes*, *Echinops*, *Calendula* and *Inula* which belong to the family Asteraceae (Christensen and Lam, 1990).

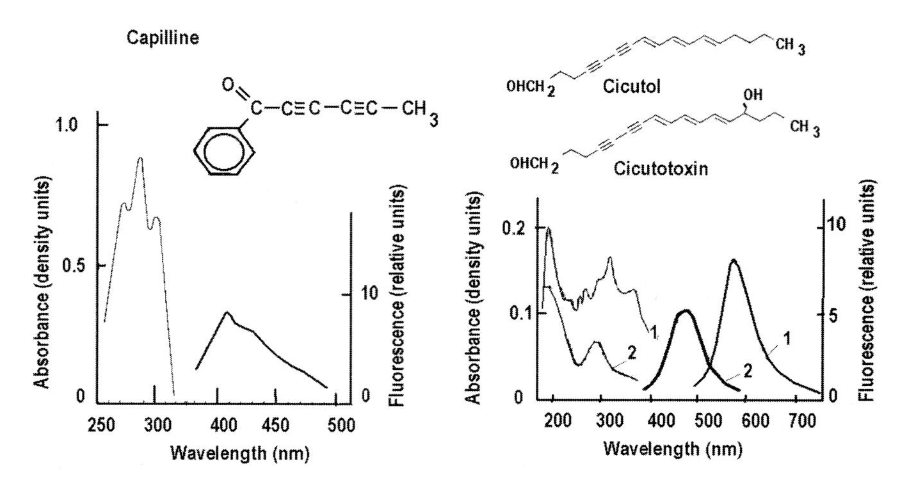

Fig. 3.14 The absorbance and fluorescence spectra of polyacetylenes. Source: Roshchina et al., 1980; and unpublished data of the author. Left – 2×10^{-4} M (in ethanol) capilline from genus *Artemisia capillaris*; Right - 1.55×10^{-8} M cicutotoxin (1) and 1.55×10^{-8} M cicutol (2) in methanol isolated from roots of *Cicuta virosa* L.

As shown in Fig. 3.14, capilline, polyacetylene from leaves of *Artemisia capillaris*, fluoresces with maxima 408 and 430 nm, while cicutotoxin and cicutol from roots of *Cicuta virosa* – with maxima 450 and 565-580 nm, relatively. Aliphatic polyacetylene cicutotoxin and its derivatives were also found in secretory roots of water hemlock *Cicuta virosa* (Anet et al., 1952; 1953; Wittstock et al., 1995). Natural and synthesized polyacetylenes fluoresce in blue-green depending on the chemical nature (Balcar et al., 2001). For instance, in *Echinops globifer (sphaerocephalus)* L. there are polyacetylenes as aliphatic and heterocyclic (Roshchina and Roshchina, 1993). The fluorescence maxima at 450-465 nm in all parts of the species are seen, whereas second maximum (or in flower petal as a shoulder) 550-560 nm may arise in leaf secretory hairs (see Chapter 4, Section 4.2.1.2).

Polyacetylenes may contribute in the spectral regions. Heterocyclic polyacetylenes fluoresced in blue at 430-450 nm (see Fig. 3.7) are also common in roots of *Tagetes* genus.

3.2.1.4 Alkaloids

Alkaloids, the heterocyclic compounds, which include nitrogen in the cycle, are often found in secretory cells of various plant species - from glandular trichomes on leaves, flowers and stems to root idioblasts and cavities (Roshchina and Roshchina, 1989; 1993). Practically all alkaloids fluoresce (Wolfbeis, 1985), except alkaloids without an aromatic or heteroaromatic ring, nicotine and imidazole derived compounds.

Some examples of the alkaloid emission, peculiar to plant alkaloid-enriched secretory structures, are in Fig. 3.15 and in Table 3.2. The excitation of the emission was 360-380 nm which is usual for a luminescence microscope. Physostigmine as indole derived alkaloid from the secretory cells of the South African fruit *Physostigma venenosum* L. fluoresces with two maxima 412 and 446 nm, while capsaicin from secretory cavities of *Capsicum annuum* L. fruit - with only one maximum 450 nm (in ethanol) or 475-480 nm (in water). Isoquinoline alkaloids sanguinarine and chelerythrine from the *Chelidonium* genus emit in orange 550-580 nm. Atropine, toluene derived alkaloid, glaucine and yohimbine, indole type alkaloid, fluoresce in the short blue spectral region, while

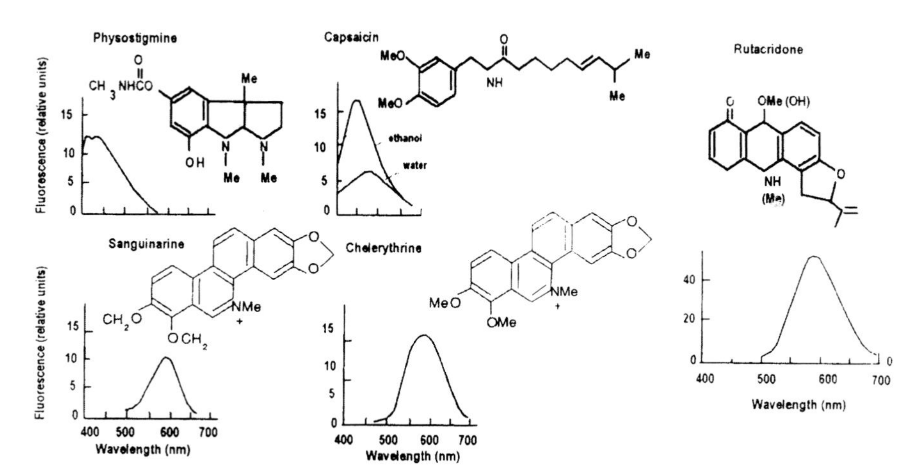

Fig. 3.15 The fluorescence spectra of some alkaloids (2 mg/ml). Physostigmine was solved in ethanol, capsaicin – in ethanol or in water, rutacridone – in chloroform, other alkaloids – in water.

isoquinoline alkaloids berberine, sanquinarine and chelerythrine as well as acridone alkaloid rutacridone – in the orange spectral region.

The emission spectra can differ strongly from each other, depending on the emitting fluorophore. There is also a dependence of alkaloid fluorescence on the solvent or polarity of the medium in the secretion. Polar solvents favour longer emission due to electron-donating substituents such as hydroxy-, methoxy-, or amino groups (Wolfbeis, 1985).

Table 3.2 The absorbance and fluorescence maxima of some alkaloids (1 mg/ml of solvent)

Compound	*The maxima wavelength (nm)*		
	absorbance	*excitation*	*fluorescence*
Atropine (in water)	270	360	415
Berberine (in ethanol)	344, 433	380	520
		480	532, 548
Capsaicin (in ethanol)	280	360	450
Casuarine (in ethanol)		360	475
Colchicine (in ethanol)		360	460
Glaucine (in water)	218-220	360	430 (weak emission)
Muscarine (in ethanol)		360	420, shoulder 440-450
Neostigmine (derivative of physostigmine)	260-270	360	415, 450
Physostigmine (in ethanol)	240,298	360	412, 446
Rutacridone (in chloroform)	395-400	360	480, 595
Sanguinarine (in water)		420	585, 670
Theophylline (in ethanol)		360	415
Yohimbine (in ethanol)	263	360	450

Wolfbeis (1985) considered peculiarities in the emission of the different alkaloid groups. Lack of an extended w-electron system in tropane and morphine alkaloids leads to the emission far in the UV-region. In morphine and its derivatives, the fluorescence maximum goes to longer wavelengths as they change to polar solvents. For phenethylamine-derived alkaloids a charge transfer interaction between fluorophore and a remote amino group can under certain conditions take place, which gives rise to a long wave exciplex emission. According to the review of Wolfbeis (1985), both quinoline and isoquinoline types of alkaloids have the fluorescence quantum yields dependent on the excitation wavelength and are highest when excitation is at the red edge of the longest-wave absorption band. Quartenary quinolinium alkaloids extracted from plants of the family Rutaceae fluoresce in the UV light, and those with a methoxy group in positions 6, 7, or 8 fluoresce in blue (maxima between 450 and 500 nm). Cationic quinolines with free hydroxy groups fluoresce even in green (515-520 nm). Tetrahydroisoquinolines and isoquinolines fluoresce in UV-region in ethanol at room temperature. The latters show long wave shift to around 450 nm after acidification, while the formers exhibit a practically unchanged fluorescence maximum. The berberines and protoberberine types of alkaloids are highly fluorescent (with maximum at 550-580 nm) depending on the wavelength of the excitation. Acridone and their derivatives usually fluoresced in green, 1-hydroxyacridone is non-fluorescent. Introduction of a methoxy group shifts the absorption and fluorescence maxima to longer wavelengths or can intensify fluorescence. The hydroxy group can have an adverse effect. Non-heteroaromatic alkaloid colchicine with a tropolone fluorophore fluoresces weakly in water with maximum near 460 nm (Wolfbeis, 1985). Heterocyclic alkaloids caffeine, theophylline (Table 3.2), and theobromine also may fluoresce at shorter wavelengths and even at UV-regions (Wolfbeis, 1985). Moreover, the alkaloid fluorescence depends on the glycosidation. For example, acridone alkaloid rutacridone with high lipophility (crystals fluoresce with maximum 600-605 nm) and solubility in chloroform (emits with maximum 590-595 nm), whereas its glycoside is solved in ethanol and demonstrates maximum in shorter wavelength emission.

3.2.1.5 Comparison of the fluorescence of intact secretory cells and secretions with the emission of individual components.

Intact secretory cells (see Chapter 2) show multicomponent fluorescence spectra, and one needs to know the chemical composition of the studied plant species and prevailing components in order to compare with the emission of individual components. The normal distribution analysis in Gaussian bell-shaped curves (the indefinite integral of the distribution function received graphically by a special computer programme) may be useful for this purpose. An attempt in this direction is first seen in Fig. 3.16 for the idioblast of intact root from pRi-transformed (inoculated by certain microorganisms) clone of *Ruta graveolens*. The clone enriched in alkaloid rutacridone (Kuzovkina et al., 1979; 1999) may serve as a model for the study. This alkaloid prevails in idioblasts because due to high concentration it may be deposited within the secretory cell in a crystalline form (see Section 3.1.1). Figure 3.16 shows the presence of the three components in the fluorescence spectrum of the intact secretory cell (the normal distribution in form of Gaussian curves). The main component is acridone alkaloid rutacridone, which occurs as both free and glycoside, but flavonoids rutin and quercetin also are found (Aliotta and Cafiero, 1999). However, unlike rutacridone emitted with maximum 595-600 nm), the

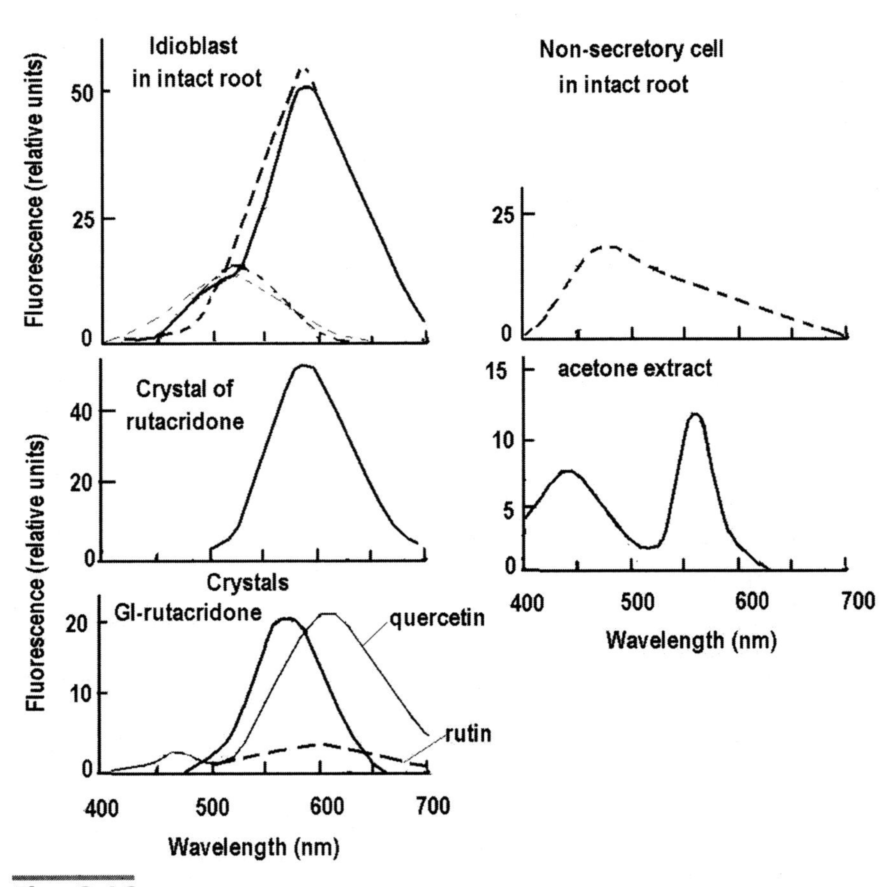

Fig. 3.16 The comparison of the fluorescence spectra from intact cells of *Ruta graveolens* and some individual compounds found in the species. The broken thin lines in idioblast are Gaussian curves related to the three components.

flavonoids have the fluorescence maxima at longer wavelengths at 605-610 nm. Moreover, crystals of rutin and quercetin demonstrated larger width of their emission maximum. Rutin is also characterized by lower fluorescence intensity. As for two minor components seen in idioblast fluorescence spectrum, they demonstrated maxima in blue (480nm) and green (525 nm) which is similar with the emission of the non-secretory cell with maximum 480 nm and shoulder at 510-520 nm. Acetone extracts (1:20) from the root contains components, which fluoresce with maxima 445 and 565 nm. Only the fluorescence spectra of the root chloroform and oil extracts (see Table 3.3) include maximum 590-595 nm. Ethanol extract may also show maximum 580-590 nm, peculiar to glycoside of rutacridone (more hydrophilic).

We cannot compare quantum yields of the compounds found in secretory cells because it is impossible to make their determinations. The data on the absorption spectra of the intact cells and the secretions are absent. Moreover, the modern concept of quantum yield is applied only to solutions of pure compounds studied. Besides, the fluorescence behaviour in the mixtures of the compounds is also not clear yet, and may be the object of the interest in future studies by physicists.

Of special interest is the analysis of contribution of some individual compounds (pigments, in particular) in plant autofluorescence *in vivo* is the interpretation of images made with the help of Laser-Scanning Confocal Microscopy (See Colour Fig. 21 in Appendix 2). The pseudocolours reflect natural emission of quercetin, rutin, azulene, chlorophyll a+b, and carotenoids.

3.2.2 Changes in the Fluorescence of Individual Compounds in Secretions

The fluorescence of the components of secretions may be changed depending on various factors: light, temperature, pH, chemicals and the composition of the secretion. This dependency is not studied especially for plant secretions, only for artificial dyes (Terenin, 1945). We need to apply some data related to individual compounds of the secretions. The changes of the fluorescence during development of the cell, in particular, and of the plant organisms as a whole, will be discussed in Chapter 4. The information about effects of external factors as well as modelling of secretory cell interior is given below.

3.2.2.1 The dependence of fluorescence on the external factors

Light. The fluorescence depends on the wavelength of the excited light. Secondary metabolites, emitted under a luminescence microscope, usually excited by ultra-violet radiation in a range 340-380 nm depending on the quartz lamp characteristics and filters. But longer wavelength excitation is also possible, for instance violet, blue, green, orange and even red, if it is necessary. The more detailed characteristics of the excitation of the compounds fluorescence in solutions by light of various wavelengths and depending on the chemical structure can be find in reviews (Baeyns, 1985; Schulman, 1985; Wolfbeis, 1985) and other special papers on Internet sites.

Phenols of secretions have different luminescence at the excitement by 313 and 365 nm wavelengths. In the first case, only phosphorescence occurs with maxima 470, 500 and 550 nm, while in the second – the fluorescence with one maximum 470-480 nm (Frolov et al., 1974). In ethanol, many flavons excited by light 365 nm fluoresce at 525-600 nm, but at 77°K the shorter wavelength fluorescence 390-510 nm also appeared (Frolov et al., 1974).

Temperature. Secretory products may fluoresce either at room temperature or at low temperature. Low temperatures such as liquid nitrogen 77°K allows one to see more peaks in the fluorescence spectra of pollen, in particular for pollen grains from two different species Japanese quince-tree *Chaenomeles japonica* and *C. sinensis* (Butkhuzi et al., 2002). There was a maximum at 586 nm in both spectra, but others differed in the blue spectral regions.

Many coumarins have no fluorescence in methanol at 20-22°C, but introduction of a hydroxyl group makes all coumarins fluorescent at room temperature, except 8-hydroxycoumarin (Wolfbeis, 1985). Fluorescence of quinoline and isoquinoline decrease with increasing temperature (Wolfbeis, 1985).

Moistening. The water and water solution of secretions quench the fluorescence of viable pollen grains of various species (Roshchina et al., 1996; 1997b; Roshchina, 2003). In some cases (see Chapter 4), water may quench the fluorescence of individual compounds.

Radiation. The intensity of fluorescence of pollen from *Hippeastrum hybridum* does not change after the γ-irradiation in doses 500-5000 Grays, although their germination was stopped at 1500 Grays (Roshchina et al., 1998b).

pH of medium. Phenols have usually green fluorescence at pH higher, than 8, but the emission is quenched or decreased at lower pH values as described in the review of Wolbeis (1985). The significant effect of pH on the fluorescence properties was demonstrated for coumarins and flavonoids. The difference between the compounds lies first in the fact that green fluorescence of 3-, 6-, and 7-hydroxycoumarins is most intensive in alkaline solutions, wherein flavones emit only weakly. Flavonols, in particular, have a greenish-yellow fluorescence after the treatment of their spots on Silicagel plates (thin layer chromatography) by 1% aquatic solution of ammonia. Various 7-hydroxycoumarins, for instance umbelliferones, have been recommended as pH indicators. Esters, ethers, and glycosides of umbelliferones are used as fluorogenic substrates. The fluorescence of 6- and 7-hydroxycoumarins is intensified with sodium carbonate or ammonia, when excited at 366 nm. Weak acids may quench the emission intensity of furanocoumarins. The indole fluorescence is quenched by protons in both aqueous and non-aqueous solutions. The intensity of fluorescence for carboline alkaloids at 470 nm in water and methanol is much higher in acidic solutions. A common feature of all nitrogen heterocycles, except indole and carbazole, is the increase in basicity upon photoexcitation. On the other hand, indole and carbazole tend to become stronger acids in their excited (fluorescent) states.

In the intact secreting cell the effects of pH may also be seen. For example, the alkaline treatment induces blue-green fluorescence of some stomata cells (Weissenböck et al., 1987). When NH_4OH was added, there was an alkaline induced bathochromic shift from maximum 470 to maximum 540 nm that was accompanied by an increase in fluorescence intensity. Moreover, a similar emission, peculiar to flavonol glycosides (kaempferol- 3,7- glycosides) was found in vacuoles of *Vicia faba* (Schnabl et al., 1986). Red fluorescence of anthocyanins (maximum 620-625 nm) is quenching by more alkaline medium (Figueiredo et al., 1990).

Solvents. The medium, mainly solvent(s), influences the fluorescence of the compounds in the secretion, for example of such a medium or solvent in secretory cells – oils, esters, ethanol, methanol as metabolites. Moreover the traces of water in organic solvent of the medium may quench the fluorescence as well as natural oxidants oxygen or reactive oxygen species such as ozone, free radicals and peroxides. As a whole, solvents may quench the fluorescence of organic compounds of secretions.

Fedorin and Georgievskii (1974b) studied the influence of solvents on the fluorescence spectra of coumarins. The authors demonstrated that umbelliferone, esculetin and scopoletin fluoresce at a longer spectral region in polar solvents (water and methanol), than in chloroform. Fluorescence intensity of furanocoumarins depressed with decreasing solvent polarity, and the maxima shift to shorter wavelengths (Wolfbeis, 1985). Peroxides stimulate the emission in the mixture water-ethanol. In particular, the fluorescence yield of azulenes in some organic non-polar solvents is constant in solvents with low polarity (Murata et al., 1972). Their fluorescence excited by ultraviolet light is measured mainly in cyclohexane from 500 to 780 nm.

The example of alkaloid rutacridone and its glycoside shows how moistening of the crystals (emitting with maxima 600 and 595 nm, relatively) by various solvents acts on the maxima positions. Aglycon and glycoside are insoluble in water and so fluorescence of their emulsion is registered. Only rutacridone glycoside may emit in blue with maxima 435 nm, while aglycon weakly fluoresces with maxima 600 nm. (Fluorescence of crystalline flavonoids quercetin and rutin are also found in the same secretory structures of rue which is quenched completely by water). Table 3.3 demonstrates the changes seen for the fluorescence spectra of rutacridone in various solvents.

Table 3.3 The fluorescence of rutacridone in various solvents

Solvent	*Fluorescence maximum (nm)*	
	Rutacridone	Rutacridone glycoside
100% acetone	456, 595	435 (weak emission)
100% ethanol (partly solved)	595(weak emission)	595
Chloroform	480, 595	No emission
Menthol oil	460, 600	595 (weak emission)

Crystals solved (0.5 mg) in 5 ml of solvent

Acetone, chloroform and menthol oil as solvents gave maxima for rutacridone both in the blue and orange spectral regions. Rutacridone glycoside fluoresces in ethanol with peak at 595 nm. The arising of blue fluorescence may show the formation of some aggregates of the compound (perhaps, dimers). This needs special analysis in the future.

Artificial ageing under UV-light and Ozone. Ageing or the effects of non-favourable factors that stimulate ageing such as UV-radiation or high concentrations of ozone change the fluorescence spectra (Brooks and Csallany, 1978; Merzlyak, 1988; 1989; Merzlyak and Pogosyan, 1988,

Merzlyak and Zhirov, 1990; Roshchina and Roshchina, 2003). This appears to be due to the formation of new fluorescent products formed as a result of lipid peroxidation. The products may be markers of cellular damage. If the plant cell or tissue is damaged or ageing, the contribution to the autofluorescence at 410-440 nm may be done by pigment lipofuscin which is a product of free radical processes in membranes (Merzlyak, 1988). In animal cells, lipofuscin fluoresces also at 550-650 nm (Eldred and Katz, 1988; Kamaukhov, 1990). Lipofuscin-like pigments are also found in chloroplasts from senescing leaves (Wilhelm and Wilhelmova, 1981; Merzlyak et al., 1984). Other products of free radical processes in living organisms can also fluoresce (Halliwell and Gutteridge, 1985). Therefore, one can not reject that free radical reactions and stable products of the processes such as peroxides also contribute to the observed autofluorescence of secreting cells. The occurrence of free radicals on the pollen surface was established earlier with the ESR technique (Dodd and Ebert, 1971). Their participation in the autofluorescence could be shown in model experiments on the secreting stigma of the pistil with noradrenaline, oxidant, generator of free radicals (superoxide anionradical and semiquinone noradrenaline radical) and antioxidants ascorbate, known as the trap of free radicals, enzyme superoxide dismutase, converting superoxide radical to hydrogen peroxide and molecular oxygen, and enzyme peroxidase, decomposing hydrogen peroxide and organic peroxides and thus breaking the free radical chains.

The contribution of peroxides and superoxide radical in the autofluorescence changes the emission from intact cells on the seedling.

For instance, some factors such as ozone stimulated ageing of carotenoid-rich pollen grains lead to the missing maxima at 530-560 nm peculiar to carotenoids (Roshchina and Karnaukhov, 1999; Roshchina, 2003). These questions will be specially discussed in Chapters 4 and 6), and the reader could also use our monograph (Roshchina and Roshchina, 2003). Here only one example for ozone will be given.

Figure 3.17 shows the example of how fumigation with ozone (0.1 μl/L) acts on the fluorescence of alkaloid rutacridone. The maximum 560 nm disappeared which demonstrated the oxidation of the compound.

Oxidants and reductants. Oxidants such as oxygen and ozone as well as their derivatives and reactive oxygen species induce the oxidation of the substances treated which leads to the changes in their fluorescence spectra

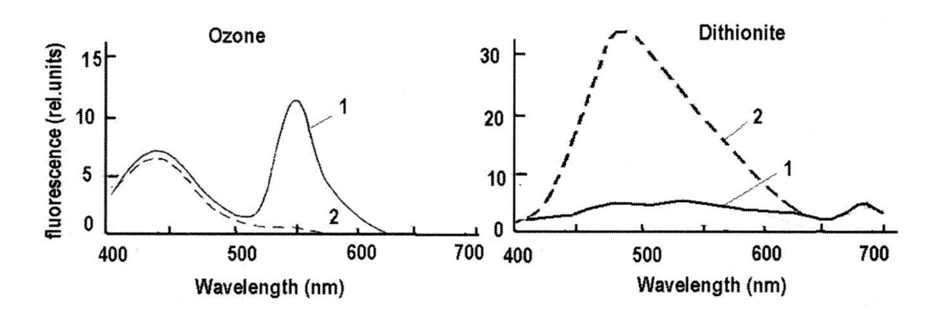

Fig. 3.17 The fluorescence spectra of some secretions such as rutacridone in ethanol treated with oxidant ozone (0.1 μL/L) and secretory cell such as vegetative microspore of horsetail *Equisetum arvense* treated reductant dithionite 10^{-5} M. 1 and 2, relatively - without the treatment and after the treatment.

as seen in Fig. 3.17. On the surface of the secretion the fluorescence depends on the formation of free radical reactions where reactive oxygen species are formed. For instance, free oxygen radicals such as superoxide anionradical, hydroxyl radical, peroxy radicals, etc, and/or peroxides (Roshchina and Roshchina, 2003). Antioxidants present in the secretions (superoxide dismutase, ascorbate, phenols, quinones and other) may interact with oxidants forming here, and new fluorescence pictures may arise. In some cases, reductant such as dithionite stimulates the green-yellow fluorescence of intact vegetative horsetail microspores, although the peak of red fluorescence is not changed (Fig. 3.17).

3.2.2.2 Modeling of the secretion

One of the approaches for understanding how fluorescence occurs in complex systems such as secretion in intact secretory cells and in cellular membranes is artificial modeling. This is a way of screening secretory products and membranous components to make a composition. This information reflects only the first steps, but the approach is important.

In intact secretory cells the fluorescence is dependent on the environment as well as the fluorescence of the individual compound in the secretions depending on the medium. The chemical nature of medium (in nature water, oil, ethanol and methanol, etc) and the composition of the secretion can define the character of the compound fluorescence. The analysis of the effects of the medium is shown in Fig. 3.18. If the secretion represents oils (non-fluorescent) with phenols such as flavonoid rutin or sesquiterpene lactone azulene or alkaloid rutacridone the emission differs

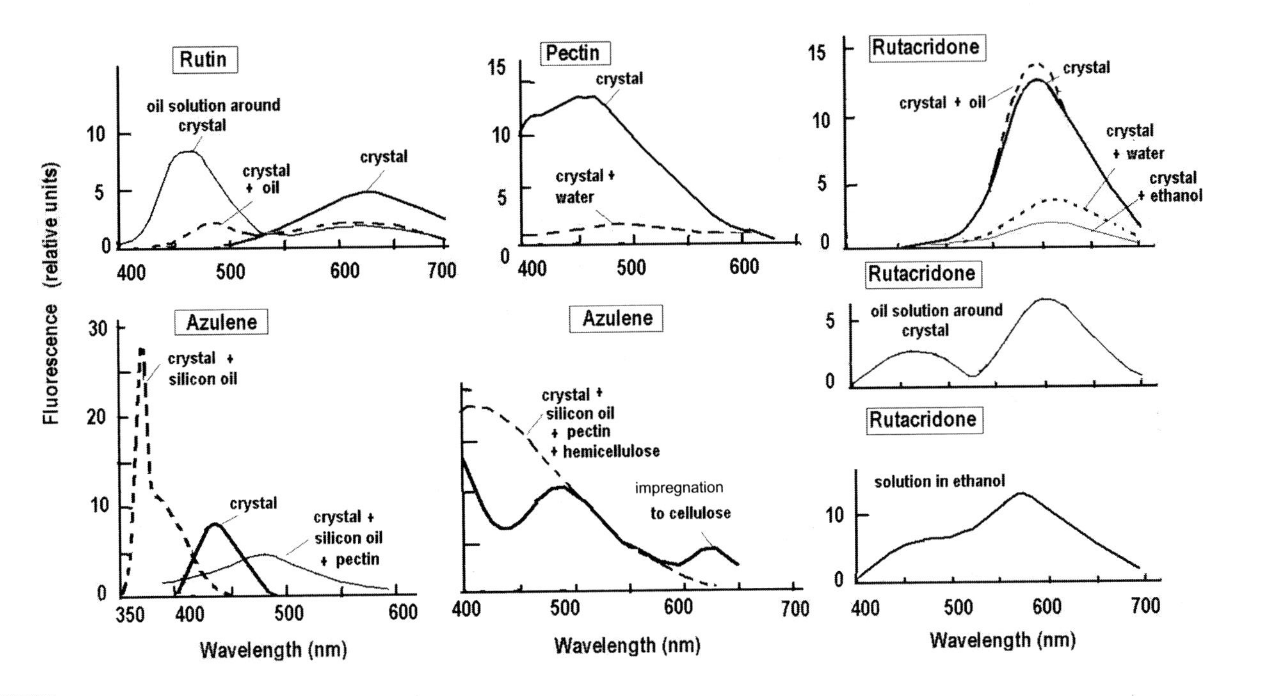

Fig. 3.18 The effects of media and different components of the media on the fluorescence spectra of some constituents of secretory cells. Oil – 2% menthol oil for the experiments with rutin and rutacridone and silicon immersion oil for those with azulene. The excitement was with light 360-380 nm. Rutacridone is solved in ethanol partly.

as can be seen by the shift of the fluorescence maxima to shorter wavelengths.

Crystal of rutin, fluorescing in orange with maximum at 595-600 nm, interacts with the menthol oil, and the oil solution of rutin in the crystal surrounding the medium fluoresces in blue with maximum 455-460 nm. The influence of cell wall fluorescence estimated on the light emission of hemicellulose and pectin as main components (pure cellulose fluoresce in blue 430-460 nm depending on the level of a purification). Crystals of the components fluoresce in blue at 408 and 450-460 nm at the excitement by light 360-380 nm. In water emulsion, hemicellulose and pectin have a weak blue fluorescence, in particular just after moistening of their crystals, water quenches the emission (Fig. 3.18). Azulenes are usually found among the essential oils of secretions (Roshchina and Roshchina, 1993), and the addition of silicon oil to their crystals leads to shifts in the fluorescence maximum to UV-emission instead of blue. Often the cell wall of secretory cells are also saturated with oils, which may include azulene, and in this case, the fluorescence in the variant: crystal + oil + pectin shifted to 490 nm. When hemicellulose was added to this mixture, all composition fluoresces in a shorter blue spectral region at 410-420 nm. If the azulene is impregnated to pure cellulose (paper filter), maxima in the blue-green and red spectral regions appeared (Fig. 3.18). Experiments with alkaloid rutacridone showed that water and ethanol quenched the orange fluorescence of the crystal, while oil induced a weak decrease, but the surrounding oil solution of the alkaloid has a new maximum 460 nm. A similar picture was observed for the fluorescent spectra of rutacridone in the ethanol solution, although maximum 460 nm converts to the shoulder.

In other cases, the oxidation factor also influences, particularly, when reactive oxygen species arise on the cell surface. For example, azulene in various solutions has undergone the influence of hydrogen peroxide (Fig. 3.19). Water quenches the fluorescence of the sesquiterpene lactone, whereas water solution of the peroxide stimulates. Although some (not so significant) stimulation at 420 nm occurs at the mixture of ethanol with H_2O_2 only in ratio 1:1 (no emission in ratios 8:1 or 9:1), but the emission intensity of the azulene solution at 420-430 nm is more, than two times higher. Moreover, the shoulder at 470 nm arises.

It is also possible to estimate the contribution of membranous fluorescence. London and Feigenson (1981) demonstrated that nitroxide spin-labelled phosphatidylcholine, the usual membranous component

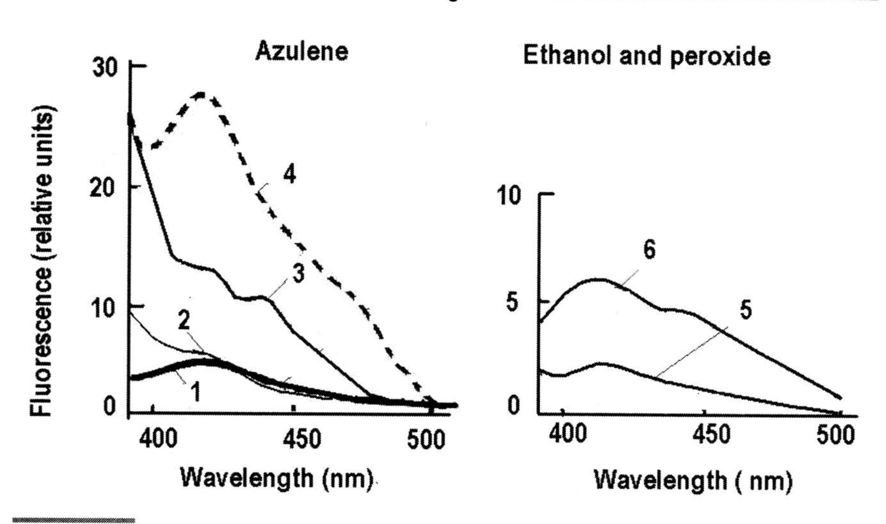

Fig. 3.19 The fluorescence spectra of azulene 10^{-4} M in ethanol without and with addition of hydrogen peroxide to final concentration 10^{-3} M. 1. azulene 10^{-4} M in 100% ethanol; 2. 1 + water (9:1 v/v); 3. 1 + water + H_2O_2 10^{-3}M (8:1:1 v/v); 4. 1 + H_2O_2 10^{-3}M (9: 1 v/v); 5. H_2O_2 10^{-3}M; 6. 100% ethanol + H_2O_2 10^{-3}M (1:1 v/v).

which also forms aqueous lipid vesicles *in vitro*, quenches fluorescence at 500 nm and shorter wavelengths (excitation by lights 340, 358, 430, 335/386 nm) in models of lipid membranes. The quenching primarily arises from the nearest spin-labelled lipid neighbours to the fluorophore.

The possible contribution of other components present in secreting cells (as both non-specialized and specialized secretory) has not been studied yet. The way compounds occur in the secretions with characteristic maxima in the absorbance and fluorescence spectra, which is favourable for the identification of the compounds in mixtures.

The analysis of the mixtures of the secretory components in natural secretions may be based on the knowledge of the components, which can quench or stimulate fluorescence. The causes, which do this, are discussed below.

3.2.2.3 Causes of the fluorescence quenching

Quenching of fluorescence occurs at the co-striking of the fluorescent molecule with other molecules. At co-striking, the excitation energy may transfer to the kinetic energy of atoms which decreases the luminescence yield. In condensed media there is a possibility of non-emitted transfer of the energy of the electron excitation into the oscillation energy and it spreads between molecules which leads to thermodynamic equilibrium. Therefore, luminescence is observed in compounds, in which the ratio of the possibilities of emitted to non-emitted energies is high, or else the energy transforms to warm energy instead of fluorescence. The theory of the fluorescence quenching has been described in monograph of Professor Lakowicz (1983) as well as an application to proteins in the review of Eftink and Ghiron (1981).

Quenching of the fluorescence is connected with the possibilities of non-emitted transfer at the increase in temperature (temperature quenching) or depends on the concentration of luminescent molecules (concentration quenching at high concentrations) or on the traces (trace quenching). π-Electron system of any unsaturated molecule may participate in the process (Parker, 1968). The fluorophore emission in some compounds may be quenched, while in other ones, it may not. In particular, the carbonyl group has a fluorescence, which is not quenched by electronic donors or proton donors in contrast with the dimethylamino group (Sauers et al., 1983). On contact, a fluorophore returns to the ground state without emission of a photon. A complex is formed between the fluorophore and the quencher, but this complex does not fluoresce (Lakowicz, 1983). In many cases of viscous liquids, when the diffusion is slow, the quenching may not be seen. Therefore, the process of quenching in a cell can reveal the diffusion rates of quenchers, when the fluorophore binds either to an individual compound (for example, protein) or to a membrane.

Quenching occurs at co-striking with other molecules, for instance with oxygen or water. In a result the fast escape of electrons occurs on the surface at the contacts. Moreover, the state of phase of compound plays a significant rolå in the quenching of the fluorescence. If the substance is in a solid state, for instance some flavonoids such as rutin and quercetin have maximum of the emission in the yellow—orange spectral region near 580 nm, but other flavonoids do not. Moreover, certain flavonoids do not demonstrate any fluorescence because as a result of internal conversion non-emitted transition of energy takes place. More rigid, well-organized systems emit in a lesser degree, than disorganized, (especially solutions) systems. In any case, an increase in a possibility of non-emitted transitions leads to a quenching of luminescence. This possibility depends on the increased temperature (temperature quenching), concentration of luminescent molecules (the concentration quenching) or admixtures (the admixture quenching). Among internal factors of quenching is the inclusion of quenching groups in aromatic compounds (Zelinskii, 1947). Inclusion of NO group with large dipole moments completely diminishes the fluorescence, whereas the inclusion of carbonyl- or carboxyl- groups only decreases the process.

There is a list of substances, which can quench fluorescence (Lakowicz, 1983). Among them, are water and oxygen as well as others. The quenching of the fluorescence by water is explained by the rapid escape of electrons from the substance located on the contact surface to water molecules. Oxygen can also quench the light emission. Singlet oxygen, hydrogen atoms in bond C-H and O-H may quench the fluorescence (Ermolaev et al., 1977). Various metabolites (histidine, cysteine, fumarate, etc) can also quench the fluorescence and many groups of chemicals (indole, carbazole and their derivatives) are sensitive to quenchers (Lakowicz, 1983). The quenching appears to occur due to the donation of an electron from the substance fluoresced to a quencher. Oxygen is a significant quencher of luminescence (Kautsky, 1939) as well as hydrogen peroxide especially studied for proteins and indoles (Eftink and Ghiron, 1981). The fluorescence of flavin adenine dinucleotide (FAD) and reduced nicotineamide adenine dinucleotide (NADH) may be quenched by adenine moiety (Lakowicz, 1983). Some ions also may quench the fluorescence (Eftink and Ghiron, 1981). Organic radicals are possible to quench the electron-excited state of many fluorescent dyes (Buchachenko et al., 1967). In particular, donors of electrons such as anthracene, fluorescein, rhodamine fluoresce, but their emission is quenched by acceptors of electrons, for instance radicals of nitric acid (nitrate radicals).

The increase in the possibility of non-emitted transition leads to the quenching of luminescence. The quenching of fluorescence in a solution may be enhanced by the increase of temperature, admixtures, changes in the pH of medium, and by an increase in the concentration of fluorescing compounds (so-called internal concentration quenching). Fluorescence of organic compounds occurs depending on the structure of the molecule and is peculiar to rigid structures which excludes a possibility of non-radiated transfer. For example, more free rotation of the part of the molecule induced the quenching of the fluorescence. The maxima of the flavonoid fluorescence shift from the blue to orange spectral region.

3.2.2.4 Increase in the light emission

Increase in the fluorescence of secretory components in mixtures is observed, but has not been especially studied yet. For example, anthocyanins stimulate the fluorescence in phenolic mixtures. The substances, which have no free hydroxylic groups in C15- position, can fluoresce in phase water-citric acid at 510, 515, 555, 570-585, shoulder 610 nm. Anthocyanins contacted with metals form complexes that stimulate the light emission. Flavons interact with anthocyanins which leads to stimulation of the anthocyanin luminescence. However, the interaction of anthocyanins with quercetin quenches the anthocyanin fluorescence (Santhanan et al., 1983). The red flavylium cation of anthocyanins forms ground-state charge transfer complexes with several naturally occurring electron donors, such as flavones (quercitrin) and benzoic and cinnamic acid derivatives. Excitation of these charge transfer complexes results in efficient static fluorescence quenching due to fast electron transfer from the copigment to the flavylium cation (Ferreira da Silva et al., 2004). The possible causes of the enhanced fluorescence in the mixtures of the secretory products *in vivo* are the electron transfer from one compound to another and forming complexes.

Conclusion

Autofluorescence of secretory cells is connected with emission of the individual secretory products which differ in their chemical nature. Fluorescence of secretions may be analyzed either to study crystals and the short time excretions by various polar and non-polar solvents or to modelling the secretions by addition of the individual natural compounds. The emission is often quenched or stimulated by other components of the secretions or medium. Among the factors influencing the fluorescence are the chemical nature of the compound, the quality of the excitation light, temperature, pH of medium, the ability of the external chemical to oxidize or reduce the fluorescent substance and others.

CHAPTER 4 Autofluorescence of Secretory Cells During their Development

Autofluorescence is changed during the development of the secretory cell, and it permits one to observe how the cells are filled with secretions and their transformation. The emission differs in unicellular and multicellular secretory structures. In this chapter, the systems in the different stages of their development will be analyzed.

4.1 DEVELOPMENT OF UNICELLULAR SYSTEMS

The unicellular secreting systems such as vegetative microspores of horsetails and ferns, belonging to spore-bearing plants, or generative spores (pollen) of seed-bearing plants have a characteristic fluorescence (see Chapter 2, Section 2.1.1). It is known that the character of their fluorescence changes during the development (Roshchina et al., 1996; 1997b; 2002; Roshchina, 1999a, b; 2003; 2004, 2005a, b). Some examples of spore-bearing and seed-bearing species will be discussed below.

4.1. 1 Development of Secretory Structures in Spore-bearing Plants

Vegetative microspores of horsetails, mosses and ferns have a weak fluorescence in a dry state, and just after moistening they start to develop. Individual cells of the microspore which can develop to multimolecular prothallium and later to thallus - gametophyte with male organs known as antheridia or/and with female organs called archegonia (Williams, 1938; Atkinson, 1938; Igura, 1956; Katoh, 1957; Bopp, 1965; Wettstein, 1965). Spores of horsetails and ferns develop depending on light quality (Stahl, 1885; Dougal, 1903; Nienburg, 1924; Wettstein, 1965).

Up to now the development cycles of some species belonging to Equisetaceae were studied by methods of the transmissent light (Beer, 1909; Castle, 1953; Nakazawa, 1956; 1958; Hurel-Py, 1959) and electron (Gulvag, 1968) microscopy. In the first case the observation of the cycles of the green-grey microspores *in vivo* needs staining with certain dyes (Nakazawa, 1956, 1958) whereas in the second case - special procedures of the fixation are required (Gulvag, 1968). Recently, the changes in autofluorescence during development of their microspores were first studied by various methods of luminescence microscopy – from the usual luminescence microscope to microspectrofluorimetry and laser-scanning confocal microscopy (Roshchina et al., 2002; 2004).

4.1.1.1 Vegetative microspores of horsetail Equisetum arvense

In the vegetative microspore of horsetail *Equisetum arvense,* as described in Chapter 2, Section 2.1.2 (Fig. 2.2), a middle part has three maxima – 460 nm, 550 nm and 680 nm, the last is peculiar to chlorophyll fluorescence. The cover of the microspore as well as elaters has no maximum 680 nm. Thus, the presence of chlorophyll is possible in dry microspores (Orange autofluorescence was observed in 1995 by Uehara and Murakami even in tapetum). But usually this peak is small in the fluorescence spectra, and the researcher sees, mainly, blue fluorescence, peculiar to the rigid coat of the microspore. The possibilities of individual substances to contribute in blue and red emission were also analyzed (Roshchina et al., 2002). Flavonoids (quercetin, kaempferol, rutin and others are found in the herb of *Equisetum arvense* (Plant Resources of Russia and Surrounded States, 1996). The compounds such as quercetin and in a small degree rutin fluoresced in the blue region (at 460-480 nm), but have second maximum at 580-600 nm (Roshchina and Melnikova, 2001). Another possibility is the presence of carotenoids in the orange region (at 530-580 nm). Azulenes out the cellulose cell wall fluoresce in the blue region (at 460 nm), but on cellulose filter it has also red (620-650 nm) emission (Roshchina et al., 1995). Similar blue pigments have been also found in the microspores, which extracts by organic solvents after acidic sedimentation were separated on Silicagel chromatograms (Roshchina et al., 2002). Similar blue pigments (separated on the Silufol-plates, showing two bands – violet-blue with Rf = 0.97 and blue with Rf = 0.91) fluoresced at 420 nm have been also found in the microspores (Roshchina et al., 2002). Total concentration of azulenes in the microspores was 0.0015 mg/mg of fresh mass.

The phenomenon of autofluorescence from vegetative microspores of the spore-breeding plant *Equisetum arvense* has been studied during the development of the cells by methods of luminescence microscopy and microspectrofluorimetry (Roshchina et al., 2002) or laser-scanning confocal microscopy (LSCM) (Roshchina et al., 2004). Autofluorescence of the developing microspore of the plant are clearly seen by using luminescence (Colour Fig. 22) and laser scanning confocal microscopy (colour Fig. 23) as well as microspectrofluorimetry (Colour Fig. 22 and Fig. 4.1). Just after moistening, the microspores start to develop as is seen from the increasing red fluorescence (Roshchina et al., 2002; 2004). Colour Fig. 22A shows that microspores are missing elaters during first 2-5 min, then the process of cell division occurs up to 2 h (for older microspores later). After 2-24 h of moistening (Colour Fig. 22A) the microspore missed the bright blue-lightening cover which fluoresce mainly with maxima 460 and 500 nm, and some liberated part of the cell contributes to the chlorophyll maximum at 680 nm, which is missing when the cover will be free completely. The liberated cell has a great maximum at 680 nm and more expressive maxima at 460 and 550 nm which is shown in Colour Fig. 22B. The shifts dealing with first time changes after moistening are also demonstrated in Fig. 4.1 in the series of experiments, using blue-red transition as a marker with the help of duable-beam microfluorimeter Radical DMF-2" (Radical, Ltd) interfaced to PC/AT compatible computer to measure fluorescence intensities at two separate wavelengths (Roshchina et al., 2002). A special programme "Microfluor" makes it possible to obtain the distribution histograms of the fluorescence intensities in the red ($I_{640-680}$) and blue ($I_{460-480}$) regions of the spectrum as well as the distribution histograms of the parameter $I = I_{red}/I_{blue}$ for 200 cells within 15-20 min with the points put on to the phasic plane, and to perform statistical treatment of the data, using Student t-test (Roshchina et al., 2002). A diffusive distribution of the elements of cover and chloroplasts could be seen more clearly. As seen in Fig. 4.1, the red fluorescence of chloroplasts strengthened just after 15 min of moistening. The cells are swollen and blue-fluorescing elaters are broken. Chloroplasts are moved from the centre to the cell wall. After 30 min moistening the amount of the cells on histograms with higher parameter I red/blue increased due to the stimulation of the red fluorescence.In the first 2-6 h of development the red fluorescence intensity is the same in the dark and in daylight (Roshchina, 2006a). Horsetails *Equisetum* cells

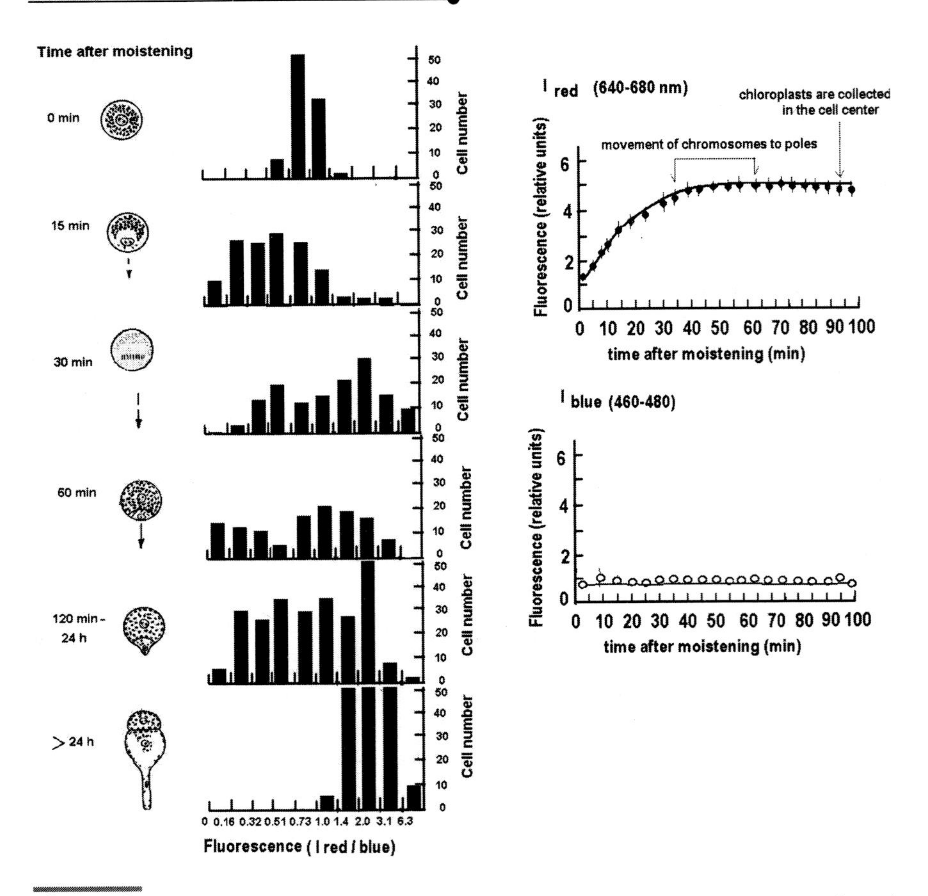

Fig. 4.1 The fluorescence intensity of the horsetail vegetative microspores. Adopted from Roshchina et al., (2002). Left panel shows schemes of microspore development as observed under a luminescent microscope (objective x 85). Middle panel shows cell frequency histograms (Student's distribution) with respect to (I) which is the ratio red emission intensity (640-680 nm): blue emission intensity (460-480 nm) measured by double beam microfluorimeter MSF. Right panel demonstrates the time kinetics of red emission intensity ($I_{640-680}$) and blue emission intensity ($I_{460-480}$) of germinating microspores.

demonstrate a polarity (Mosebach, 1943; Haupt, 1957; Bloch, 1965) which is seen after 30 min moistening. In this stage chloroplasts are clearly seen as red globules through the transparent cell cover (colour Fig. 23). They are concentrated around the blue-lightening nucleus like Nienburg (1924) saw it. After 60 min of development the chloroplasts concentrated on one side of the cell, the cell started to divide and there was back movement of red-fluorescing chloroplasts in the middle of the cell. According to Nakazawa (1958), it indicates the beginning of the cell division. Nienburg (1924) has shown the induction of polarity in microspores of *Equisetum*:

1. Microspore (one cell) with chloroplasts around nucleus →
2. Then chloroplasts are concentrated on one side and the nucleus on the other side →
3. The division of nucleus, meiosis →
4. Two cells with separated nuclei are formed (the big cell with majority of chloroplasts and the small cell with a small amount of chloroplasts) →
5. The big cell is divided again, and multicellular thallus develops;
6. The small cell forms growing a rhizoid cell.

120 min (sometimes to 24-72 hours) after moistening, the cell missed off its blue-fluorescing rigid cover (Colour Fig. 23), and the new second cell is clearly seen. The cell without rigid cover divides, forming two cells, from one of which will form a multicellular prothallium and then the thallus (Colour Fig. 23 and Fig. 4.1). The other cell will be a rhizoid, which anchors the thallus to a substrate (a soil) and where chloroplasts will be decreased and, then completely, reduced. This new cell becomes a rhizoid cell later, and its chloroplasts are missing. The other cell (prothallium-maded or prothallial) starts to divide, representing the beginning of the multicellular thallium (gametophyte). So after 24 h of the development from microspores multicellular thallium with non-fluoresced rhizoid arose. These data are correlated with histochemical investigations and with staining by artificial dyes (Beer, 1909; Nienburg, 1924; Nakazawa, 1956; 1958) and biochemical analysis of the content of chlorophyll and azulenes (Roshchina et al., 2002). The chlorophyll fluorescence reflects the first stages of the microspores development and serves as their marker. Dry blue-fluorescing microspores showed the maxima in their fluorescence spectra – major at 460 nm and small at 680 nm, which demonstrated the presence of chlorophyll in undeveloped single cells of Equisetaceae as more primitive taxon.

The comparison of the changes in the fluorescence and the changes in the pigment composition of the spores can be seen in Table 4.1. The parameter I red/blue was used as a characteristic of synthetic activity of cells as the band at 640-680 nm in the red region of spectra is due to mainly chlorophyll and, perhaps, also some other pigments, such as azulenes

Table 4.1 The changes in the fluorescence intensity (parameter I red/blue - the average values from 3 experiments) and the pigment content at the first stages of development of microspores.

Time after moistening	*Fluorescence intensity parameter I red/blue*	*Pigment content, mg/g of fresh mass*	
		Chlorophyll a + b	*Carotenoids*
0 min	1.01 ± 0.1	0.47 ± 0.03	0.20 ± 0.2
15 min	0.82 ± 0.3	0.39 ± 0.1	0.22 ± 0.01
30 min	1.20 ± 0.2	0.55 ± 0.03	0.24 ± 0.05
60 min	2.20 ± 0.2	1.00 ± 0.07	0.21 ± 0.04
120 min	2.70 ± 0.3	1.10 ± 0.04	0.35 ± 0.08
24 hours	3.21 ± 0.3	1.24 ± 0.01	0.27 ± 0.03

(Roshchina et al., 1995; 2002), whereas the band at 460 nm in the blue region is associated with cover compounds, mainly phenols and azulenes.

On the object glasses placed on Petri dishes the development of the microspore lasted not more than three d. Red cells developed and after three d showed non-fluorescing rhizoid (seen in transparent light of a usual microscope). The following development of vegetative microspores was observed 1.on the object glasses covered with parafilm and put on the wet camera of Petri dishes (short experiments - first wk) and 2. on the paper filters on the wet camera of Petri dishes (long experiments – more than 2 months). The nutrition medium was water or Knop solution. Just after 15-min of moistening the amount of rose- and then red-fluorescing cells arose in the blue-fluorescing microspores population which demonstrated the beginning of the development. After 24 h the fluorescence maximum at 680 nm grew more than 20 fold. Red cells developed on paper filters strengthening their fluorescence at 680 nm due to their division and following growth as the prothallium body. After 25-40 d of the development red gametophyte fluorescing mainly at 680 nm formed antheridia with blue-fluorescing spermatozoids (male cells) and archegonia with the egg cell (female cell) – blue fluorescing without the chlorophyll maximum. The microspectrofluorimetry allowed one to observe the main cycle of the *Equisetum* development, which is not usually clearly seen under the transparent light of a microscope.

According to Elenevskii et al., (2000), the viability of the vegetative microspores is not more than 2-3 wk, but our experiments show that it may be 2-4 months in dry form storing. The development of the spores is dependent on environmental conditions. The dark does not influence during the 1st d of development, only longer dark conditions lead to the depression (Colour Fig. 22) (Stahl, 1885). Figure 4.2 demonstrates the stages of development of prothallium and then thallus, where sexual organs are forming. After 1 wk of the germination on the paper filtres, the red fluorescence intensity increased which shows the intensive chlorophyll synthesis in the forming thallus. The picture is similar in the first two wk. Although spermatogenous tissue of antheridia (male organ) may be seen as it has bright blue or blue-green fluorescence with maxima at 450-460 nm. After 21d, one can see the differentiation of the thallus more clearly. Unicellular lamina plate converts to multicellular (rhizoids on the lower side). On the upper side, vertical laminar branches, from which the sexual organs are formed after 4-6 wk. Many neighbouring spores also affect the germination and following development, in particular in dense sowings – green thread instead laminar plate of gametophyte. Heart-like young gametophyte, intensively fluorescing in the blue, yellow and red spectral

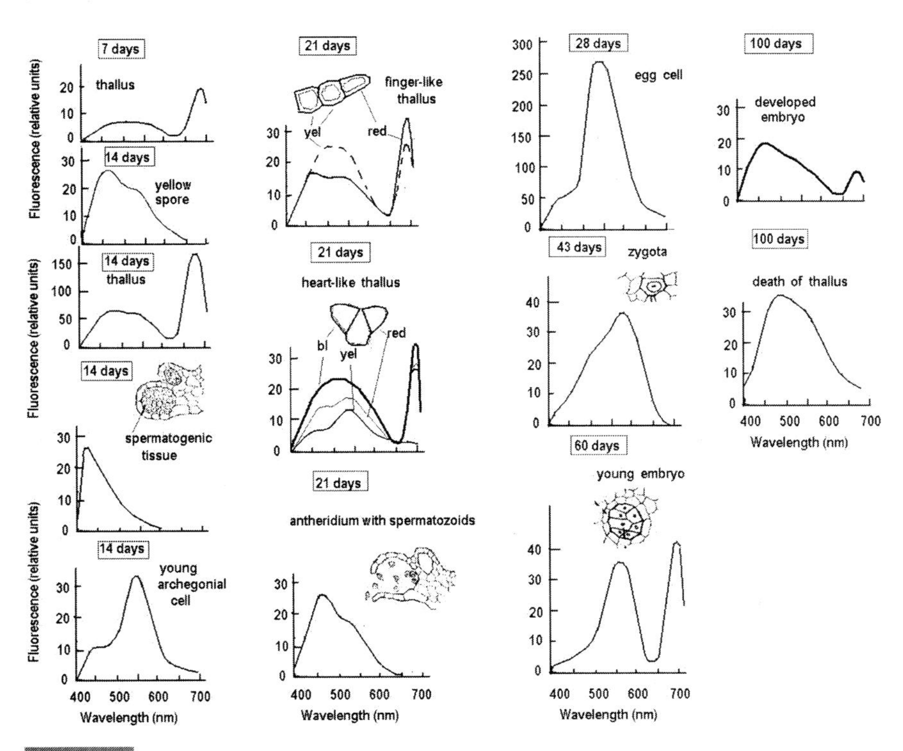

Fig. 4.2 The fluorescence spectra of the *Equisetum arvense* surface during the development of multicellular gametophyte with generative organs

regions, differs from the finger-like rhizoid (which has a weak blue emission if any). After 28 d non-germinated spores emit in green-yellow (maxima 475 and 680 nm as well as a shoulder 525 nm), while the gametophytes with similar spectrum arise. A green-yellow egg cell is seen. The structures have enhanced red fluorescence which identifies it among non-generative cells. Figure 4.2 shows the formation of the multicellular thallium (gametophyte) with developed male organs – antheridia and female organs – archegonia. Spermatogenous tissue and individual spermium fluoresce at 530-570 nm without chlorophyll maximum (see Fig. 2.2 in Chapter 2), whereas young and old gametophytes (7 and 8 wk, relatively) stored the same character of the spectra as unicellular microspore, only chlorophyll fluorescence intensity at 680 nm became much more – 6 times. The antheridium differs in the fluorescence between cells, perhaps due to the emission of carotenoids at 520-565 nm. Hauke and Thompson (1971) have shown the presence of carotenoid pigments associated with antheridial formation in *Equisetum* gametophyte. The egg cell was lightening in the blue (shoulder 500 nm) and orange (maximum 595 nm) regions, whereas surrounded cell – only in the red region

There are differences between female or male or germaphrodite gametophytes. Female or bisexual gametophytes have sizes from 1 mm to 2-3 cm, while with male organs are smaller. Spermatogenous tissues of the male organ antheridia form many (up to 100) spermatozoids after 21-25 d of moistening. The female organs are formed later, and every bottle-like archegonia (see Chapter 2, Section 2.1.2.) has one egg cell, to which the spermatozoid fuses, and the zygote arises. The fertilized egg is divided transversely, forming, then multicellular embryo.

Thus, the autofluorescence of the *Equisetum arvense* clearly distinguishes the stages of the development *in vivo* without any dyes. Enhanced red fluorescence serves as a marker for the development to start in microspores. Undeveloped microspores look like blue- lightening, and the maximum 680 nm are missed. Any scientist without special training can distinguish the antheridia and archegonia on the thallium due to their difference in the fluorescence.

4.1.1.2 Development of moss

The fluorescence of vegetative microspores from mosses and ferns after moistening has not studied before this publication. The prothallium and thallus develop just after the germination and first division of the microspores.

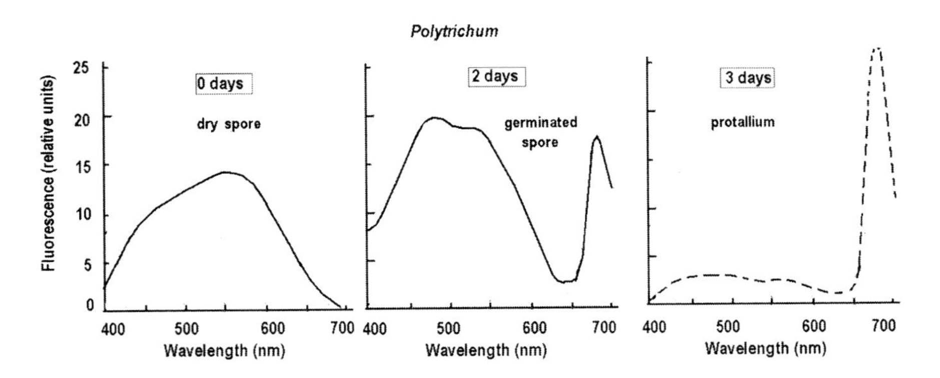

Fig. 4.3 The fluorescence spectrum of the germinated vegetative microspores of moss *Polytrichum* sp

Figure 4.3 demonstrates how vegetative microspore of moss *Polytrichum* germinate (has no the chlorophyll maximum in the dry state) and fluoresces in blue-yellow with maxima 460 (or shoulder) and 580 nm. After two d of germination, the red fluorescence at 680 nm appeared in the spectrum then the divided cell is converted to prothallium with bright fluorescence, peculiar to chlorophyll.

4.1.2 Development of Seed-bearing Plants

The fluorescence of secretory cells from seed-bearing species can be changed during plant development (Roshchina et al., 1996; 1997a, b; Kovaleva and Roshchina, 1999; Roshchina and Melnikova, 1999). Some examples of similar changes will be discussed below.

4.1.2.1 Pollen (generative microspores)

Mature and immature pollen. Development of generative (pollen) microspores in seed-breeding plants begins in the tapetum of the forming anther in the flower. According to Stanley and Linskens (1974), the forming pollen in tapetum contain chloroplasts, which are missed after maturing.

Immature pollen differs in the fluorescence spectra in comparison with the matured one, for instance as shown for the microspores of *Philadelphus grandiflorus* and *Tussilago farfara* (Fig. 4.4). In both species non-matured pollen has maximum 660-680 nm in the red spectral region. Non-matured pollen grains of first species have maxima in the blue (465-

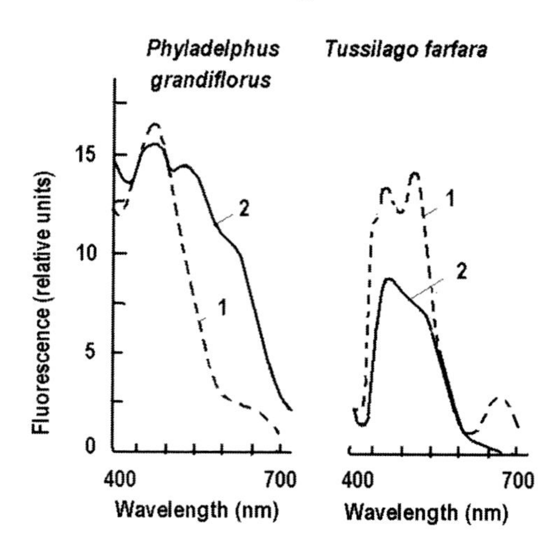

Fig. 4.4 The fluorescence spectrum of non-matured (1) and matured pollen (2) (Adopted from Roshchina et al., 1997b).

470 nm) and in red (665 nm) spectral regions, whereas matured pollen grains also demonstrated new maxima in green (510-520 nm) and shoulder 620 nm. Pollen grains of second species showed peaks in blue-green (shoulder 440 nm, maxima 465 and 518 nm) and red (maximum 680 nm), while the matured one – in blue with maximum at 470-480 nm and in yellow (maximum 550 nm). Table 4.2 shows the corelation between the appearance of the fluorescence maxima and content of some pigments such as carotenoids and azulenes.

At maturing of pollen from *Philadelphus grandiflorus*, the character of the fluorescence spectra changes significantly at the transfer from non-matured state (lightly yellow colour) to the matured one (orange colour). Both forms had maximum in the blue spectral region 465-470 nm, and the non-matured form also – small shoulder 665 nm. The differences were observed in the orange and red regions, where only matured pollen grains have several maxima at 510-520 nm (yellow emission) and at 620 nm (red emission). Since in this species, pollen lacks anthocyanins, accordingly the maxima may belong to flavins and carotenoids emitted at 500-550 nm and azulenes, which in a cell wall fluoresce at 660-620 nm (Roshchina et al., 1995). Table 4.2 shows that matured pollen of *Philadelphus grandiflorus* contain a smaller amount (approximately 10 times) of carotenoids than matured pollen grains which corelated with a lack of expressed maxima in the yellow-orange spectral region, but there is only one maximum in blue (465-480 nm). In matured pollen xanthophylls are formed. The lack of azulenes was also peculiar to non-matured pollen which corelates with the absence of fluorescence maximum in the red spectral region, there is shoulder at 665 nm, which is related to the traces of chlorophyll which disappeared in a mature form. Thus, a transfer of pollen from *Philadelphus grandiflorus* from the non-matured to matured state is accompanied by the enhancement of biosynthesis of xanthophylls and azulenes.

Table 4.2 The position of maxima in the fluorescence spectra of immature and mature pollens of *Philadelphus grandiflorus* and contents of some pigments in their exine (Roshchina et al., 1997b)

Pollen	*Maxima of pollen fluorescence*	*Total pigment contents, ng/mg fr wt*		
		Carotenoids		*Azulenes*
		total	*xanthophylls*	
Immature	465-480, shoulder 665	55.9 ± 0.05	6.1 ± 0.03	traces
Mature	465-480, 510-520, 620	72.3 ± 0.06	62.8 ± 0.03	6.0 ± 0.02

Small concentrations of carotenoids, especially xanthophylls, confirm the supposition that the pigment contributes in the fluorescence of mature pollen. Their deficit is in immature microspores.

While for *Tussilago farfara*, it was seen that non-matured greenish pollen evolved from anthers had three maxima of fluorescence – 465-470, 518 and 680 nm (Fig. 4.4). The latter maximum belonged to chlorophyll and disappeared in matured pollen, which spill out of anthers (Roshchina et al., 1995; 1997). There were other changes in the fluorescence spectra. Smaller maximum at 480 nm was observed for matured pollen that appears relate to different ways of metabolism in comparison with non-matured pollen. The latter maximum is likely to have a higher content of components such as flavonoids and NAD(P)H, which fluoresce in blue. In matured pollen, the maximum shifted from yellow to yellow – orange region (535-550 nm) and the peak decreased. This may be related to the changes in the composition of the carotenoids at maturation.

According to Willemse (1971), the fluorescence of pollen *Pinus sylvestris* changes significantly during pollen formation in the sporangium of the conifer plant. In sporangium, the main amount of pollen grains are in the centre of the structure, surrounded by tapetum (tissue, which serves for the nutrition of developing spores). Forming pollen grains develops through a stage of tetrad. Tetrad is a group from four gaploid spores, which are formed from the mother cell of the spore as a result of meiosis. At the tetrad stage, pollen fluoresces in green with maximum 500 nm, or in green-yellow (maximum more 500 nm). In anther these pollen spores, both tetrad and the wings, show a small light emission only. During the development, after the tetrad stage, a shift occurs to bluish green (about 490 nm). Probably, more flavonoids and flavins are formed along the maturing.

Pollen and pistils of *Petunia hybrida* from self-compatible and self-incompatible clones demonstrated different fluorescence spectra during development (Kovaleva and Roshchina, 1999). Pollen of the flower bud from both clones were identical (maxima 495 and 640 nm), while in an open flower there were differences. In a self -compatible clone there are two maxima at 460-470 and 520 nm and shoulder 640 nm, related, perhaps, to azulene or flavonoids rutin or quercetin. Unlike the self-compatible clone, in the self-incompatible clone pollen has no similar shoulder. The pistil of the self-compatible clone had maximum in green at 530 nm (flavins) whereas those from the self-incompatible – in green at 440 nm (terpenoids) in the flower bud before the flower opened. In a completely open flower, the pistil stigma of the former clone demonstrated maxima 480, 510 and 680 nm, but in the latter clone – 480, 510 and 595 nm. It is likely that the self-incompatible clone produced pollen, which lack chlorophyll (maximum at 680 nm). It appears to be redox reactions on the surface of pollen at the interaction "pollen-pistil". They include reduction-oxidation of the surface components, which differ in self-compatible and self-incompatible clone. The difference in the fluorescence spectra of the clones could be related also to genetic peculiarities, in particular with the activity of S-gene.

In some cases the development of the pollen in flower buds at the stage of a completely open flower and closing of the flower was observed, in particular for *Epiphyllum hybridum* (Roshchina et al., 1997b). Anthers, which were cut from an open flower, fluoresced presumably in blue with maximum 460-480 nm. In the yellow-orange spectral region, wide maximum 510-550 nm was seen. At ageing and drying of the anther for several days this maximum became smoother.

The fluorescence intensity is also changed at pollen maturing due to large water loss from the anther tissue. The pollen grain is therefore a dehydrated entity, and its emission enhances.

Pollen germination. Pollen may germinate either on the pistil stigma of a flower or in artificial nutrient medium (sometimes in water only). Certain pollens of the plant undergo meiosis giving clusters of cells with haploid chromosome content. During the first mitotic division of the microspore, two unequal cells are produced. One is the generative cell which ultimately gives rise to two sperm cells and which is embedded in the other cell to be produced at this first mitosis, the vegetative cell. The pollen grain after anthesis is therefore correctly termed as a multicellular male gametophyte. The two male sperm cells are formed in the pollen grain or in the pollen tube produced from the pollen grain at the time of germination of pollen. In angiosperms the pollen grain alights on the receptive surface or stigma of the female portion of the flower and germinates there (Mascarenhas, 1975; Cresti et al., 1992; McCormick, 1993).

The changes in the fluorescence occur at the development of pollen. Among models for the study of pollen germination on artificial nutrient media are knights' star *Hippeastrum hybridum*, mock-orange *Philadelphus grandiflorus*, clivia *Clivia* sp. and petunia *Petunia hybrida* (Roshchina et al., 1996, 1997b; Roshchina, 2001a, b; 2004a; 2005a, b, c; 2006a, b, c). Besides, pollen grains and pistils of intact laboratory model herbaceous plants *Hippeastrum hybridum* (Amaryllidaceae) is a suitable object for the experiments with the pollen development *in vivo* (on pistil stigma) because it has large pollen grains and the longest pistil. Moreover, *Hippeastrum* pollen germinates directly in water and does not need additional nutrition (Johri and Vasil, 1961).

The fluorescence spectra has been studied during the first steps of development after moistening of pollen grains of *Hippeastrum hybridium* by the nutrient medium in some earlier papers (Roshchina et al., 1996; 1997b). As shown in Fig. 4.5, two main types of fluorescent grains, differing in their intensity are seen: weak lightening which are viable and brightly lightening, usually non-viable (Roshchina et al., 1997a; Roshchina, 2003). The non-viable pollens were swollen and maintained their emission, but did not germinate during 24 h and do not form pollen tubes at all. The average intensity of the luminescence was 2-3 – fold as high as that for germinating pollen grains. In contrast, the viable pollen grains decreased their autofluorescence during germination, and after 60-120 min, where pollen tubes arose, the emission was quenched completely. Water quenches the emission, and, as a result, pollen with forming pollen tube has no or weak fluorescence in first hours after moistening. (Later in some species one can see the blue or blue-green emission of tube cell walls). Thus autofluorescence may be an indicator of pollen viability during the first 10-

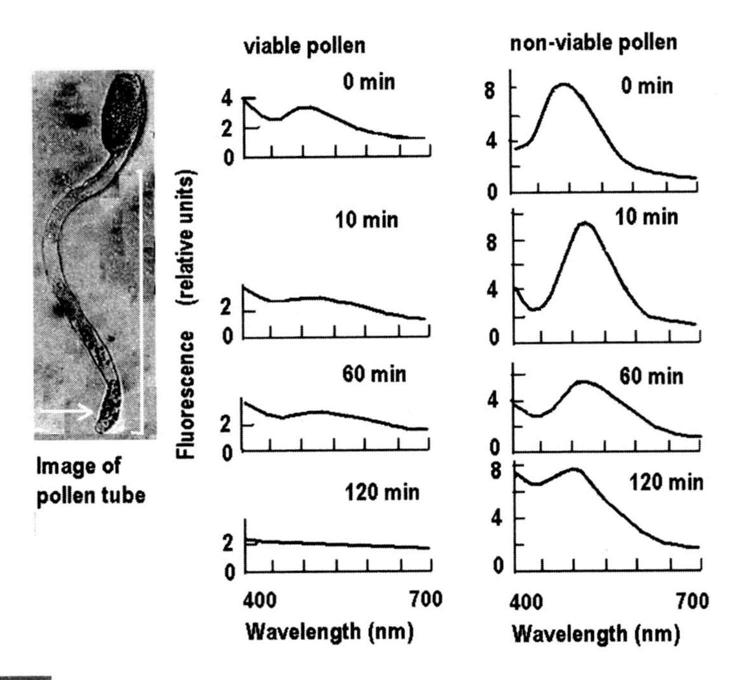

Fig. 4.5 The image of developed pollen tube (left) and fluorescence spectra of viable and not-viable pollen from *Hippeastrum hybridum* after moistening by nutrient medium (right). Adopted from Roshchina et al., (1996; 1997b) and Roshchina (2004a). Bar = 200 mm. Arrow shows the spermium on the tip of the pollen tube.

30 min after moistening. The phenomenon has been confirmed by experiments on pollen from other species *Aesculus hippocastanum*, *Sorbus aucuparia*, and *Betula verrucosa*.

Ageing pollen. We can also distinguish the fluorescence of fresh and ageing pollen (Fig. 4.6). Old pollen from *Hippeastrum hybridum* has two maxima in blue (465-470 nm) and in yellow-orange (545-580 nm) of the emission, instead of one (480 nm) in fresh. Three new peaks (455, 540 and 675 nm) in the fluorescence spectra also appeared for ageing microspores of *Papaver orientale* whereas for fresh-collected ones there was only one maximum 470-480 nm. This demonstrated the formation of new compounds, perhaps products of the lipid peroxidation and oxidation.

4.1.2.2 Development of leaf hair (emergence)

The leaf and stem hairs (emergencies) of the *Urtica* genus are unicellular systems, which also change their fluorescence during the development. The

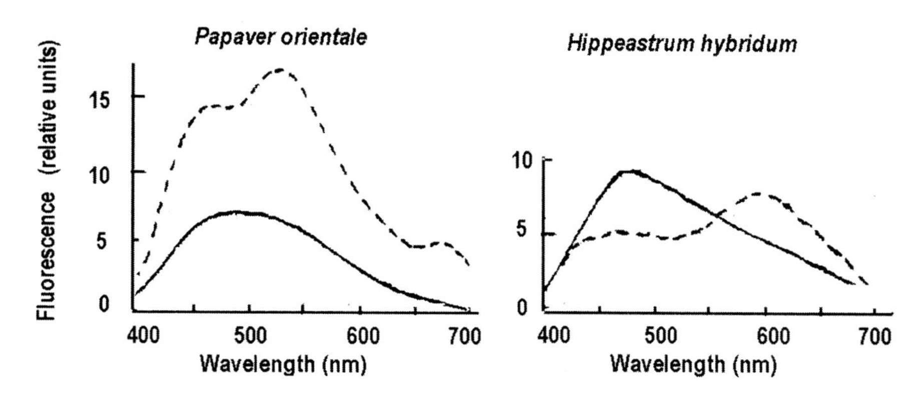

Fig. 4.6 The fluorescence spectra of fresh and ageing pollen. Unbroken line – fresh-collected pollen; broken line – pollen, which is stored for 1 y.

examples for two species *Urtica dioica* and *Urtica urens* may be analyzed (Figs. 4.7 and 4.8). The changes in the intensity of fluorescence are observed along the stinging hair of the common nettle *Urtica dioica* (Fig. 4.7). At the top of the hair where there is intensive evacuation of the secretion, the intensity of fluorescence is minimal whereas at the base of the emergence of the hair is highest. After one week of germination the *Urtica dioica* seedling secretory hairs arose and fluoresced (basis of the emergence is lightening) in green-yellow with maxima 525 and 585 nm as well as in red with maximum 680 nm, peculiar to chlorophyll. Secretion released out from the hair had only one peak 475 nm of the emission which is usually seen for the stinging emergencies from the adult plant leaf. Some crystals of the secretions are observed on the leaf surface and emit in blue-green with maximum 480 nm. After two wk of the seedling development the maxima shifted to the shorter wavelength region - 455 and 515-525 nm.

As for *Urtica urens*, secretory hairs appeared on one wk old seedlings as well as crystals of the secretions located both on the upper and lower side of the leaf (Fig. 4.8). Secretory hairs and the crystals have similar fluorescence spectra with blue-fluorescing (in lower leaf side) or and yellow –fluorescing (in upper leaf side) hairs have blue-lightening crystals at the base of the cell maximum 450-470 nm, but in some cases on the upper side of the leaf, the emergence also demonstrated a shoulder 550 nm. Among the forming hairs, one can not distinguish stinging or non-stinging emergencies; it may be done only after 2 wk of seedling development. In the latter case, stinging hairs were formed, while non-stinging emergencies

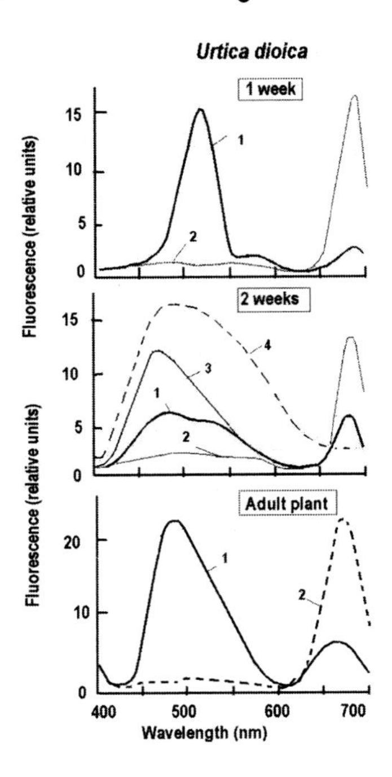

Fig. 4.7 The fluorescence spectra of *Urtica dioica* secretory hairs during 2 wk after the beginning of the development. 1 – secretory hair (blue fluorescence), 2 – non-secretory cell; 3 – secretion from the hair; 4 – crystal on the surface of the seedling.

started to form and weakly fluoresce. Adult stinging hairs fluoresced mainly in blue- or in blue-green with maximum 475-480 nm, and chlorophyll also contributed in red emission with maximum 680 nm. Crystals at the tip of the emergencies fluoresced with one maximum in blue - 460-470 nm. The stinging hairs on the upper side of the leaf emitted more intensively, than those on the lower side, and moreover, also have shoulder 550 nm in the yellow spectral region. The green-yellow emission was significant as the basis of emergencies.

4.2 DEVELOPMENT OF COMPLEX SECRETORY STRUCTURES

Fluorescence of complex secretory structures such as oil cells and ducts as well as glandular trichomes is also changed during their development. This reflects the alterations in their chemical composition.

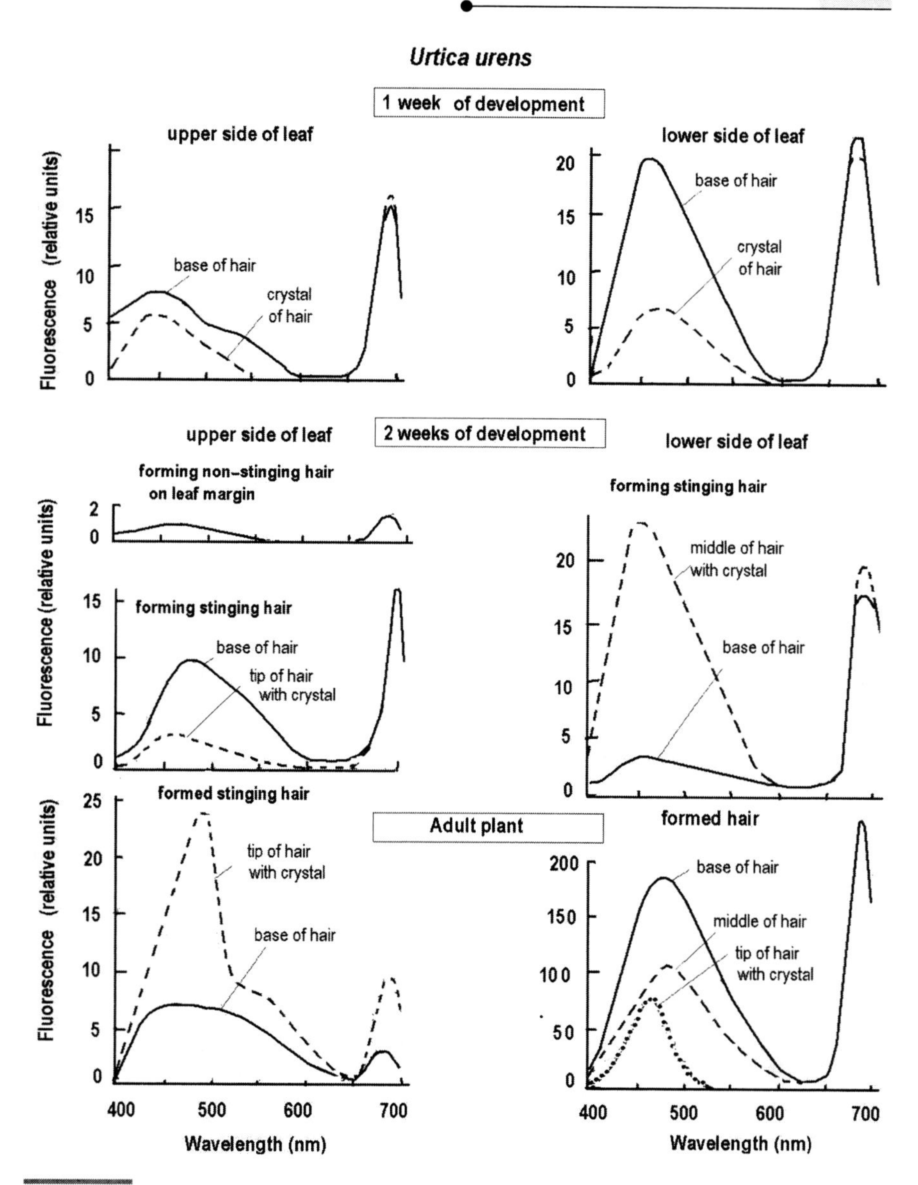

Fig. 4.8 The fluorescence spectra of *Urtica urens* secretory hairs during 2 wk

4.2.1 Glandular Structures

Changes in the light emission of glands or glandular structures should be considered depending on the secretion filled in the cell.

4.2.1.1 Terpenoid-enriched secretory structures

Some examples of terpenoid-enriched secretory structures such as oil cells and ducts as well as glandular trichomes, in which light emission is changed during their development are discussed below.

Oil structures. Essential oils are often found in secretory oil cells of *Heracleum* genus (Roshchina and Roshchina, 1993). Similar oil-containing secretory structures are located in stems (near the red anthocyanin-containing non-secretory cells) and flowers of *Heracleum sibiricum,* which also include furanocoumarins psoralens, besides terpenoids. Figure 4.9 shows the changes in the fluorescence of the stem oil cells during development. We can see the enhancement of the blue-green fluorescence with maximum at 480-500 nm of secretory cell, approximately 5 times for the stem after 10 days (July 18) from the beginning of the experiment on July 8, 2003. On July 28, besides maximum 480-500 nm, the shoulder at 540 nm was often seen on the ageing stem. Perhaps, secretory cells contain psoralens and coumarins, which fluoresce in blue-green and can be accumulated during development.

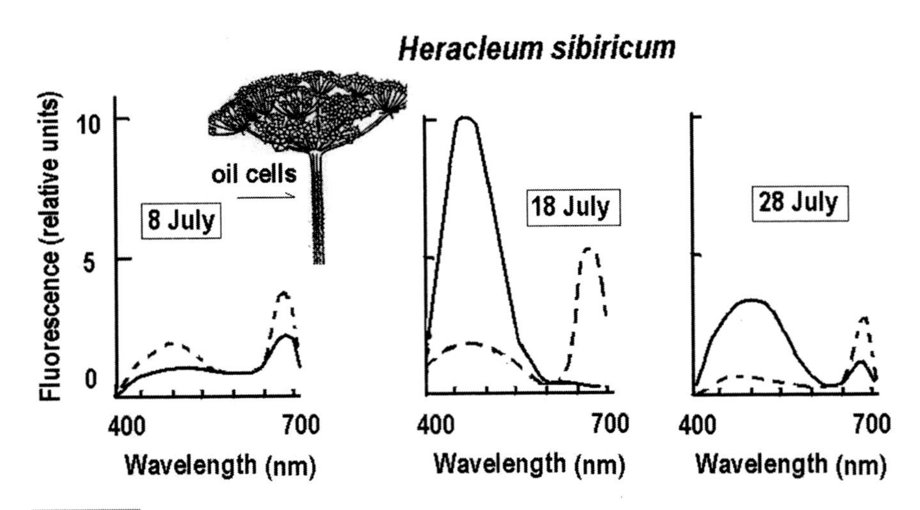

Fig. 4.9 The fluorescence spectra of oil cells on the stem from *Heracleum sibiricum* during their development. Unbroken line – secretory cell with oil secretions, broken line – non-secretory cell.

Glands and glandular hairs. The fluorescence spectra of glandular trichomes on the leaves from seedlings belonging to *Calendula officinalis* are shown in Fig. 4.10. Secretory hairs appeared just 7-10 d of the seed

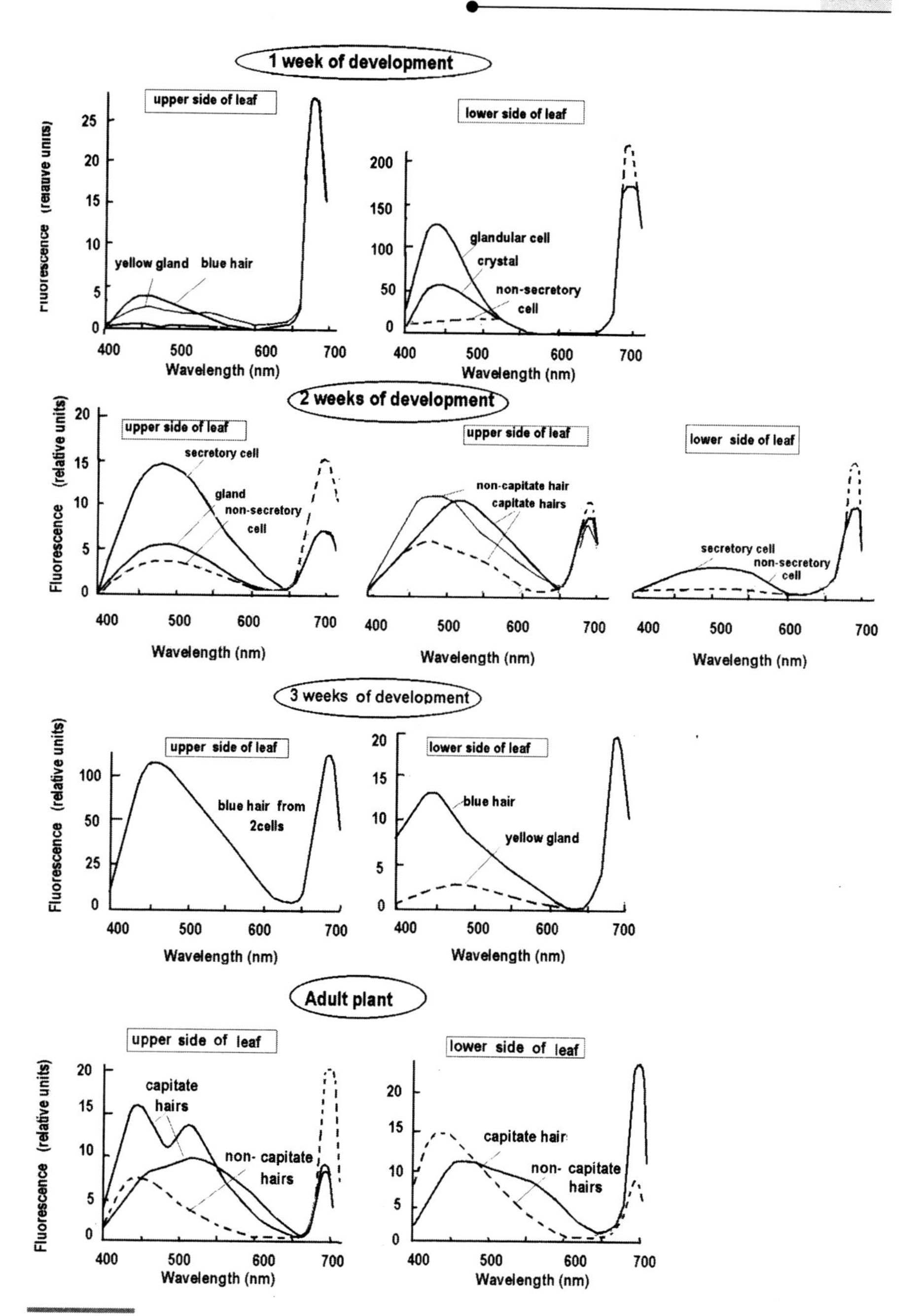

Fig. 4.10 The fluorescence spectra of leaves from seedlings belonging to *Calendula officinalis*.

germination (Fig. 4.10, Table 4.3). Glands were seen in just 7-d old plants: yellow-fluorescing glands with maxima of the emission 450 and 550 nm for the upper side of the leaf and 450 nm for the lower side as well as blue-fluorescing hairs with maximum 450 nm only for the upper side of the leaf. The emission intensity of the gland and the crystal of the secretion within the same gland on the upper side of the leaf was 5 times higher, than on the lower side. Two wk later, all secretory cells emit in green-yellow (maximum 500-530 nm), and the beginning of capitated hair was seen. But only after 3 wk of development, one can clearly see the appearance of multicellular glands and glandular hairs fluorescing in green-yellow and blue. The adult plant contains differentiated glands and trichomes. Secretory capitate hairs of the upper leaf side have three maxima in the fluorescence spectra 450 (480), 525-530 and 680 nm, while non-capitated hairs of both leaf sides - maxima 450 (480) and 680 nm.

Table 4.3 The fluorescence intensity of the glandular hairs of *Callendula officinalis* measured by double beam microspectrofluorimeter in green-yellow ($I_{520-540}$) and in red ($I_{640-680}$) part of the spectrum during the seedling development. Excitation 420-436 nm.

Variant	$I_{520-540}$	$I_{640-680}$
2-3 d old seedling No hairs. Red-fluorescing	0	4.2 ± 0.3
5-7 d old seedling Glandular hair, which appeared only on lower part of first leaf	0.5 ± 0.016	5.73 ± 0.7
5-7 d old seedling Part of first leaf, free of hairs	0.1 ± 0.02	5.9 ± 0.2
10 d old seedling Green-fluorescing hair of lower side of leaf	0.9 ± 0.009	10 ± 0.003
10 d old seedling Part of lower leaf side, free of hairs	0.68 ± 0.2	9.8 ± 0.01

The chemical composition of the secretory hair changes, and this is reflected by the changes in fluorescence observed during seedling development. The emission intensity of forming secretory hairs also changes during leaf development (Table 4.3). Better development of secretory hairs is seen for the lower side of the primary leaf. Green fluorescence intensity increases, approximately upto 9 fold, in the sites of the glandular trichome formation, depending on the accumulation of the secretion enriched in terpenoids. The part of the leaf, which is free of the glandular structures, has no significant green fluorescence. In this part and in secretory hairs, red fluorescence at 640-680 nm also increased during the development of the seedling.

Using the luminescence technique, the glandular hairs of *Sigesbeckia jorullensis* Kunth (Asteraceae) fluorescence and composition of essential oil at the flower and leaf development were studied (Heinrich et al., 2002). The hairs showed a subcuticular space filled mainly with essential oil, yellow coloured cells of the upper layer, green cells of the lower layers and a long stalk. Lipophilic droplets are visible on the cuticle of the subcuticular space. Blue fluorescence was observed for lipophilic materials under UV light (365/395/420). Red autofluorescence of chloroplasts was also seen both in the lower layers of the head and stalk cells. Even in young hairs the apical layer did not contain chlorophyll, and the bright blue autofluorescence indicates lipophilic material. The upper head cells stained more intensely with Sudan black B than the lower ones. Yellow secondary fluorescence indicated the presence of flavonoids in the head cells, and of flavonoid aglycones in the secretion products on the surface of the head and stalk cells. Flavonoids were present in vacuoles, whereas the inner part of the cells, where the nucleus is surrounded by plastids, is free of flavonoids, as indicated by the lack of secondary fluorescence.

4.2.1.2 Phenol-, alkaloid- and polyacetylene-containing structures

Phenol-, alkaloid and polyacetylene-containing structures are known for various plant taxa. In such plants there is a difference in the development of the secretory cells on the upper and lower sides of the first leaf forming from seedlings of herbaceous or from leaf buds of woody plants. The differences are also seen in secretory structures of mature and immature flowers as well as of developing roots. Some examples will be presented below.

Seedling development. The example of the fluorescence in forming of hairs during the development of the *Lavatera* sp. seedling shows a difference between secretory structures of the lower and upper sides of the first leave (Fig. 4.11). On the upper side glandular hair and gland fluoresce with maxima in blue (maximum 470 nm) and in yellow (maximum 550 nm) and there is also red emission at 680 nm related to chlorophyll. However, at the lower side we can see blue-fluorescing hairs with only one maximum 475 nm and yellow-fluorescing gland – with peaks 580 and 680 nm. Two-week old seedlings demonstrate a similar fluorescence

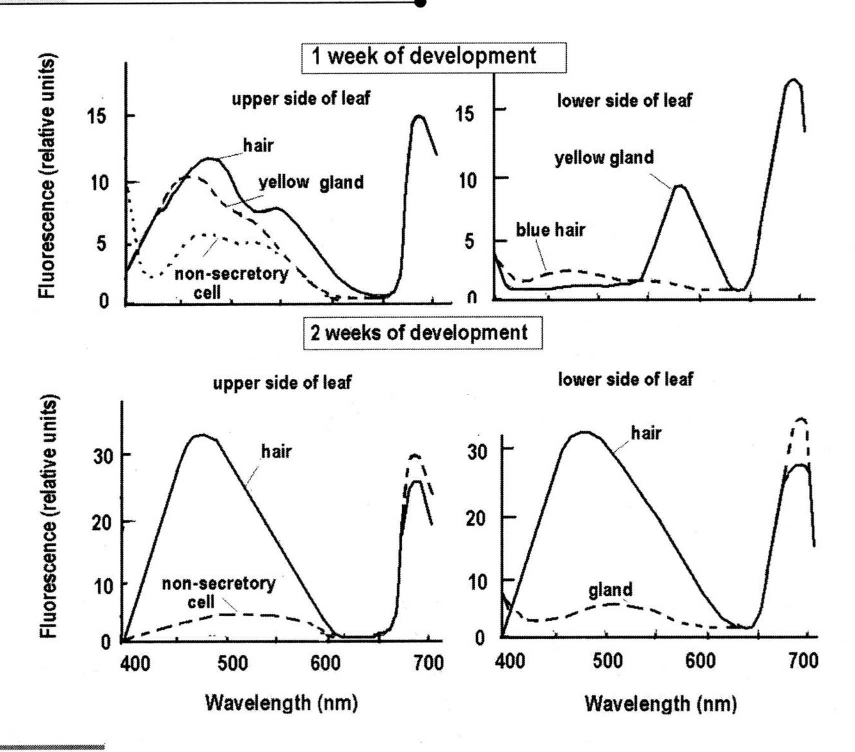

Fig. 4.11 The fluorescence spectra of the leaf secretory cells from *Lavatera* sp. during development.

spectrum of secretory hairs with those of the adult plant (See Fig. 2.35. in Chapter 2). Only maxima in the blue region are seen.

Bud development. In *Swida alba*, the first leaf of buds have three maxima 445, 555 and 680 nm for the lower side and 445, 530 and small 680 nm for the upper side (Fig. 4.12). Glands and glandular cells on the lower side in the young leaf and old leaf fluoresced only with two maxima – in blue 445 nm and in red 680 nm, although yellow-fluoresceng hair was seen in the old leaf. The first leaf of the bud has intensive emission of the green-yellow-fluorescing hairs (maximum 530 nm), which differed from blue-fluorescing hairs (maximum 420-430 nm) that shows the presence of different secretory products in the glandular structures.

Flavonoids aglycons in the mixture with terpenoids contained in the bud scales of many woody species of the genera *Aesculus, Alnus, Populus* and *Betula* (Fahn, 1978; Wollenweber et al., 1987; 1991). In spring the filling

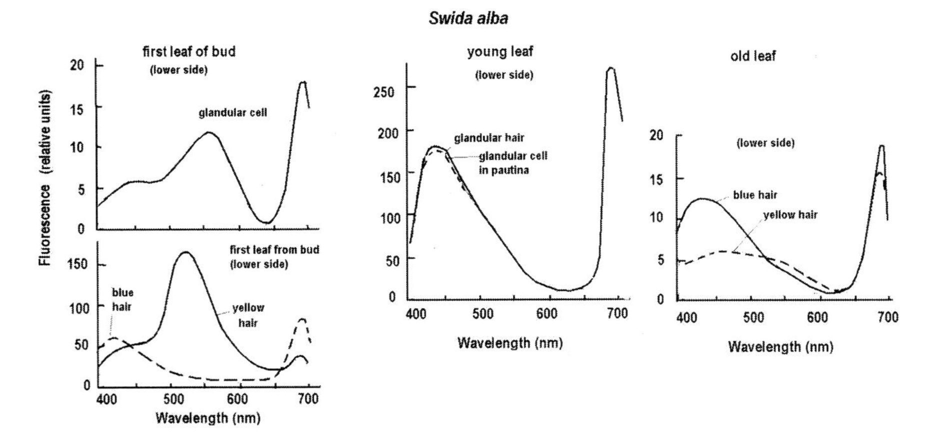

Fig. 4.12 The fluorescence spectra of secretory structures of *Swida alba* during the development of leaf buds.

of secretory glands (colleters) of the scales by resin-like secretion leads to the bright luminescence, mainly in the blue-yellow region of spectra and, except *Aesculus*, in the red region at 650-690 nm which is peculiar to chlorophyll (Fig. 4.13). The form of fluorescence spectra of *Betula* and *Populus* changed during this time period, unlike the *Aesculus* bud scales where only the height of the peak at 500-510 nm changed. In *Betula* spectra there was shoulder at 480 nm, maximum at 520-530 nm, and a chlorophyll maximum at 680 nm on April 1, then on April 14, the shoulder disappeared and the height of the 520-530 nm maximum sharply decreased, and at last on April 21, all maxima decreased. Unlike *Betula*, in *Populus* spectra on April 14, shoulder at 480 nm was absent, but the maximum at 510 nm was high, and on April 21, the decrease in the entire blue region fluorescence was observed, whereas marked chlorophyll maximum had not disappeared yet.

The intensity of emitted energy as the secretion is accumulating, increased up to a peak after 2 weeks in April, and decreased to zero at the end of April when the bud scale dropped and the young leaf was opening. At the beginning of May the leaves are unrolled from the buds. The secretory function serves for the defence of the primordial leaf from damage factors such as pests and late frosts. In the bud secretions phenolic compounds, mainly flavonoids and esters of aromatic acids as well as terpenoids prevail (Wollenweber et al., 1987; 1991; Roshchina and

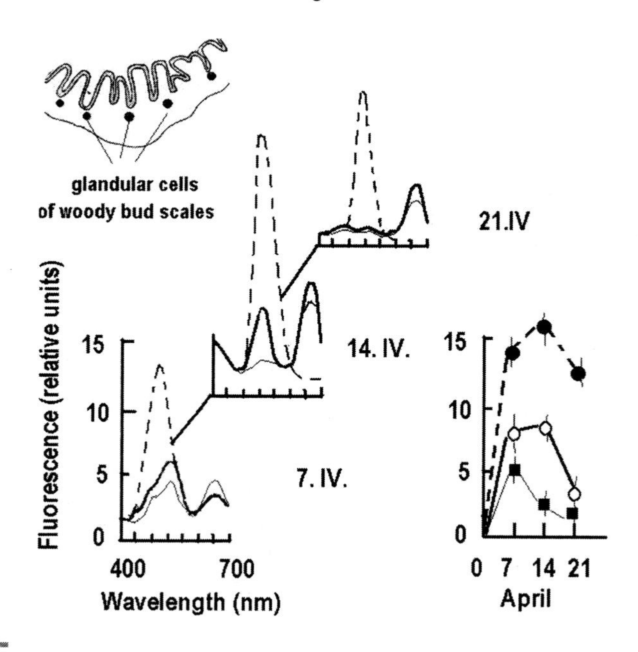

Fig. 4.13 The scheme of location, the fluorescence spectra (middle) and kinetics of the light emission intensity in the maximum (right) of developing secretory cells in scales of woody buds through April 7, 14 and 21. Adopted from Roshchina et al., (1998a). Thick unbroken line - *Populus balsamifera*; thin unbroken line – *Betula verrucosa*; broken line - *Aesculus hippocastanum*. At the start of the experiment, fluorescence was not observed.

Roshchina, 1989). In the bud exudates of *Populus deltoides* similar components are found (Greenaway et al., 1990).

Mature and immature flowers. The flower secretory cells enriched in polyacetylenes, in particular in *Echinops* genus (Golovkin et al., 2001), differ in their fluorescence in matured red and immature green ligulate flowers as well as from leaf secretory hairs (Fig. 4.14). Although both of them have fluorescence with maxima in blue at 450-460 nm, but the contribution of yellow emission of shoulder at 540-550 nm is clear only in the mature flower secretory cell. But unlike the mature flower and leaf, yellow fluorescence at 530-550 nm is seen in pollen of the immature flower.

Developing roots and root culture. Alkaloid-containing plants may also demonstrate alteration in the fluorescence of secretory cells. As seen from Fig. 4.15, the intensity of light emission at 580-600 nm from components of excretions in the intact root (acridone alkaloids, in particular rutacridone) depends on the age of the organ. (The main population of root

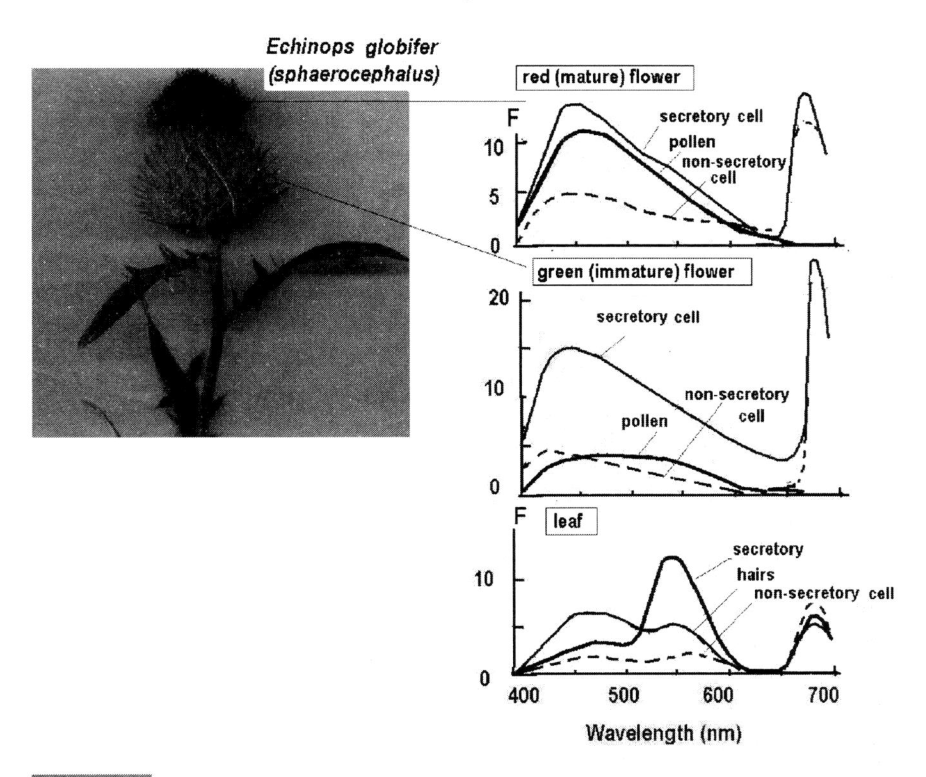

Fig. 4.14 The fluorescence spectra of various secretory cells in petals of mature and immature flowers from *Echinops globifer (sphaerocephalus)* in a comparison with leaf secretory cells.

cell fluorescence in blue at 470-480 nm). In the young cell the orange emission of tip increases approximately 5 times in a comparison with the lightening before secretion accumulated, while the single crystals and secretory hairs have low fluorescence intensity. In the old root, the secretory process decreases as a whole, and no difference can be seen in the tip fluorescence between cells before and at the secretion accumulation. Nevertheless, some crystals emit brighter, than in the young root. Unlike intact roots, the root culture, which include orange-fluorescing (maximum 590-600 nm) idioblasts, also demonstrate brightly yellow-fluorescing (maximum 550-560 nm) cells filled with the secondary metabolites, which differ from acridone alkaloids.

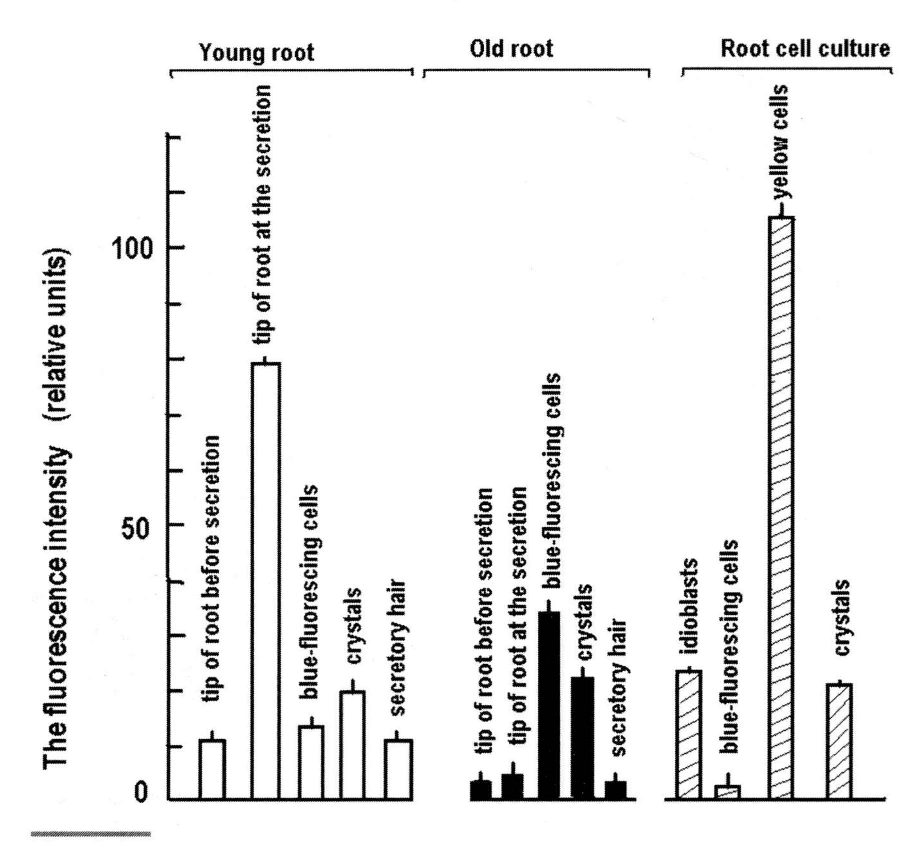

Fig. 4.15 The fluorescence intensity at 580-600 nm (acridone alkaloids) of the various root parts and of root cell culture in *Ruta graveolens.* Young root – 22 d old; Old root – 44 d-old. Excitation 360-380 nm.

4.2.1.3 Resin-containing cells

Resin-containing cells of conifers have a bright fluorescence, and this phenomenon is useful for the analysis of the development of similar structures as will be demonstrated in Figs. 4.16 and 4.17.

Seedlings. The developing 3-d old seedlings of *Pinus sylvestris* shows a cholorophyll-less part of the developing root tip (hypocotyl start) and chlorophyll-enriched part of developing leaf meristem (see the upper side of Fig. 4.16 for *Pinus sylvestris*) which is seen from the fluorescence spectra. Figure 4.16 shows that the tip and future root meristem of 2-3 d old seedling fluoresce 1. in the blue-yellow region with maxima 470 and

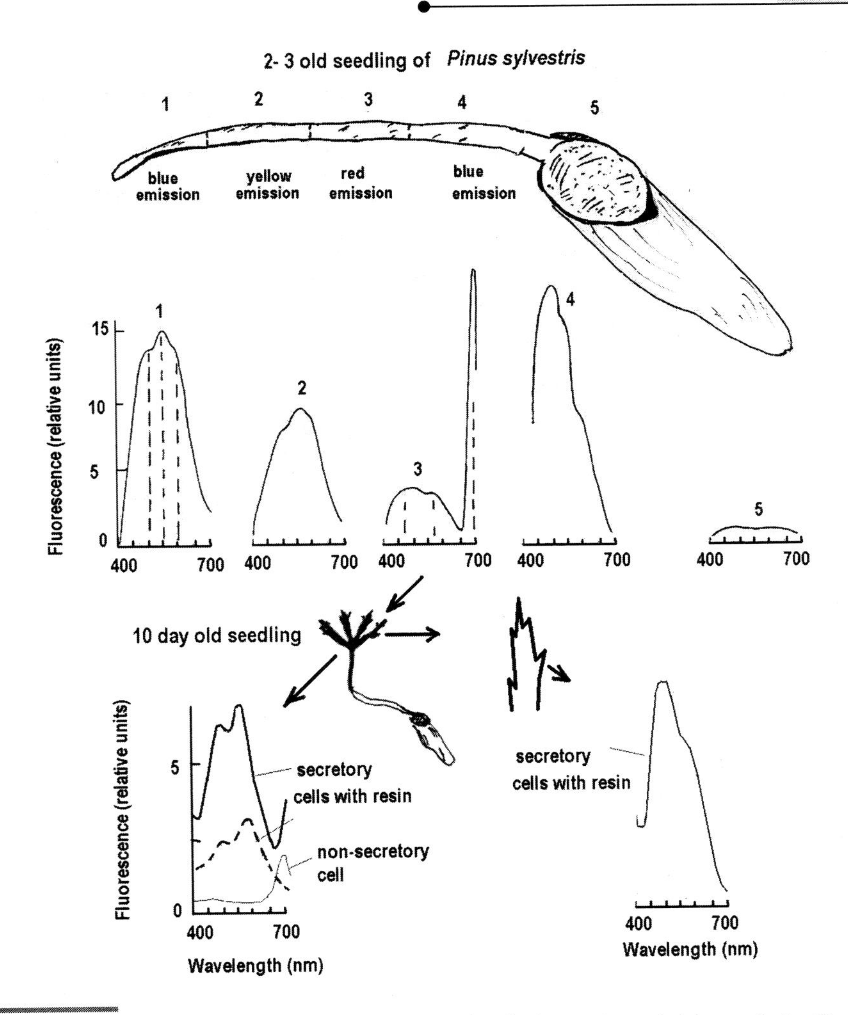

Fig. 4.16 The fluorescence spectra of the developing resin-containing cells in *Pinus sylvestris* in 2-3 or 10 d old seedlings.

550 nm, then 2. yellow emission occurs with maxima 450 and 550 nm; 3. Red emission with maxima 470, 570 and 680 nm; 4. Blue emission with maxima 475, 530 and 600 nm; 5. Small emission of the seed cover, if any. 10 d old seedlings have formed leaf plates with secretory resinous cells fluorescing with maxima 490 and 570 nm or 500 and 600 nm whereas surrounded non-secretory cells emit in red with maximum 680 nm, peculiar to chlorophyll. Primary needles have resinous cells emitting with

maxima 525 and 600 nm. The maximum 450 nm, peculiar to the mature adult needles of the same species (Table 2.1, Chapter 2), is absent at the stage of development. As in a mature leaf, common maxima in the yellow-orange spectral region (525-566 nm) are seen.

Cones. Cones of *Larix decidua* fluoresce due to the presence of resin-containing secretory cells (Fig. 4.17). In the green cone, the emission is with maxima 475 and 550 nm, whereas the cone became red when the fluorescence of resin excreted has only one maxima 460-470 nm and small shoulder at 550-575 nm as well as secretory cells. Mature yellow-brownish cone demonstrated only maxima 530 nm.

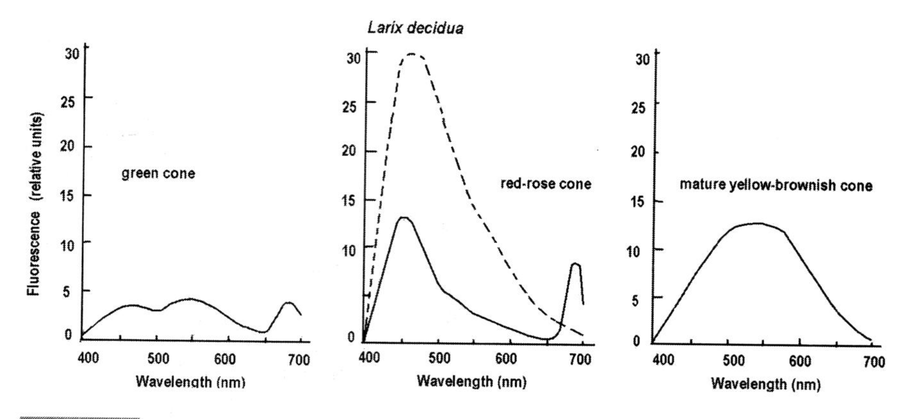

Fig. 4.17 The fluorescence spectra of secretory cells of developing cone from *Larix decidua*. Unbroken line – secreting cell; broken line – released resin on the cone surface

4.2.1.4 Salt-containing cells

Salt-containing glands of *Chenopodium album* include crystals, which fluorescence in yellow, being excited by light with maximum 420 nm, or in blue at the UV (360-380 nm) excitation (see Chapter 2, Section 2.2.2.4). Seeds of the plant also contain green-yellow fluorescing crystals, but there are no any glandular structures on the surface. 1, 3, 5, and 8-d-old seedlings of the plant were developed from seeds moistened with water or the salt solutions, and their fluorescence was analyzed. Each seedling has green-yellow fluorescence at a tip (developing root meristem) and red fluorescence - in part near the seed (developing leaf meristem in primary leaf forming from cotyledon). Only after 8 d of the seed germination in the red fluorescing part of the seedlings, which later transform into a primary leaf, and glands with salt crystals are seen. There are no such crystals in green-yellow fluorescing tip of the seedling, from which the primary root will arise. The nature of the salts, which form the salt gland, has been studied in the experiments with the seed germination in various salt media – from water to 10^{-4}-10^{-1}M salt solutions of sodium sulphate, sodium carbonate, potassium chloride and sodium chloride. As shown in Table 4.4, in the red-fluorescing part of the seedling the green-fluorescing spots arose in seedlings developed in sodium sulphate salts. Other salts used were not effective. Only 10^{-1}M sodium sulphate stimulated the salt gland formation. The green-yellow fluorescence of salt crystals in the glands was seen in the red-fluorescing part of the seedling earlier, when similar glandular structures arise in a control (after 8 d of the seed germination). The highest concentration of sodium sulphate 10^{-1} M is effective for the salt gland arising. After 8 d of the development, the fluorescence intensity in salt glands formed on the lower side of the primary leaf strongly enhanced and was about 210% of control, while red fluorescence decreased up to 45-47% of control due to the formation of the glandular structures. Thus, fluorescence of salt crystals in forming salt glands may be a marker for the study of plant seed development among halophytes.

Table 4.4 The fluorescence intensity (relative units) of the developing 3 d old seedlings from *Chenopodium album.* Excitation 420-436 nm

Variant	*Green fluorescence at 520-530 nm* ($I_{520\text{-}530}$)	*Red fluorescence at 640-680 nm* ($I_{640\text{-}680}$)
Control without salts (no salt glands yet)	0.05 ± 0.01	0.90 ± 0.09
Sodium sulphate 10^{-4} - 10^{-2} M (no salt glands yet)	0.05 ± 0.005	0.92 ± 0.01
Sodium sulfate 10^{-1} M (developing salt glands). Fluorescent crystal in gland	0.11 ± 0.02	0.84 ± 0.10
Sodium sulphate 10^{-1} M (developing salt glands). Part of seedling without salt gland	0.04 ± 0.01	0.77 ± 0.09

Conclusion

Autofluorescence of plant secretory structures changes during their development which reflects the alterations in a composition or/and redox state of secretory products accumulated. The phenomenon could be an indicator of the structures' formation, based on the appearance of their secretions at the earlier stage. Finally, light emission may be useful for the analysis of pollen maturing or its self-incompatibility as well as ageing of the microspores.

CHAPTER 5

Fluorescence of Living Cells at Intercellular Contacts

The alterations in cellular autofluorescence occur at interactions between cells, for example between sexual cells of one and the same plant species or between cells of various species. Contact with neighbours at so-called "effects of neighbour" takes place during any chemical relations between species known as allelopathic interactions (allelopathy). This includes relations in biocenosis (plant-plant, plant-animal, plant-microorganism) or at interaction pest- organism. Chemical interactions such as at fertilization of sexual cells or allelopathic relationships may also be considered as biogenic chemical influence. The earlier diagnostics helps to design more suitable phytocenosis in agriculture and the forest economy.

At cell-cell contacts one cell-acceptor with the help of its own excretion and sensor systems receives a chemosignal from a secretion excreted by the other cell, which is known as a cell-donor. There is recognition of the signal and the response to the signalling (Knox et al., 1976; Murphy, 1992). The fluorescence of both contacting cells may change. Therefore, the fluorescence could be considered as a response on the external chemosignal and, simultaneously, as an indicator of the signalling.

Secretions play an essential role as chemical signals or/and as medium for the reception of the signals (Roshchina, 1999a, b; 2001a, b). The cell-donor for chemosignal (fluorescent signalling substance in its excretion) and the cell-acceptor of the same chemosignal (recognizing systems in its excretion) interact via their excretions. The secretory products are located in surface (0.01–0.02 μm in diameter) channels of the cell wall, for example in the cuticle of nectary cells (Koteeva, 2005), or in pollen exine (Rowley et al., 1959; Roshchina et al., 1998d) as well in the secretory ducts of pistil stigma (Fahn, 1978). The changes in the fluorescence of the cell-donor and cell-acceptor may be observed under a luminescence microscope or with a help of microspectrofluorimetry (Roshchina et al., 1996; Roshchina and Melnikova, 1996; 1998a, b; 1999; Roshchina, 2003; 2005a). Microspectrofluorimetry and the new Laser Scanning Confocal Microscopes from Leica can be used for recording the fluorescence spectra in several ways: 1) the diagnostics of secretory cells containing allelochemicals; and 2) dynamics of chemical interaction between secreting cells.

Some examples of the fluorescence at the cell-cell contacts and some of the problems will be discussed below.

5.1 POLLEN-PISTIL INTERACTIONS. CONTACTS AT THE FERTILIZATION

The fluorescence of both contacting cells may be changed, for instance at fertilization after interaction between sexual cells – female and male gametophytes such as pollen and pistil in a flower) or chemical relations between cells of different species (pollen of own and foreign species usually have contacts, when added on one and the same pistil). The fluorescence spectra will be considered with examples of secretory cells from non-generative and generative organs. In the latter cases "pollen-pistil" and "pollen-pollen" interactions will be analyzed in connection with fertility and pollen allelopathy.

5.1.1 Interactions at the Contact "Pollen-pistil"

As described in Chapter 2, pollen and pistil fluoresce under ultra-violet light. During fertilization of plants, the generative male microspore known as pollen, is added to the surface of the pistil by wind, insects or by the human hand in the selection practice. Cell-cell interactions of pollen with a cell of the pistil stigma at their contact result in changes in the fluorescence spectra of cells contacted (Fig. 5.1). The fluorescence of both contacting cells are changed if the cells belong to one and the same plant species (Roshchina et al., 1996; 1997b; Roshchina and Melnikova, 1999). In the opposite case, there were no changes (see also Section 5.2.1). Before pollen addition only weak green-yellow emission of the pistil stigma papillae was seen. Visual increase in blue fluorescence of the stigma papillae, when pollen was put down on the pistil surface was observed especially at cross-pollination (Fig. 5.1). The events occurring upon cross- and self-pollination

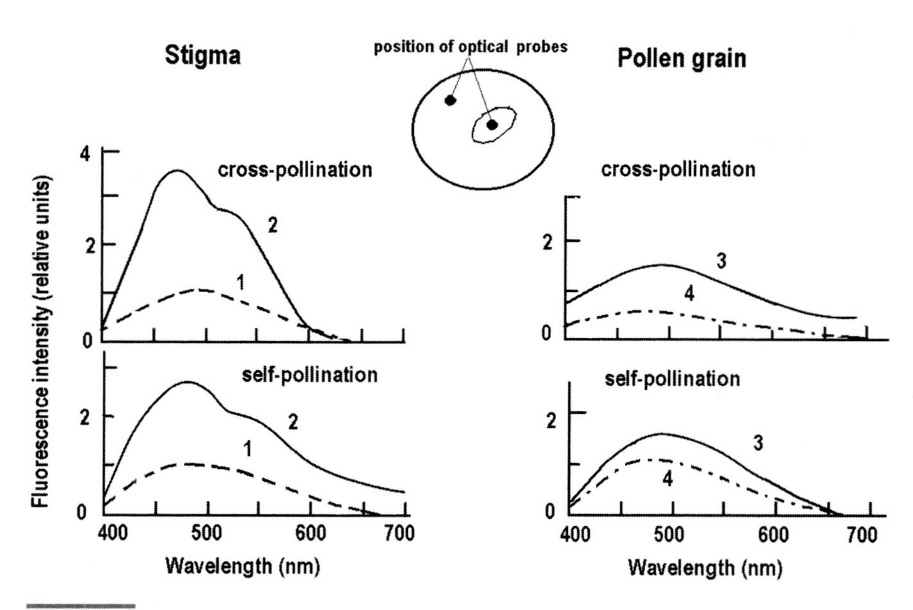

Fig. 5.1 The fluorescence spectra of pollen and pistil stigmas (stigmata) of *Hippeastrum hybridum* upon self-pollination and cross-pollination. Adopted from Roshchina et al., (1996; 1997b). 1. Pistil without pollen; 2. Pistil at the contact "pollen-pistil"; 3. Pollen lying on pistil stigma; 4. Pollen out pistil.

seem to be more complicated. The pollen grains of diverse plant species are transferred to the stigma surface, but only the pollen of "own" species is capable of germination. Probably, the liquid excreted by the stigma contains inhibitors and stimulators of pollen germination, including plant-specific alkaloids, phenols, terpenoids, and other fluorescent substances. These compounds may also be involved in free-radical reactions occurring on both contacting surfaces; thus, they may rapidly trigger pollen germination or participate in germination as a component of the signalling system.

With the help of fluorescent cytodiagnostics, a researcher can know whether pistil cells recognize, pollen is "own" or "foreign". Figure 5.2 shows the changes in fluorescence of the pistil with pollen from different species, which are added on the stigma. So the pistil refuses to receive the "foreign" pollen, does not to permit it to germinate, and then to react with an egg cell. Moreover, pollen can germinate only on the pistil of the same plant species. Possible contacts between "own" and "foreign" pollen grains are also demonstrated out pistil, modelling their interactions on the pistil stigma (Roshchina and Melnikova, 1996; 1999). This is interpreted (see

Section 5.2.1.1) as models of allelopathic relations between pollen of different plant species which leads to the inhibition or the stimulation of their ability to grow on the pistil stigma (Roshchina, 2001b). The knowledge is necessary for the specialists in the plant selection and the genetic practice. As shown in Fig. 5.2, the fluorescence of "foreign" pollen from *Dactylis glomerata* leads to small changes of the pistil stigma fluorescence if any, while the addition of "own" pollen induces the appearance of the response, which is similar for those seen in the Fig. 5.1.

Fundamental studies of fluorescence in cell-cell contacts under a luminescence microscope may be done so that the plant is not damaged. For instance, there was similar manipulation with the pistil, which was pollinated, but was not separated from the flower of a whole plant *Hippeastrum hybridum* (Roshchina et al., 1998a; Roshchina and Melnikova,

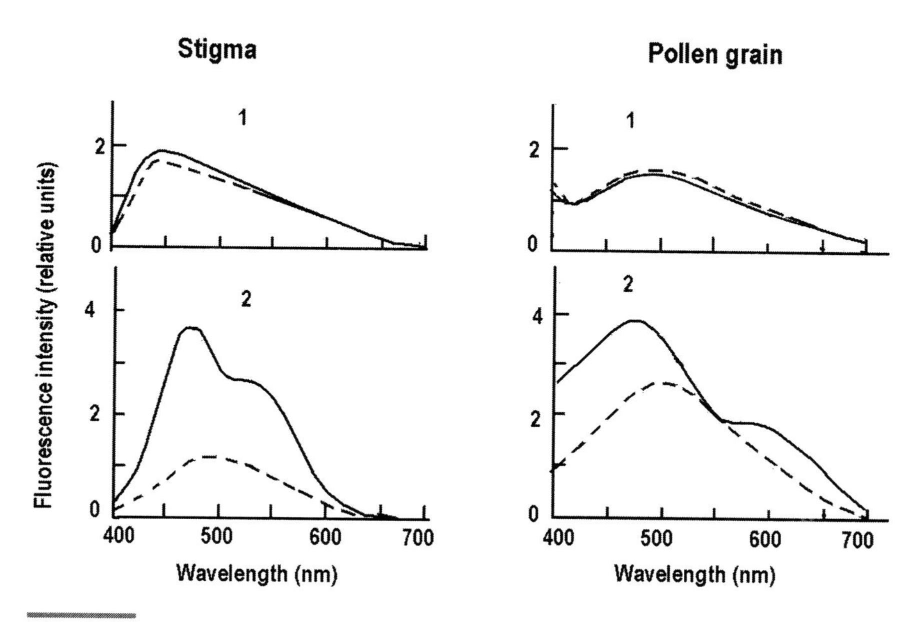

Fig. 5.2 Changes in the fluorescence spectra at pollen-pistil contacts. Position of optical probe is shown on Fig. 5.1 for the pistil stigma and pollen added on the pistil surface. Adopted from Roshchina et al., (1996). Left. The fluorescence spectra of the pistil stigma of *Hippeastrum hybridum* without (---) and with (___) pollen grain added. 1. Foreign pollen of *Dactylis glomerata*; 2. Self-pollen of *Hippeastrum hybridum.* Right. The fluorescence spectra of pollen with (_) and without (---) interaction with the pistil stigma of *Hippeastrum hybridum.* 1. Foreign pollen of *Dactylis glomerata*; 2. Self-pollen of *Hippeastrum hybridum.* All changes are observed only at the contact of pistil with self-pollen - pollen of the same plant species.

1998a). In this case, after pollination and the fluorescence measurement with microspectrofluorimeter the plant was studied, following its growth and development up to the formation of normal fruits and viable seeds.

The difference in the form of the spectra and position of their maxima both for the pistils and pollen grains depends on the plant species and the composition of the surface of the pollen grains (Roshchina et al., 1998a; Roshchina and Melnikova, 1999). Table 2.1 summarized the fluorescence maxima of the studied pistil and pollen from many species and shows the variety of peaks, dealing with the different components of their surface and excreta. Their fluorescence spectra show maxima at 460-490 nm (blue), 510-550 nm (green-yellow) and 620-680 nm (orange- red). Blue and blue-green fluorescence could be related to phenolic compounds, green-yellow-with carotenoids and orange-red – with chlorophyll (Wolfbeis, 1985) or azulenes (Roshchina et al., 1995). Depending on the plant species, the pistil stigmas also contain phenols in the exudates (Roshchina et al., 1998a; Roshchina and Melnikova, 1999), which can fluoresce both in the blue and yellow spectral region.

5.1.2 Mechanisms of the Fluorescence Changes at Pollination

5.1.2.1 Effects of individual components of pollen secretions on the pistil fluorescence.

The search for possible mechanisms of the changes in the above-mentioned fluorescence at the contacts "pollen-pistil" should include recognition of specific signal-stimuli in plant excretions by the acceptor cell. Similar chemosignals may be acetylcholine and biogenic amines (dopamine, noradrenaline, serotonin and histamine) known in animal cells as neurotransmitters and also found in plant cells, in particular in pollen (Roshchina, 2001a, b). Another group of chemosignals is known as reactive oxygen species – ozone, free radicals and peroxides (Gamaley and Klyubin, 1996; 1999; Roshchina and Roshchina, 2003).

Acetylcholine and biogenic amines. The treatment of the pistil stigma with acetylcholine and histamine as well as their agonists (imitators) and antagonists (blockers of membrane receptors) bound in one and the same part of the membrane also induced the shifts in the fluorescence spectra and intensity (Roshchina and Melnikova, 1998a, b; Roshchina, 2001a, b). As illustrated in Fig. 5.3, acetylcholine almost doubled the fluorescence intensity in maximum 475 nm and shoulder 530 nm. However, after the

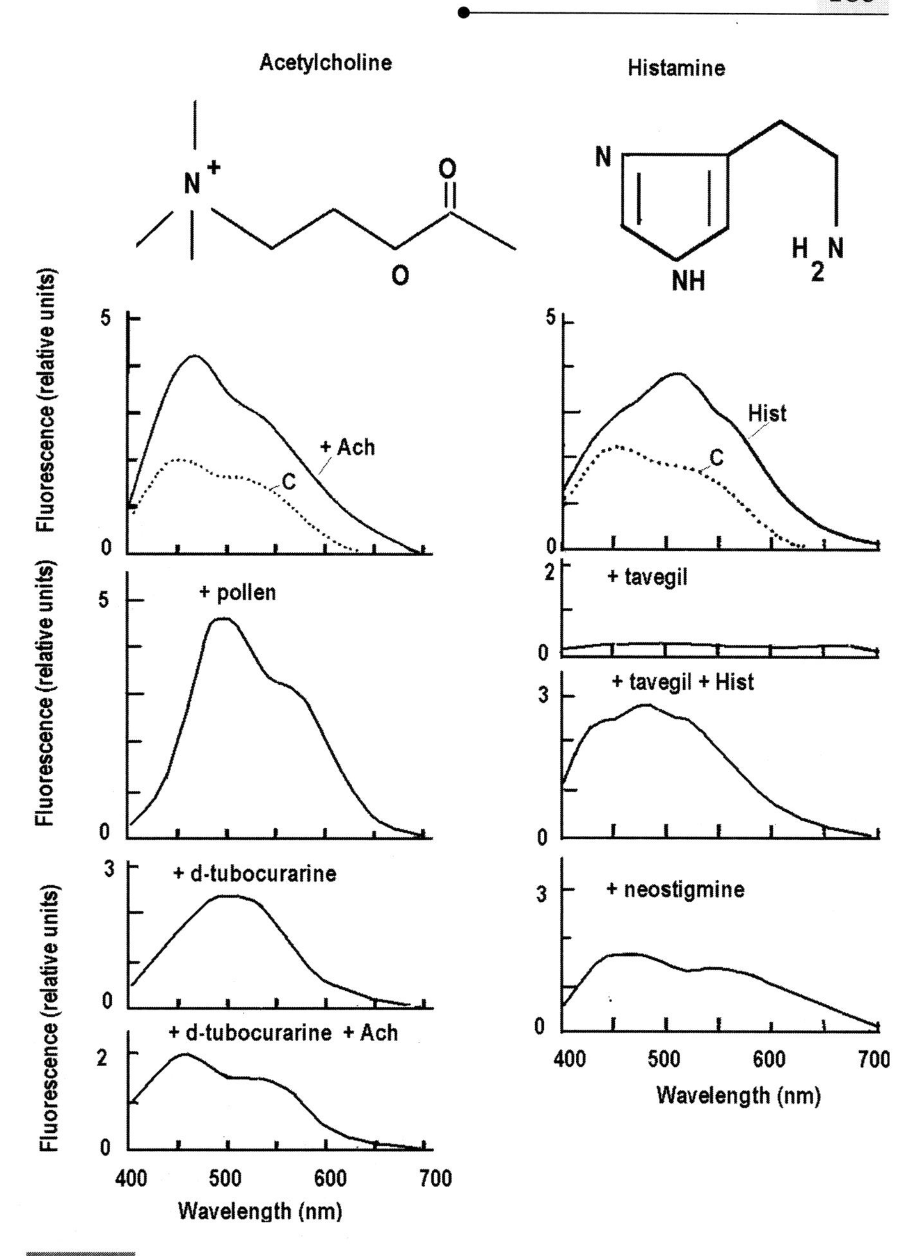

Fig. 5.3 The effects of 10^{-5} M neurotransmitters acetylcholine (Ach), histamine (Hist) their antagonists, and anticholinesterase agent neostigmine on the fluorescence spectra of the pistil stigma in *Hippeastrum hybridum* flower C (control). Source: Roshchina, 2001a.

preliminary treatment with its antagonist d-tubocurarine, which is a blocker of animal cholinoreceptor, the effects were not observed. A similar picture was observed for the antagonist atropine (Roshchina, 2001a, b). It showed the presence of the receptors of acetylcholine on the stigma surface.

Histamine also stimulated the autofluorescence of pistil stigma (Fig. 5.3) and unlike acetylcholine, induced maximum at 510 nm, and two shoulders 450 and 560 nm. The fluorescence intensity was also stimulated, but after preliminary treatment with tavegyl (clemastine), an antagonist of histamine, the emission decreased in a comparison with histamine alone. This showed the presence of structures like histamine receptors on the stigma surface. Another surface sensor on the pistil is enzyme cholinesterase, present in the stigma excretions (Roshchina and Semenova, 1995). The treatment of the pistil stigma with neostigmine, an inhibitor of cholinesterase also smoothens the fluorescence spectra of the structure (Fig. 5.3). It was seen that acetylcholine and histamine induced the changes in the pistil stigma fluorescence, as is observed at pollen addition (Roshchina, 2001a). Moreover, this response was absent if the pistil stigma was first treated with antagonists of acetylcholine or histamine such as d-tubocurarine or clemastin (tavegyl) which links receptors on the cellular surface. Thus, autofluorescence of the cells at contact with pollen and pistil appears to be a biosensor reaction for neurotransmitters and the antitransmitter substances.

Reactive oxygen species. Reactive oxygen species –ozone and its derivatives free radicals and peroxides are constantly formed in atmosphere and water (Bailey, 1958; 1973; 1978; Baird, 1995). The formation and decomposition of oxygen radicals and peroxides occur in living cells according to the scheme:

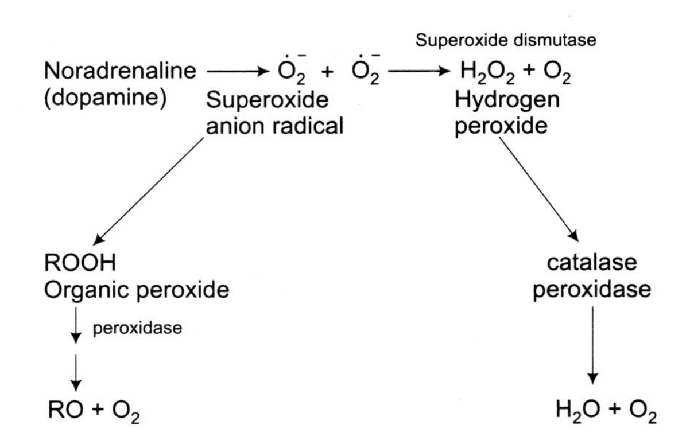

Reactive oxygen species such as free radicals, in particular superoxide anion radical, and peroxides arise both in pollen and pistil. As shown in Fig. 5.4, the fluorescence spectra of the pistil stigma changed during the first 22-300 sec after addition of the species (Rochchina et al., 1998a). Light emission increased after the addition of noradrenaline which generates superoxide anion radical $\dot{O}_2^-$. The shift of maximum from 500 to 530 nm and shoulder 530 nm occurs after the treatment with hydrogen peroxide (this reagent itself has a weak fluorescence), but water has no effect. Another character of the fluorescence spectra was observed under the treatments with antioxidants, constantly found on the cellular surface. Low-molecular antioxidant ascorbate forms shoulder 600 nm in the emission spectrum, which is transformed to maximum at 560-600 nm. It may act as a stabilizing agent for the fluorescence of aquatic solutions (Baeyens, 1985). Antioxidant enzyme superoxide dismutase (catalyzer of the superoxide radical dismutation which leads to the formation of hydrogen peroxide)

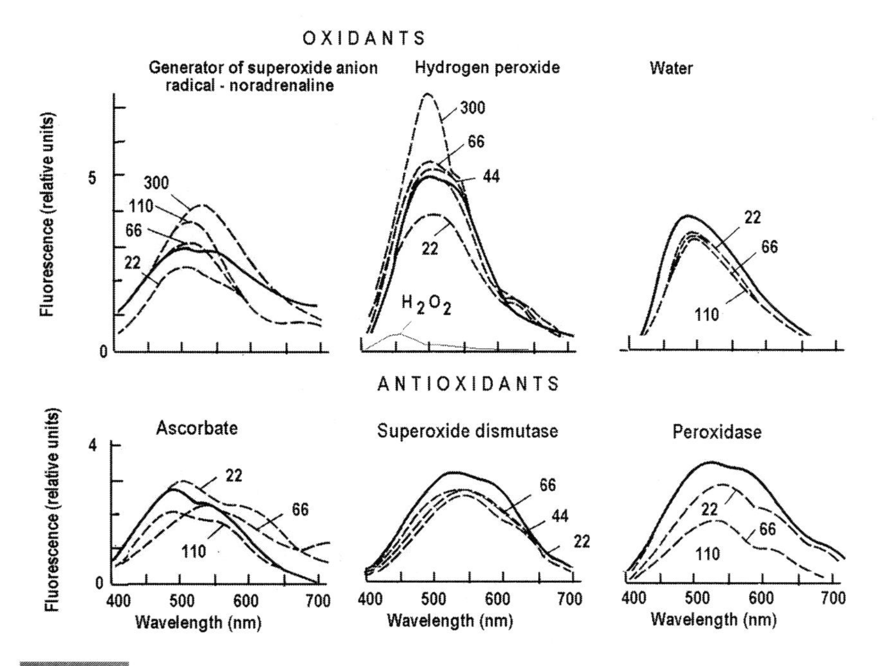

Fig. 5.4 The effects of oxidants and antioxidants on the fluorescence of the pistil stigma in *Hippeastrum hybridum*. Adopted from Roshchina et al., (1998a). Unbroken line–without treatment, broken line – 22, 66, 110 and 300 s after the addition of reagent: noradrenaline 10^{-4} M, hydrogen peroxide10^{-5} M, ascorbate 10^{-4} M, superoxide dismutase (0.6 mg/ml) and peroxidase (0.6 mg/ml). H_2O_2 – emission of hydrogen peroxide out cell.

induces the arising of new maximum 580 nm, and peroxidase (catalyzer of the peroxides decomposition) maxima 530 and 600 nm. Maximum 530 nm may be related to oxidized flavins, and a longer wavelength maxima - to flavonoids or azulenes in a complex with the cell wall.

Blue pigment of azulene present in pollen exine (Roshchina et al., 1995) is also an antioxidant and also can be the origin of free radical (Barriero et al., 1992). If it is added on the pistil stigma the fluorescence of the cellular surface increases, and smooth maxima 500 and 600 nm become clear and 2.5 fold higher during 22-110 sec, when the emission achieves a control level (Fig. 5.5).

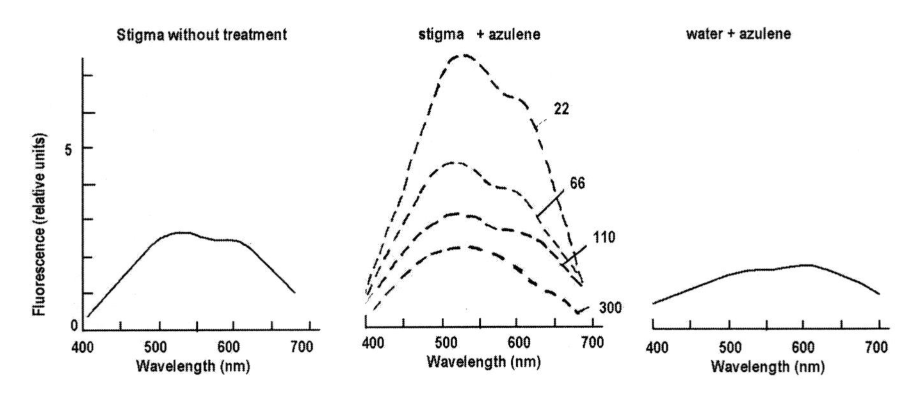

Fig. 5.5 The effects of 10^{-5} M azulene on the fluorescence of pistil stigma in *Hippeastrum hybridum.* Unbroken line – without treatment, broken line – 22, 66, 110 and 300 s after the addition of reagent.

5.1.2.2 Effects of individual components of pistil and pollen secretions on the pollen fluorescence

Fluorescence of pollen grains changes on contact with secretion of pistil or secretions of other plant species, which can be carried on the same stigma by wind or insects. Among of the substances in the secretions of both pistil and pollen are reactive oxygen species (Roshchina et al., 2003a, b), while azulenes and carotenoids are released by exine of pollens (Roshchina et al., 1995). As seen in Fig. 5.6, the experiments lasted $<$ 400 sec with superoxide dismutase added on the pollen surface result in short time changes in the fluorescence spectra – instead maximum 490 nm, two maxima 500 nm and 580 or 600 nm arose. But after 5-10 min the picture cannot be seen. Superoxide anion radical participates in the contact and is

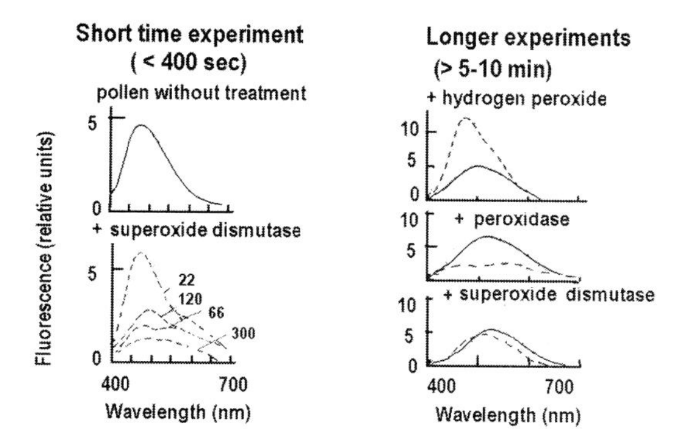

Fig. 5.6 The effects of oxidants and antioxidants on the fluorescence of pollen from *Hippeastrum hybridum* in short – and long-time experiments. Left. Unbroken line – without treatment, broken line – 22, 66, 110 and 300 s after the addition of superoxide dismutase (0.1 mg/ml). Right. Unbroken line – without treatment, broken line – 5-10 min after the addition of hydrogen peroxide 10^{-4} M, peroxidase (0.1 mg/ml) and superoxide dismutase (0.1 mg/ml).

diminished by superoxide peroxidase, converting the radical into hydrogen peroxide which can be seen from the quenching of the pollen fluorescence. In longer experiments, only effects of hydrogen peroxide (stimulation of the emission intensity and appearance of shoulder 530 nm) occur, and when peroxidase was added, decreased the emission. Antioxidant azulene decreases the emission intensity of pollen just after the addition of the substances. However, it is able to penetrate into a cell and fluoresce in various compartments (see Chapter 7).

Unlike the effects of neurotransmitters, which acts on the pistil emission, acetylcholine has no significant influence on pollen fluorescence, but noradrenaline and dopamine as generators of free radicals (superoxide anion radical) in low concentration stimulate the process (Roshchina, 2001a, b). The components are present in secretions of both pistil and pollen. But effects of acetylcholine on pollen fluorescence may be indirect because inhibitors of cholinesterase, enzyme, which hydrolyzes the neurotransmitter, such as neostigmine and tetramethylammonium stimulates the light emission of pollen from *Hippeastrum hybridum* (Roshchina and Melnikova, 1998a). Another possible mechanism of fluorescence changes may be similar with those described by Erokhova et al., (2005), when acetylcholine alter the membranous environment of carotenoids.

5.2 INTERACTIONS AT THE ALLELOPATHY. CONTACTS BETWEEN CELLS FROM DIFFERENT PLANT SPECIES

Chemical relations between species based on the sensitivity of cells to excretions of other cells is known as the phenomenon of allelopathy or "neighbour effects". Biologically active compounds known as allelochemicals are able to regulate the pollen and seed germination found among the fluorescing secretions. Connections between the secretory function and chemosignalling in allelopathic relations have been analyzed on model systems in such cell-cell contacts as "pollen-pistil", "root-seed (seedling)" and others, where alterations in the cellular fluorescence occur.

5.2.1 Pollen-pollen Interactions

Foreign pollen, which can be carried by wind or insects on the surface of pistil stigma, interacts with pollen of "own" species which may lead to antagonistic or favourable effects on pollen germination. Early in the 1930's and 1940's the ability of pollen mixed from different species to react with each other, stimulating or inhibiting pollen tube growth, was reported in the papers of Branscheidt (1930), Zanoni (1930) and Golubinskii (1946) and analyzed in Grümmer's monograph (1955). The authors analyzed both cultural lawn species from the genera *Lilium*, *Tulipa*, *Narcissus* and species from natural phytocenosis such as *Corylus avellana*, *Linaria vulgaris*, *Papaver somniferum*, etc. Beginning in the 1980's, experiments connected with allelopathy were conducted on the basis of pollen germination *in vivo* (Char, 1977; Sukhada and Jayachandra, 1980 a; Thomson et al., 1982; Murphy, 1992; 1999), where fertilization by mixtures of "own" and "foreign" pollen grains was observed.

Pollen autofluorescence of many allelopathically active species was described earlier (Chapter 2), and the phenomenon was recently considered in context of allelopathy (Roshchina and Melnikova, 1996; 1998a, 1999; Roshchina, 2001b). Rapid changes in the fluorescence on the mixtures of pollen from different species may be an indicator of chemical relations. The responses are obviously dependent on both the composition of the excreta of plant-donor pollen and the chemosensory peculiarities of the surface component of the plant-acceptor pollen. The data were received in experiments, involving pollen of 1) meadow, wind-pollinated species, 2) forest-living, wind-pollinated species, 3) meadow, wind-pollinated and insect-pollinated species, grown in flower gardens; 4) insect-pollinated species, grown in flower gardens and weeds; 5) weeds.

5.2.1.1 Fluorescence of pollen mixtures

The fluorescence spectra of the pollen mixtures may reflect first quick responses in pollen allelopathy. One example is shown in Fig. 5.7, where the fluorescence spectra of mixtures given for herbaceous meadow plants such as meadow foxtail *Alopecurus pratensis* L. and orchard grass *Dactylis glomerata* L. (Graminae), which are wind-pollinated, and from cultural flower garden species poppy *Papaver orientale* L. (Papaveraceae) and day lily *Hemerocallis fulva* L. (Liliaceae). The changes are seen, mainly, in fluorescence intensity. The value is 1.5 fold higher for *Alopecurus* in a mixture of *Alopecurus* and *Papaver* whereas poppy has no visible shifts. In contrast, meadow foxtail decreases its autofluorescence intensity and shifts the main maximum to the long wavelength region when mixed with day lily. The latter has no changes in its autofluorescence. When pollen grains of *Hemerocallis* and *Dactylis* are mixed, orchard grass undergoes significant changes. A new maximum or shoulder at 480 nm arises, major maximum at 525 nm become flatter, and red fluorescence with maximum at 650 nm is more obvious, while the intensity of fluorescence in maxima decreases more than two fold. As for day lily pollen, its autofluorescence also

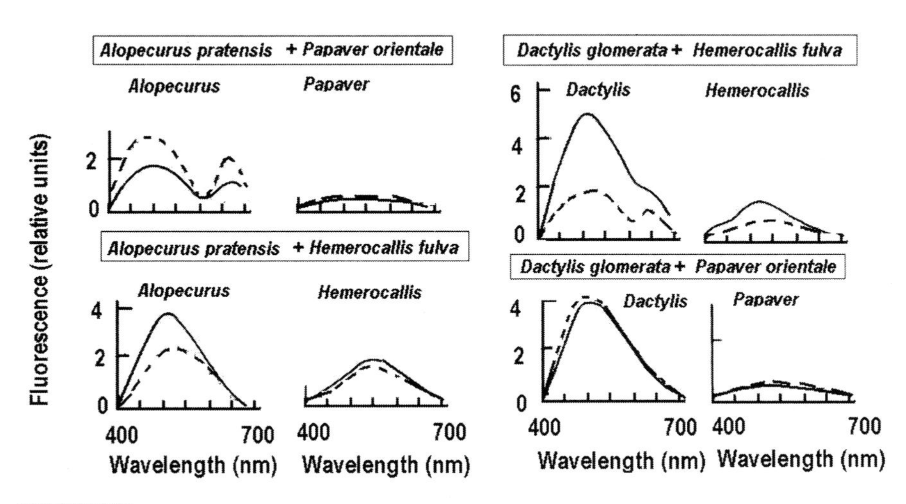

Fig. 5.7 The fluorescence spectra of the pollen mixtures from various species (in ratio 2 mg: 2 mg of pollen grains). Unbroken and broken lines – without and after addition of foreign pollen, relatively.

decreased. If pollen of orchard grass and poppy are mixed, the fluorescence of both is not changed.

There were are also rapid changes (1-2 min) in fluorescence intensity which reflected the effects of volatile excreta of the pollen grains from weed species *Artemisia vulgaris* and *Urtica dioica* and day lily *Hemerocallis fulva* known for flower gardens or parterres, as well as of wind-pollinated, woody species *Larix decidua* and *Betula verrucosa* (Table 5.1). When the microspores of weed species or woody species interacted, in both cases a 60% drop in light emission was observed. By contrast, day lily increased its fluorescence by 45-80% in the presence of weed pollen grains. According to Stanley and Liskens (1974), the pollen has a noticeable smell. Terpenoids prevail among the volatile excreta (Egorov and Egofarova, 1971) and may influence the foreign pollen grains as well as attract insects.

Table 5.1 Fast observation of chemical interactions of dry pollen grains in mixtures estimated by the fluorescence intensity.

Species of pollen-acceptor of volatile excreta	*Species of pollen-donor of volatile excreta*	*Fluorescence intensity of pollen acceptor at 475 nm (% of control)*
Urtica dioica L.	*Artemisia vulgaris* L.	40 ± 1.0
Hemerocallis fulva L.	*Artemisia vulgaris* L.	181 ± 2.3.
Hemerocallis fulva L.	*Urtica dioica* L.	145 ± 5.1
Larix decidua L.	*Betula verrucosa* Ehrh.	45 ± 2.4.

Pollen-pollen antagonism, estimated on the ability to germinate, is known, mainly, for cultural plants of parterres (Ortega et al., 1988; Anaya et al., 1992; Murphy, 1999a, b) or, whereas a mutual stimulation was first found for natural field plants by Golubinskii (1946). Unlike the experiments needed for 1-24 h for the pollen tube growth, fluorescence allowed testing the chemosensitivity of different pollen species in mixtures for 1-2 min which may indicate compatible or incompatible species in phytocenosis. The ability to depress or stimulate fertilization by foreign pollen could be one of the determinants of mutual existence of the species. Certainly, pollen of *Papaver orientale* and *Hemerocallis fulva*, as parterre species, were either insensitive or weakly sensitive to pollen of field-grown, wind-pollinated *Alopecurus pratensis* and *Dactylis glomerata*. Only pollen of weeds *Artemisia vulgaris* and *Urtica dioica* stimulated the fluorescence of Hemerocallis pollen. In contrast, the pollen grains of weeds and woody species, wind pollinated, were extremely sensitive to each other, decreasing their autofluorescence. The sensitivity between pollen of meadow, wind-pollinated plants *Alopecurus* and *Dactylis* was marked, but without a strong correlation with their germination.

Experiments with the mixtures of woody plants such as *Larix decidua* L., *Betula verrucosa* Ehrh. and *Populus balsamifera* L. was also carried out (Roshchina and Melnikova, 1996). Dry pollen of *Larix* (maximum 460 nm) decreased the emission of dry pollen of *Populus*, but not of *Betula* pollen (maximum 550 nm). Moreover, pollen of *Betula* stimulated the emission of *Larix* pollen grains. Water extracts of the species may stimulate or inhibit the fluorescence of *Larix* and *Betula* pollens. In these cases, the emission was seen only for non-germinated microspores. The fluorescence of *Populus* grains (maxima 460, 520 and 580 nm) is not changed significantly after the addition of dry pollen from other species or their water extracts, but in last case, the non-viable pollen emitted.

Murphy (2007) has applied this mode to pollen of *Hieracium canadense* and its transfer to the sympatric, confamilial, and concurrently flowering *Sonchus arvensis* and *Sonchus oleraceus*. When an excitation wavelength of 435 nm is used, *Hieracium canadense* pollen may emit in yellow-queen wavelength, (maximum 520 nm) while pollen from both *Sonchus* species do not. Similar results have been received for *Phleum pratense* when its pollen alights on stigmas of *Elytrigia repens*, *Danthonia spicata*, and *Danthonia compressa* as it too emits at 520 nm whereas pollen from the other species does not emit at this wavelength. Images of pollen mixtures were received by laser scanning confocal microscopy (Roshchina et al., 2007b; Yashina et al., 2007). Colour Fig.25 shows that the red-fluorescing excretions from pollens of *Artemisia absinthium* or *Solidago virgaurea* concentrated on the green-fluorescing surface of *Knautia arvensis*. In variant *Knautia- Artemisia*, the germination of pollen from *Knautia* completely blocked whereas development of pollen from *Artemisia* decreased to 33 % of control. On the contrary, in pollen mixture *Knautia- Solidago*, the amount of germinated pollens of *Knautia* achieved 209 % of control, and LSCM-images demonstrated the red-fluorescing excretions from *Solidago* pollen on *Knautia* pollen tubes.

5.2.1.2 Possible contribution of allelochemicals in the fluorescence of pollen mixtures

Fast processes observed as quick shifts in the fluorescence spectra or intensity should be associated with 1. metabolic processes, such as redox changes of NAD(P)H and flavins (Karnaukhov, 1978); 2. photodynamic changes of some photosensitive pigments of excreta or (and) structures of cellular surface (Aucoin et al., 1992); 3. free radical reactions, both in excreta and on the cellular surface (Roshchina, 1999a, b; 2001b). Moreover, allelochemicals may also fluoresce and contribute in the emission alterations. The oxidation leads to formation of hydrogen peroxide, peroxides (ozonides) of allelochemicals, free radicals such as superoxide anion radical, hydroxyl radical and radicals of allelochemicals. These active oxygen species can participate in allelopathic effects. Alkaloid sanguinarine, monoterpenes, some sesquiterpene lactones and peroxides are considered as triggers of pollen germination whereas proazulenes austricine and grosshemine – as blockers. Since low concentrations of the studied allelochemicals induced their effects, it could consider them as participators in a chemosignalling on contacting plant surfaces.

Phenols, in particular flavonoids, may contribute significantly in the blue light emission of pollen grains. Stanley and Linskens (1974) reviewed the following information for pollen. For instance, herbaceous species such as *Alopecurus pratensis* included approximately equal amounts of quercetin, kaempferol and isorhamnetin whereas *Dactylis glomerata* – quercetin, isorhamnetin and traces of kampferol, the compounds, except kaempferol, have significant fluorescence. In pollen of woody species such as *Larix decidua* kaempferol and naringenin are present (0.04-0.06% per pollen dry weight), in pollen of *Populus, Salix* and *Betula* genera - quercetin and kaempferol. According Murphy (2007), phenols, such as flavonoids may participate in the allelopathic relations between pollen of different species. This author isolated and identified high concentrations of isorhamnetin from pollen of *Phleum pratense* (fluoresces at 520 nm) using high-speed counter-current chromatography. Blue and green fluorescence of pollen may be related to phenols such as anthocyanins, which at excitation wavelengths 355-360 nm or 410 nm have maxima 450 or 518-530 nm, relatively (Drabent et al., 1999; Figueiredo et al., 1990; Sighicelli et al., 2005). As seen on Fig.3.11 (Chapter 3), anthraquinone hypericin from plants of *Hypericum* genus fluoresce in red (600-660 nm) and can be released from anther and pollen. Colour Fig. 26 shows how this red-emitting pigment stained the blue/green-fluorescing pollen of *Plantago major*. Blue-green fluorescence of pollen grains may be due to nicotine-amide nucleotides (Karnaukhov, 1988; Morales et al., 1994), or blue pigments azulenes (Roshchina et al., 1995) as well as phenylacetic acid, which are found, in particular in corn pollen (Anaya et al., 1992). The fluorescence maxima in the red region (at 610-680 nm) of the spectra may be associated with anthocyanins , azulenes and chlorophyll (Drabent et al., 1999; Roshchina et al., 1995; 1998a).

Among volatile compounds, emitted from pollen monoterpenes, for instance, in μg/weight linalool 0.66-1.08; nerol 3.49-0.84; geraniol 1.46-0.52 in *Vitis* pollen (Egorov and Egofarova, 1971). Monoterpenes, weak fluorescing at 420-430 nm, react with the cell - acceptor surface in a few seconds, inducing redox reactions or acting on enzyme systems connected with fluorescence. Lipophilic components of pollen such as carotenoids, chlorophyll and sesquiterpene lactones also may contribute in the fluorescence.

5.2.2 Microorganism-plant or Fungi-plant Relations

The microbial cell, which contacts with the plant cell may induce the synthesis of new fluorescent compounds here. The examples are flavonoids arising in cortical parts of white clover root during nodule initiation after infection by *Rhizobium leguminosarum* (Mathesius et al., 1998). The plant cell begins to luminescence in blue, having maximum 450 nm in the fluorescence spectra. Among the fluorescent compounds, which may serve as markers of the microbe-plant relations, are 7, 4' –dihydroxyflavon and, probably, isoflavonoid formononetin. Microbial interactions with plants may play an important role in biocenosis (Schmidt and Ley, 1999).

Many pathogenic fungi have been found to autofluoresce when exposed to ultraviolet light (Rohringer et al., 1977; Mann, 1983; Rost, 1995; Elston, 2001). These fungi include Blastomyces, Cryptococcus, Candida, Aspergillus, Coccidioides, and occasionally, Histoplasma. No autofluorescence was observed with Mucor. Fluorescent are fungal spores and hyphae. The pest invasion induced the formation of fluorescent leaf lesions that look blue-light (Gray et al., 2002). Similar bright autofluorescence as a sign of the blast fungus infection is seen in cells of leaf sheaths in economically important grass species – rice, barley, onion, etc (Koga, 1994). It has emerged as a model system to study fungal–plant interactions (Park et al., 2004). Many fungi produce specialized infection cells (in which germ tubes produced from conidia differentiate) called appressoria that adhere tightly to the plant surface with appressorium mucilage. In the appressoria regions, one often can see a significant autofluorescence that may be as one of the response to the infection, beginning from earlier stages of the fungal penetration (Park et al., 2004). Here are high concentrations of actin filaments and the visiable autofluorescence may be related to this process. Moreover, in hyphae of fungi of *Gigospora gigantea*, the fluorescence was observed only in living cytoplasm and has been absent in empty, dead hyphal segments (Sejalon-Delmas and Magnier, 1998). The fluorescent (unknown) material secreted in cytoplasm moved via cytoplasmic streaming of the fungi that permited to recommend autofluorescence as natural vital marker. Unlike *Gigaspora* species, autofluorescence of *Glomus* was related to wall chitin (Jabaji-Hare et al., 1984). The advantages of fluorescent microscopy for fungal screening include: (1) no special staining procedures required; (2) no time delay, as involved with special stains; (3) the ability to scan sections at a relatively low power; and (4) the ability to tentatively identify the fungus. Cytoplasmic autofluorescence of an arbuscular mycorrhizal fungus *Gigaspora gigantea* is observed in plants (Sejalon-Delmas et al., 1998). Similar responses of the non-mycotrophic plant *Salsola kali* (Chenopodianceae) and the mycotrophic grass *Agropyron dasystachyum* were found in the invasion by vesicular-arbuscular mycorrhizal fungi mixture of *Glomus* spp. and by *Gigaspora margarita.* In *A. dasystachyum* normal mycorrhizal development occurred and no root browning or autofluorescence was observed, indicating a compatible reaction, while in *S. kali*, the roots reacted in the fungal invasion by autofuorescing bright yellow, suggesting lignification. Yellow autofluorescence is associated with necrosis in barley infected to *Puccinia hordei* (Hoogkamp et al., 1998).

5.2.3 Fluorescence at Modelling of Allelopathic Relations

In nature there are several types of allelopathic relations among non-generative cells: "root-seed (seedling)", "root- root", "seed-seed", "vegetative microspore-root", "vegetative microspore-seed" or "microorganism- plant cell" "pollen-pollen" contacts. We have no direct data about fluorescence changes at similar relations. If the secretory cell fluoresces, their emission can be changed in such relations which could be observed in model experiments.

Root-seed (seedling) relations. Root-seed (seedling) contacts occur, when the seed or root are not far from each other and may know of the neighbouring seed or root via their secretions. One model example is described in Fig. 5.8 for the seedlings of conifer plants, which show the sensitivity to free radicals and peroxides as secretory products. The root tip is known to release active oxygen forms (Cruz-Ortega and Anaya, 2007). It can be seen from the fluorescence (Fig. 5.8) of the developing 3-d old seedlings of *Larix sibirica* and *Picea pungens* which have (1) a cholorophyll-less part of the developing root tip (hypocotyl start and forming root meristem) and (2) chlorophyll-enriched part of the developing leaf (leaf meristem). Leaf meristem fluoresces mainly in red, and has weak fluorescence reaction on the antioxidant additions, while meristem in root tip emits in blue and is sensitive to similar treatments. Root tip of the first object shows the emission maxima 445, 460 and shoulder 500 nm whereas those of the second – only maximum 450 nm. If superoxide anion radical

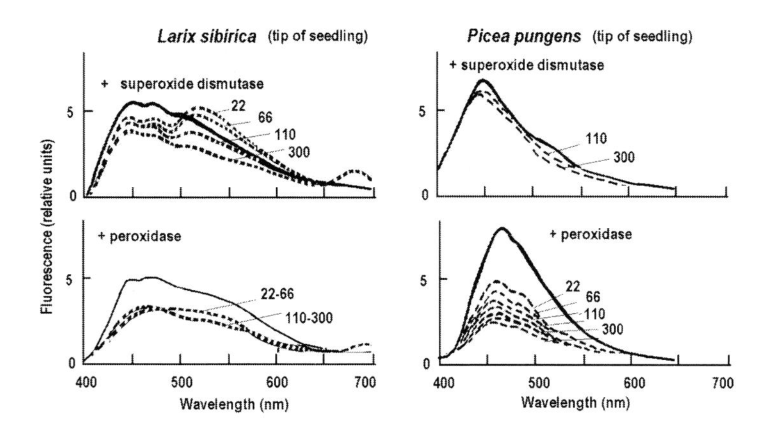

Fig. 5.8 The effects of antioxidant enzymes (0.1 mg/ml) on fluorescence of the seedlings from *Larix sibirica* and *Picea pungens*. Unbroken line – without treatment, broken line – 22, 66, 110 and 300 s after the addition of superoxide dismutase or peroxidase.

is diminished by superoxide dismutase, one can see the appearance of maximum 525 nm for *Larix*, instead of shoulder 500 nm and no similar effects - for *Picea*. In our model experiments, dopamine neurotransmitter found in the root, seed or microspore excretions and able to form reactive oxygen species $\dot{O}_2^-$ (Roshchina, 2001a), increased 10 times the yellowish fluorescence of germinating radish seed. But there is visible decrease of autofluorescence in the maximum for both species after the addition of peroxidase, catalyzing the peroxides' decomposition. New maximum 525 nm in *Larix* may belong to oxidized flavins. Thus, roots excreted reactive oxygen species may change the fluorescence of contacting roots or seeds of other plant species.

Vegetative microspore-seed or microspore-microspore relations. Vegetative microspores of horsetails, mosses and ferns also participate in the allelopathic relations with the same microspores of other species as well as with roots and seeds. The excretions of the microspores may fluoresce, as can be seen for *Equisetum arvense* (Fig. 5.9). The yellow pigment with absorbance maximum at 415-420 nm fluoresces in the orange-red spectral region. In the fluorescence spectrum we see maxima 475 and 575-580 nm as well as shoulders 620 and 675 nm. In antheridial bodies red pigment related to carotenoid was found (Hauke and Thompson, 1973), and it does not excluded that is the same compound. The extract may contain allelochemicals, which, depending on the concentration, was able to depress or stimulate the germination of *Raphanus sativus* seeds (Roshchina, unpublished data).

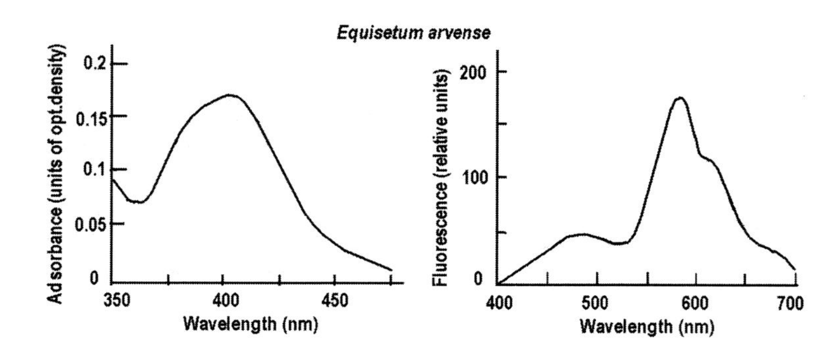

Fig. 5.9 The absorbance and fluorescence spectra of red colour excretion on white paper filter from moistened vegetative microspore of *Equisetum arvense.*

In our model experiments on pollen allelopathy connected with the fluorescence changes in mixtures of generative microspores, orange-red fluorescence (640-680nm) of pollen from *Plantago major* (earlier had only blue emission) was induced by the presence of pollen from *Calendula* officinalis, which released carotenoid-enriched secretion. The ethanol extract from the *Calendula* pollen (1:20) dryed and then wixed with water (weak-fluoresced) also induced similar orange-red emission.(208 ± 10 % of control). Like this, carotenoid-enriched pollen from *Hemerocallis fulva* stimulated both orange and green fluorescence of pollen from *Alopecurus pratensis* or *Dactylis glomerata* (Fig.5.7). Carotenoid bodies are usual in many pollens (Stanley and Linskens, 1974) and to be excreted appear to act as antioxidants in redox reaction on the pollen surface. As seen on Colour Fig.25, orange-fluorescing unknown components are released from pollen of *Artemisia absinthium* and *Solidago virgaurea*. Phenolic components of the excretions may be participate in the emission. In red-orange spectral region fluoresce anthocyanins and anthraquinones, in particular hypericin. Colour Fig.26 shows the appearance of the red fluorescence in pollen of *Plantago major* (1057 ± 8 % of control) in the mixture with pollen from *Hypericum perforatum* contained hypericin. The addition of this water-solved pigment purified by thin-layer chromatographic method also induced the same lightening picture (stimulation of red emission to 870 ± 15 % of control).

5.2.4 Mechanisms of the Fluorescence Changes

In biocenosis, allelochemicals are found in the fluorescing secretions where individual components are able to regulate the seed germination and development of seedlings and roots. The physiological response as visible changes of fluorescence may also depend on a ability of the acting substance to penetrate through the plasmalemma (Roshchina, 2005a). If the compound is not practically permeable (mainly hydrophilic components), an initiation of pollen or seed development occurs indirectly, outside the cell interior, the mechanism of chemical signalling, which includes an excitation of the sensor in the plasmalemma and a cascade of the messenger informational reactions within the cytoplasm to nucleus, such as, cyclic nucleotides, inositol triphosphate, Ca^{2+}-ions which may serve as secondary messengers (Roshchina, 1991; Franklin et al., 1994; Roshchina et al., 1998c). On the other hand, substances (mainly lipophilic) which penetrate within a cell have a direct contact with nucleus and so initiate or retard the beginning of germination (Roshchina et al., 1998b). The fluorescent compounds which penetrates into a cell may serve as fluorescent markers and probes due to the selective binding with cellular compartments (see Chapter 7).

Most rapid are free radical reactions of allelochemicals. They are seen in plant excreta containing allelochemicals (Yurin et al., 1972), especially in contact with ozone of air (Roshchina, 1996; Roshchina and Roshchina, 2003), on the pollen surface (Dodd and Ebert, 1971), on the surface of seeds and leaves tested on final products,and seen as spots of fluorescent pigment. The oxidation of allelochemicals leads to the formation of hydrogen peroxide, peroxides (ozonides) of allelochemicals, free radicals such as superoxide anion radical, hydroxyl radical and radicals of allelochemicals. Peroxides may stimulate plant growth reactions (Roshchina and Roshchina, 2003).

Low concentrations of allelochemicals may act as participators in a chemosignalling on the contacting plant surfaces. Reactive oxygen intermediates such as superoxide radical and OH-radicals, as well as hydrogen peroxide, can play a role of mediators, activating the transcription factor (Schreck et al., 1991). The oxidants which are released, such as, ozone, paraquate and noradrenaline decreased the index of the seed germination (Roshchina, 1991; Roshchina et al., 1998d; Roshchina and Roshchina, 2003). It is possible to rupture the free radical chain by use of antioxidants such as phenols of the cell-acceptor excretions or antioxidant enzymes such as peroxidase or superoxide dismutase (Roshchina and Mel'nikova, 1998a). Antioxidants alkaloid rutacridone or flavonoids rutin and quercetin appear to be blockers of seed germination (Roshchina, 2001b; 2005d). Known photodynamic processes are connected with the antioxidants-allelochemicals excreted on the cellular surface such as furanocoumarins (Ceska et al., 1986), or located in surface glands, such as hypericin and juglone and its derivatives (Aucoin et al., 1992). Antioxidant enzymes peroxidases are excreted into the extracellular space of the vegetative cells and on the first steps of pollen moistening. Peroxidase activity is found in excretions. The enzyme decomposes H_2O_2 and prevents the free radical from spreading. Endogenous peroxidase decreases the index of pollen germination but in a smaller degree than superoxide radical generators, paraquate and noradrenaline (Roshchina and Mel'nikova, 1998a).

The quick changes in fluorescence at 430-470 nm from a plant surface may be due to NAD(P)H fluorescence. For instance, for 25 s the fluorescence at 460 nm in ascitic tumours decreased by 50% in the presence of fatty acids (Shwartsburd and Aslanidi, 1991). Chappele et al. (1990) analyzing blue fluorescence (460-490 nm) of normal green leaves, concluded that the emission of water-soluble compounds deals with NADPH (maximum at 460 nm), flavins FMN and FAD (maximum at 440 nm), oxidized FMN has maximum at 525-535 nm (Hastings, 1986). Significant addition to the emission at 440-480 nm may be also done by phenolic substances, especially in a complex with metals (Inglet, 1959). Fluorescence in the region 500-520 nm may be due to riboflavin (525 nm) and some carotenoids (500-525 nm) (Chappelle et al., 1990). Carotenoids of pollen may break a spreading of free radical chain of lipid peroxidation (Roshchina and Karnaukhov, 1999) and so retard main oxidative processes that seen from fluorescence alterations. Anthocyanins released from pollen of *Papaver orientale* may stimulate fluorescence of contacting pollen from *Alopecurus pratensis* in the green and red spectral regions (Fig.5.7). These flavonoids may also serve as antioxidants. In many cases, the fluorescence at allelopathic relations is determinated by balance between oxidants and antioxidants, mainly accumulation of enzymes superoxide dismutase and peroxidases or low-molecular compounds such as ascorbate, phenols, carotenoids and azulenes.

Blue-green emission of excretions from fall bud scales of trees may be connected with phenols (Popravko et al., 1969; 1982; Wollenweber et al., 1987; 1991; Feucht and Treutter 1990). Fluorescing phenols of secretions has been recommended for the study of chemical interaction between plants (Shapovalov, 1973). Fluorescence of intact cells at 600-650 nm, differing from chlorophyll are supposed to connect with blue pigments azulenes, occurring in oil cells of leaves and flowers (Roshchina and Roshchina, 1993) or (and) with lipofuscin, pigment of ageing, found both in the retinal pigment epithelia (Katz and Robinson, 1986) and the plant tissue (Merzlyak, 1988).

Concentrating on above-menthined aspects of proposed mechanisms of visible changes in emission, it should keep in mind that physical base of the alteration is out of our description. This may be of interests specially of physists and chemists.

The position of fluorescence as an indicator reaction in allelopathy becomes clear from the common conception on the cellular level, which is as follows. Possible mechanisms of the allelochemical action deal with a common basis of chemosensitive reactions for many living organisms in cell-cell communication (Roshchina, 1999a): 1. receptor-sensor mechanisms, or 2. free radical mechanisms of a signalling in cell-cell communication. A response to the interaction between contacting cells includes several events: 1. a recognition of specific signal-stimulus in plant excreta by cell-receptor; 2. spreading chemical information within cell-acceptor; 3. a formation of characteristic response of cell-acceptor – a germination of a pollen or a seed or the regulation of the seedling or root growth. Recognition of signal-stimulus occurs via a secretion as a primary medium for a chemosignal spreading from the surface of a plant donor of allelochemical to the plasmalemma of the plant-acceptor. Secretion of the recipient cover may serve as a recognizing and transporting liquid such as olfactory slime in animals for chemosignal (Roshchina, 1991; 2001a). The secretion contains proteins, lipids, and low-molecular substances. The components may interact with the allelochemical as a redox agent or free radical precursors (Roshchina, 1996; 1999a, b). The alteration in visible fluorescence at allelochemical relations deals with the participation of the above-mentioned compounds either as secretions of the cell-donor or located on the surface of the cell-acceptor as the secretory chemosignal.

Conclusion

Intercellular relations may be accompanied by the alterations in the autofluorescence of contacting cells. Secretions released by the cell-donor serves as a chemical signal for the cell-acceptor which receives the fluorescent secretory product, recognizes and responds to the physiological changes, which are seen as shifts in their fluorescence spectra and the light emission intensity. Among cell-cell contacts, whose interactions accompanied by the changes in their fluorescence are pollen-pistil (own or foreign) recognition and allelopathic relationships such as pollen-pollen, root-seeds and others. The mechanism of the autofluorescence during the cellular interactions deals with the presence of known biologically active compounds in the secretions, in particular neurotransmitters, oxidants and antioxidants.

CHAPTER

6 Autofluorescence in Cellular Diagnostics

Autofluorescence is a phenomenon which could be applied to cellular diagnostics. In perspective, it is used for the search of biosensors and as an indication of some of the reactions which occur in living cells.

The autofluorescence application in a special method for Herbarium analysis in ultraviolet has been given by Eisner et al., (1973) and is recommended for the practice of botanical studies of flowers. Today fluorescent microscopy may be widely used in all kinds of botanical investigation and for education as can be seen from Chapters 2, 4, and 5. Depending on the purpose, vital images of fluorescing pollen grains are also applied to their allergen-coupled analysis and determination in any material, especially in the air and soil samples (Driessen et al., 1989). As observed in the review of Roshchina (2003), autofluorescence of secreting cells could be used: 1. in express-microanalysis of the accumulation of secondary metabolites in secretory cells during this development without long-term biochemical procedures; 2. in diagnostics of cellular damage; 3. in analysis of cell-cell interactions; 4. for special indicatory reactions of living cells; 5. as a biosensor reaction. These possibilities will be discussed below.

6.1 EXPRESS-MICROANALYSIS OF THE STATE AND ACCUMULATION OF SECONDARY METABOLITES

During plant development secretory cells may change both the composition of the fluorescing secretions and their amount that can be measured, estimating their characteristic fluorescence spectra and the fluorescence intensity (Roshchina et al., 1997a, b; 1998a; Roshchina, 2006a; 2007b).

6.1.1 Secretory Cells in Medicinal and Economic Plants

Most medicinal and economical plants have secretory cells, where pharmacologically-valuable secondary metabolites are accumulated (Roshchina and Roshchina, 1993; Roshchina et al., 1998a). For instance, the accumulation of different fluorescing components in *Humulus lupulus* L. (Fig. 6.1) is seen from the fluorescence spectra. The developing glands changes the fluorescence from blue with maxima 465 (similar with blue-fluorescing secretory hair with maximum 465-470 nm) and 550 nm to green-yellow with one maximum 535-540 nm (in a matured cone). The cones of *Humulus lupulus* contain sesquiterpene lactones such as E-β-farnesene (Roshchina and Roshchina, 1993). Experiments with leaves and flowers of some plants, which belong to the family Asteraceae and contain both sesquiterpene lactones and their derivatives such as azulenes, show the increase of green-yellow or yellow-orange fluorescence of these intact cells during their development (Roshchina et al., 1998a; Roshchina, 1999a; Roshchina and Melnikova, 1999; Roshchina, 2003, 2005b). For instance, such emission is observed only after the appearance of formed secretory cells on the petals of flower from the medicinal plant *Achilea millefolium* (Roshchina, 2003). The immature cells fluoresce in blue-green, and when they are developed, also – in the red spectral region. Azulenes are absent in immature glands and appear in developed secretory structures. In a free

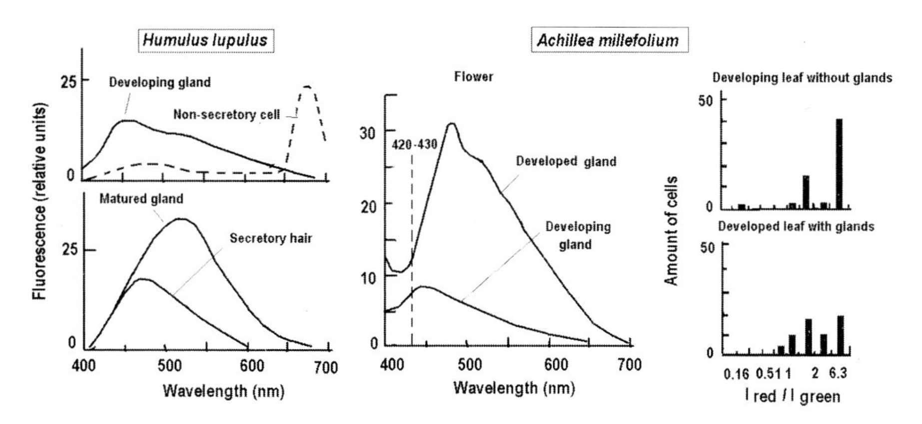

Fig. 6.1 The fluorescence spectra of non-matured and matured cone secretory cells of *Humulus lupulus.* (Left) and *Achillea millefolium* (Middle). Right. Histograms (Student distribution) of the leaf cell fluorescence intensity (Ired/I green) of *Achillea millefolium.* The excitation by UV-light 360-380 nm. I red = $I_{640\text{-}680\ nm}$; I green = $I_{520\text{-}530\ nm}$.

state they may fluoresce in blue at 420-430 nm, but when impregnated to the cell wall – in red (Roshchina et al., 1995). As seen in Fig. 6.1 (middle), red fluorescence of the glandular cells in flower Achillea arises only in the mature gland. Using a double-beam microspecrofluorimeter (see Chapter 1, Section 1.2), the ratio of the fluorescence intensities I red 640-680 nm (maximum of emission at 640 nm): I green (maximum of emission at 530 nm) was measured for leaf glands (Fig. 6.1. Right). The histogram, which demonstrates the distribution of the fluorescing cells, shows the appearance of cells with increased green fluorescence. But red fluorescence of chlorophyll becomes smaller due to the increase of secretory cells on the leaf surface. This ratio decreases in petals, which are covered with matured secretory cells, in comparison with petals of the flower bud, where there are only a few or no similar cells. This occurs due to a filling of matured secretory cells with green-fluorescing secretion. Table 6.1 demonstrates the fluorescence from secretory cells in matured organs of medicinal plants from genera *Achillea solidago* and Hypericum. The fluorescence intensity in the green and red spectral regions for fresh pharmacological materials from *Achillea millefolium*, whose flowers have rose-coloured petals, contain blue-green fluorescing glandular cells filled with the components that differed from those of white petals (Table 6.1). Blue-green glands of white petals emitted 20 fold more intensively, and it was the same with the emission of glandular hairs of leaves. Table 6.1 shows the difference in the

Table 6.1 The fluorecence intensity (relative units) of the terpenoid-accumulated cells of medicinal plant species. Excitation 420-436 nm.

Plant	*Organ*	*Cell*	*Common* $I_{640/530}$	*Glandular cell* I_{530}	*Out glandular structure* I_{530}
Achillea millefolium (Asteraceae)	Flower (Petal)	Glandular cell of rose-coloured petal	20.3 ± 3.0	0.01 ± 3.0	
		Glandular cell of white petal	2.97± 0.16	0.21 ± 0.04	0.06 ± 0.007
	Leaf	Glandular hair	8.4 ± 0.9	0.26 ± 0.04	0.09 ± 0.03
Solidago canadensis (Asteraceae)	Leaf	Oil reservoir	69.5 ± 3.0	0.26 ± 0.01	0.02 ± 0.013
		Secretory hair	17.4 ± 5.0	0.17 ± 0.06	
Hypericum perforatum (Hypericaceae)	Flower (Petal)	Light gland	27.0 ± 2.5	0.12 ± 0.01	0.12 ± 0.008
		Blue-coloured gland		15.0 ± 4.0	no

intensity of various secretory cells on the surface of the other medicinal plants. This permits a quick determination of the maturity of the plant and of pharmacologically valuable material. Moreover, in future, pharmacologists may be able to prepare special scales of the fluorescence for determining calibration by using a fluorescein scale. In particular, for microspectofluorimeter MSF (Karnaukhov et al., 1982; 1983) 1 cm (1 relative unit) is equivalent to 7×10^{-9} M of fluorescein (Roshchina et al., 1998a). The lightening of artificial dye fluorescein may be a blank for comparison with known components of secretory structures, for example azulenes and carotenoids, whose concentrations are changed during maturing of the secretory cells (Roshchina et al., 1998b).

The filling of a secretion may be different for various plant secretory cells among the organs. One of the examples is for *Rubus odoratus* (Table 6.2). Even the accumulation of the secretions in various secretory hairs of *Rubus odoratus* differs as seen from the fluorescence intensity at 530 and 640-680 nm. The main concentration of the green fluorescing secretions was for the secretory hair of the stem. In the capitate hairs, the head with small amounts of chloroplasts and the stalk enriched in the chlorophyll-containing plastids have equal intensity in the green spectral region, but different in red fluorescence. The stalk demonstrated more intensive (approximately two fold) red fluorescence.

Table 6.2 The fluorescence intensity (relative units) of secretory hairs of *Rubus odoratus*. Excitation 360-380 nm.

Secretory cell	I_{530}	$I_{640\text{-}680}$
Hair of leaf lower side	0.05 ± 0.009	1.69 ± 0.2
Hair of stem	0.58 ± 0.03	2.9 ± 0.7
Hair of flower sepal head	0.02 ± 0.009	2.5 ± 0.7
stalk	0.03 ± 0.009	4.2 ± 0.9
Hair of petal	0.04 ± 0.008	0.47 ± 0.03

When the secretory cell accumulates secretory products in higher concentrations, we can see crystals arise within secretory structures or excreted on the cell surface, as it can be seen for tomato (Chapter 3, Section 3.1). The emission maximum of the secretory hair interior was in blue 450-455 nm and of the secretion crystal lying within the secretory cell – 465-470 nm, while the crystal excreted on the leaf surface has another maximum 500 nm (in the blue-green region). There is a maximum in the accumulation of secretory products since the crystallization is observed. Moreover, the composition of the crystals located within and out of secretory hairs differs. In the first case, terpenes prevail whereas in the second–phenols (solved in the essential oils) may be also released.

Autofluorescence of secretory cells may be useful for the express - analysis of medicinal and economic plants whose secretory cells of vegetative tissues contain valuable components as an active matter.

6.1.2 Pollen State

The state of pollen grains which are generative microspores with male gametes is important for humans for many reasons. First of all, it defines normal breeding of forests, economic and medicinal plant species.The most significant is the maturing and fertility of pollen as the male gametophyte for the fertilization of female gametophyte which leads to the formation of fruits and seeds (see Chapter 2, Section 2.2.1). Moreover, fluorescence of pollen could be useful for distinguishing self-compatible and a self-incompatible pollen of the same species. The pollen viability and self-compatibility may be identified from their autofluorescence. The maturing and fertility of pollen as the male gametophyte for the fertilization of female gametophyte which leads to the formation of fruits and seeds is most significant (see Chapters 4 and 5). The fluorescence of pollen grains is also useful in meteorological and environmental monitoring, especially for the determination of seasonal peaks in the allergic reactions in humans. Moreover, pollen fluorescence is a useful indicator in palaeontology and criminalistics (van Gijzel, 1961; 1967; Roshchina et al., 1997b; Yeloff and Hunt, 2005).

6.1.2.1 Analysis of pollen maturing and fertility

The registration of the fluorescence spectra of pollen grains of some plants collected in natural conditions and the male sexual cells of a mouse demonstrates remarkable differences between functionally related structures of plants and animals (Roshchina et al., 1996; 1997b; Roshchina and Melnikova, 1996; 1999). The difference is in the higher intensity of light emission in pollen and in the position of maxima in the fluorescence spectra. Table 2.2 (Chapter 2, Section 2.2.1.) summarizes the absorbance and fluorescence maxima of studied pollen species and shows the variety of peaks dealing with the different components of their surface and excreta. Among them are mono-, di- and three component systems.

The absorbance spectra of pollen have from one to three maxima at 380-390 nm, 410-450 nm, and 760-840 nm, while their fluorescence spectra have maxima - at 460-490 nm, 510-550 nm and 620-680 nm. The first could be connected with phenolic compounds, the second – with carotenoids, and the third with chlorophyll or azulenes (Wolfbeis, 1985; Chappele et al., 1990; Roshchina et al., 1995).

Unlike the sperm of a mouse, the male gametophyte known as pollen shows a clearly seen maxima in the blue-green region at 480-500 nm (corb scabious, wild heliotrope, catch weed, hawthorn) and/or (and) second maxima in the yellow-orange region (celandine poppy, delphinium, apple tree). Sometimes a third maximum in the red region at 600-650 nm arises, as can be seen for an apple tree. As shown in Chapter 2 (Section 2.2.1.), there are from one to three maxima at 460-490 nm, 510-550 nm, and 620-680 nm in pollen fluorescence spectra of various species. It depends on the substances included in the sporopollenin and/or exuded on the pollen surface. Van Gijzel (1961; 1967; 1971), Willemse (1971) and Driessen et al., (1989) recorded pollen fluorescence spectra of forest and herb trees and observed one or two maxima in them, mainly in the blue and orange-red regions. The fluorescence of 21 grass species may change (20-40 nm shifts to longer wavelengths) during the first 10-30 s after the UV-light excitation was switched on (Driessen et al., 1989). In particular, within 30 sec the pollen cytoplasm of *Bromus hordeaceus* became yellow-orange, instead yellow. Pore emission in pollen grains of *Festuca arundinacea* and *Holcus mollis* was yellow, while in other species – white-blue. Among 21 studied grass species, maxima in blue (460-490 nm) were peculiar to Alopecurus *pratensis, Arrhenatherum elatius, Dactylis glomerata, Elymus repens, Holcus lanatus, Lolium perenne, Molinea caerulea, Phalaris arundinacea, Phleum pratensis* and *Poa trivialis,* while *Poa pratensis, Agrostis stolonifera,* and *Alopecurus pratensis,* fluoresce in green (500-530 nm), and *Bromus hordeaceus, Cynosurus cristatus, Festuca arundinacea,* and *Holcus mollis* – in green-yellow or yellow (540-575 nm). The fluorescence of pollen may be useful for the estimation of the cell state and viability (see below Section 6.2.1) as well as self-incompatibility (see Section 6.2.2).

When pollen grains are mature, they often have no maxima in the red region whereas in immature pollen grains of catch weed the maximum at 680 nm peculiar to chlorophyll is observed. Fluorescence of dry non-matured and matured pollen differs. For instance, non-matured pollen of *Tussilago farfara* demonstrated green fluorescence with maxima 465, 518 and 680 nm in the fluorescence spectra whereas matured pollen, which is lack of chlorophyll, showed, mainly, yellow fluorescence with maxima 465, 520 and 535-540 nm (Roshchina et al., 1997a; 1998a). Blue fluorescence (maxima at 465-470 nm and small shoulder at 665 nm) of non-matured pollen of *Philadelphus grandiflorus* changed to yellow-orange emission (maxima at 465, 510-520 and 620 nm) in matured pollen grains (Roshchina et al., 1997a; 1998a) which occurred due to the synthesis of carotenoids (maximum 520 nm) and azulenes (maximum 620-640 nm) during the development of the generative microspores (Roshchina et al., 1995; 1997a; 1998a). Luminescent and Laser Scanning Confocal Microscopy allow one to distinguish mature (without chlorophyll) and immature fluorescing in red due to the chlorophyll. One of the examples is shown on Colour Fig. 24 (Appendix 2)

The fluorescence spectra of mature and immature pollen grains of green-house and outdoor plants have been shown in an earlier paper (Roshchina et al., 1998a). These differ in intensity of light emission and in the position of maxima of the fluorescence spectra. Mature pollen grains often have no 680 nm maximum, whereas in immature pollen grains of *Hymenocallis* 680 nm maximum, typical chlorophyll is observed. Pollen mainly demonstrates clearly seen maxima in the blue-green region (480-500 nm). Sometimes the third maximum in the orange (550-570 nm) or red region (600-650 nm) arises depends on the substances contained in the sporopollenin or/and exuded on the pollen surface. For instance, p-coumaric acid, a monomer in the sporopollenin skeleton as in the pollen of *Pinus mugo* Turra. (Wehling et al., 1989), and it may emit in blue.

Flavonoids such as kaempferol fluoresce in blue, and their deficit leads to a decrease in pollen fertility (Mo et al., 1992; Vogt et al., 1994). Several mmoles of kaempferol to the petunia pollen are enough to restore its normal fertility (Vogt et al., 1994). Pollination or wound-induced kaempferol accumulation in petunia stigmas enhances seed production.

Thus, the colour of fluorescence, fluorescence spectra and intensivity of the emission in maxima may be useful for analysis of mature and immature pollen.

6.1.2.2 Pollen in atmosphere as an ecological and allergic factor

The pollen accumulation of some primary and secondary products, which can be released and are able to induce allergy, has been widely discussed. Pollen is a seasonal problem for meteorologists, who analyze weather changes, and for millions of people around the world, who suffer from allergenic reactions to the antigens embedded on the outer casing of these microscopic grains (Knox, 1979; Leuschner, 1993). Tiny grains of pollen are released into the atmosphere by a wide spectrum of flowers, trees, weeds, grasses and other plants that reproduce seasonally. Although most of these fertilizing gametes seldom reach their destinations, many find their way into the noses and throats of unsuspecting humans where they illicit an allergenic reaction. Commonly referred to as “hay fever”, pollen allergies induce rhinitis and other effects that often cause sneezing, coughing, red and watery eyes, and related symptoms. Smaller numbers of people are more seriously affected and must undergo a more aggressive treatment. In dangerous periods of the spreading of pollens, which induce allergy, it is possible to determine the presence of the microspores in atmosphere based on their fluorescence. Pollen samples are usually collected in Nature by ecologists and meteorologists in order to analysis of the pollen-allergen concentration for profilactic human information. The bright autofluorescence (from blue to red) is observed for many dry pollens, but water quenches the emission. The image presented below reveals green autofluorescence of tiny pollen grains visualized utilizing wide field fluorescence microscopy. Incident light fluorescence microscopy is growing rapidly in importance as an investigational tool in the fields of medical and biological research. All photomicrographs in this gallery were taken with Olympus microscopes employing UIS optics and a PM-30 automatic camera system.

Pollen autofluorescence is analyzed by the estimation of the formation of the antioxidants (Bors, 2005) and the messenger substances with proinflammatore and immune-modulating capacities, peculiar to allergic inflammation such as polysaturated fatty acids (Traidi-Hoffmann et al., 2005). These lipophilic components participate in lipid peroxidation and various fluorescent products arise.

6.1.2.3 The diagnostics of honey and bee products – pollen loads and perga

The analysis of honey and other bee products is also useful on the basis of autofluorescence of their components (Roshchina et al., 1997c). These

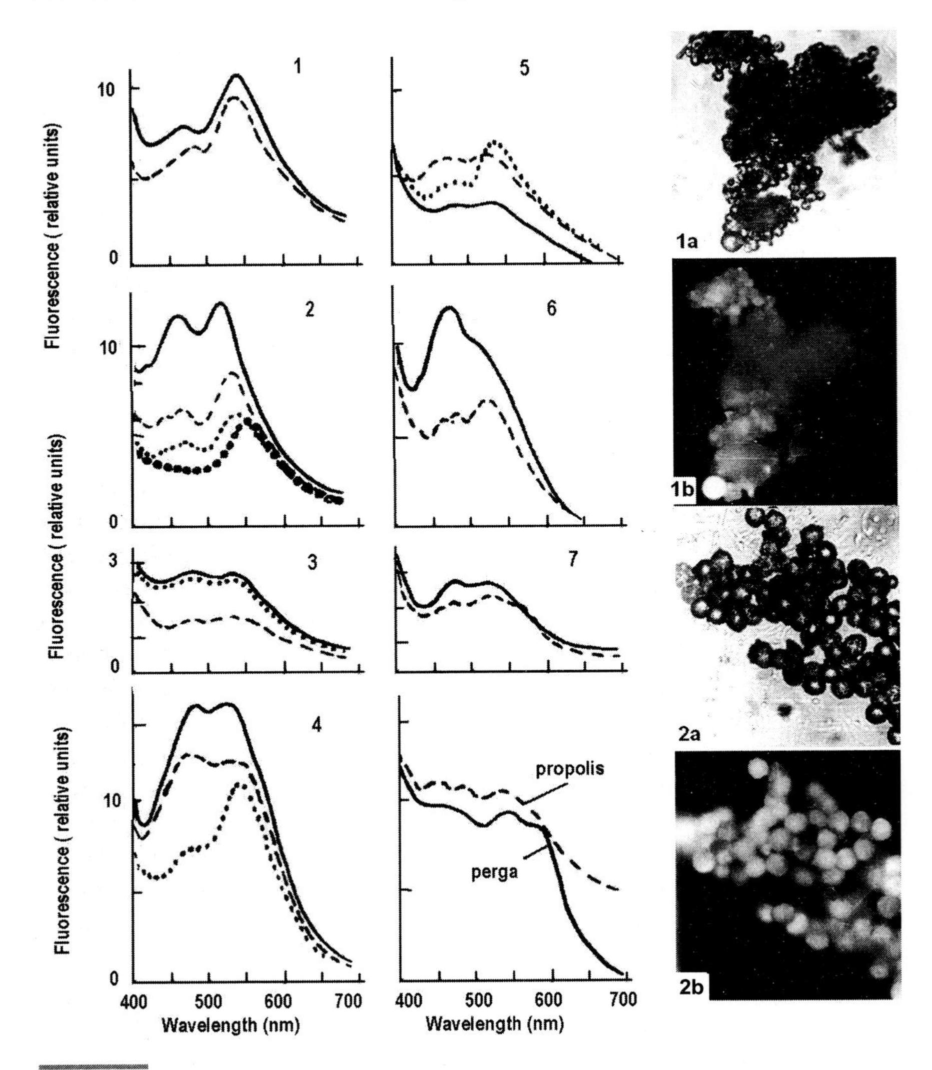

Fig. 6.2 The fluorescence of pollen loads and perga. Left – the fluorescence spectra of various samples of pollen load (1-7), perga (8) and propolis (9). Unbroken line – the type of pollen prevailing in the sample. Right -the images of pollen loads under transmission (1a, 2a) and luminescent (1b, 2b) microscope. Different fluorescence intensity in the pollen grains is seen.

components are pollen in honey, pollen load (pollen collected by bees), perga (transformed pollen load) and propolis (bee glue). As seen in the Fig. 6.2, the samples of bee products show that the composition of the pollen among the various pollens differs from each other. They fluoresce in the blue-orange spectral region. The maxima at 430-470 nm are peculiar to pigments such as flavonoids and azulene, whereas maxima at 500-560 nm – carotenoids and coumarins. In some samples, both types of maxima are present. The comparison of water and ethanolic extracts from the samples shows that probes 1, 2 and 4 are enriched in carotenoids while probes 2 and 4 also contain flavonoids. Other samples are not concentrated in these pigments. As for perga (pollen-load which pass through the bee nutrient system), it included only transformed pigments which results in non- expressed maxima in their fluorescence spectra. The same picture was seen for propolis (bee glue) enriched with phenolic compounds (Melnikova et al., 1997b).

In some cases, depending on the season of the pollen load collection we can determine what pollen is included in pollen load. For instance, maximum 520-530 nm is peculiar to *Trifolium repens*, *T. pratense* and *Salix caprea*. Maximum 550 nm is seen in the fluorescence spectra of *Taraxacum officinale*, *Medicago falcata* and *Cerasus vulgaris* (Fig. 6.3). The plants, from which bees collect pollen and nectar for honey, includes meadow species such as clovers, dandelion or willows flowering in early spring. Some of them are similar in their fluorescence spectra with pollen loads collected by bees (Fig. 6.2). As it is demonstrated in Fig. 6.2, pollen loads from 7 samples were multicomponent, although there are no prevailing components among the fluorescence spectra. In pollen loads 1 and 4 one maximum is seen, mainly, in green-yellowish (520-530 nm), while in pollen loads 2 and 6 the same maximum was characteristic only for small amount of pollen grains. This peak is usually correlated with the increased content of carotenoids (Roshchina et al., 1997c). Comparing the fluorescence spectra of pollen received from clovers, both *Trifolium pratense* and *T. arvense*, we should mark the same maximum 520-530 nm that are characteristic for the meadow plants (Fig. 6.3), as well as for pollen from dandelion *Taraxacum officinale* and other species. Non-matured pollen of clovers also have maximum at 680 nm, peculiar to chlorophyll whereas in pollen loads studied – they do not. In the blue region of the spectra pollen grains from samples of pollen loads 2, 4 and 6 fluoresce with maximum 460-480 nm and the same peak is characteristic for willow *Salix caprea*. Depending of the time when the pollen load was collected and from analysis of the fluorescence spectra or a colour of fluorescence under a luminescence microscope, one can establish, which pollen is present in the bee collection. Moreover, maxima of prevailing pollen grains indicate which

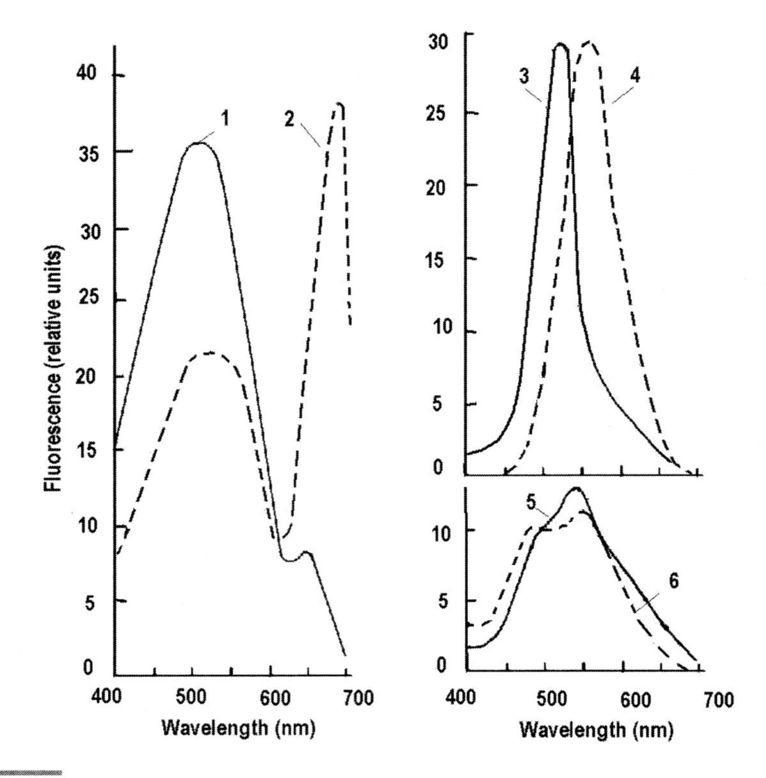

Fig. 6.3 The fluorescence spectra of main types of pollen collected by bees. 1. *Trifolium pratense*; 2. *Trifolium repens*; 3. *Salix caprea;* 4. *Taraxacum officinale;* 5. *Medicago falcata*; 6. *Cerasus vulgaris*

components are present, for an example in green-yellow - carotenoids, while in blue - flavonoids and some terpenes (azulenes). Red fluorescence may show the presence of chlorophyll, anthocyanins and, in some cases, azulenes. Thus, autofluorescence of pollen loads may serve as an indicator in express–analysis of pollen loads without long lasting biochemical procedures. As for the products of pollen such as transformed products of bees – perga (pollen food) or propolis (bee glue served for the defence of the bee hive against diseases), their fluorescence spectra have smoothed maxima and, therefore, are not as informative as the fluorescence spectra of pollen loads (Fig. 6.2). Pollens are also found in various samples of honey, and the fluorescence spectra of the samples may give information about the plant species visited by bees, and its nectar may be the base of the honey sample.

6.2 CELL VIABILITY AND COMPATIBILITY

The cells of pollen and pistil may be viable or non-viable, and there could be a complete lack of fertility after being damaged by any factor or by the inability of some plant clone for self-pollination. This knowledge is important for genetics and selectors as well as for gene engineering. Autofluorescence of the cells could be informative in the analysis of the fertility.

6.2.1 Pollen Viability

It is possible to know cell viability by measuring fluorescence of pollen moistened with nutrient medium (Roshchina et al., 1997b). We can see the changes in the fluorescence intensity of the untreated cells (autofluorescence) or of the treated with natural dye rutacridone (Roshchina, 2002) as shown in Table 6.3. In both cases the fluorescence enhances. But in a larger degree, the intensity of autofluorescence increases, approximately three fold in non-viable pollen. This fact was established not only for pollen of *Hippeastrum hybridum*, but also for pollen of *Philadelphus grandiflorus*, *Plantago major*, *Betula verrucosa* and other species.

Table 6.3 The fluorescence intensity (relative units) of viable and non-viable pollen of *Hippeastrum hybridum* after 30 min of moistening

Pollen	*Autoluorescence at 460-475 nm*	*The fluorescence at 530 nm after the treatment with rutacridone*
Viable	1.2 ± 0.1	1.23 ± 0.09
Non-viable	3.3 ± 0.2	1.7 ± 0.1

Pollen tubes of germinating microspore usually has no or weak fluorescence, but sometimes their cell walls may fluoresce in blue-green (*Hippeastrum hybridum*) or even in red (*Hypericum perforatum*).

6.2.2 Self-incompatibility of Pollen

The fluorescence of pollen could be useful for distinguishing self-compatible and a self-incompatible pollen of the same species, as can be seen for *Petunia* clones (Roshchina et al., 1997b; Kovaleva and Roshchina, 1999). In self-incompatible clones, although pollen is given from a flower of the same species as the pistil, self-pollination is impossible. The differences in the fluorescence spectra of pollen and pistil stigma from self-compatible and self-incompatible clones of *Petunia hybrida* are shown in Fig. 6.4. This reflects the alterations in the composition of fluorescent metabolites. In the self-incompatible clone there was no emission at 620-640 nm in the pollen fluorescence spectra which correlated with the absence of azulenes (Table 2.3) and with a decrease in the chlorophyll synthesis for the pistil stigma. The yellow fluorescence at 530-560 nm was also small due to the lower concentration of carotenoids (Roshchina et al., 1997a; 1998a).

Pollen and pistils of *Petunia hybrida* from self-compatible and self-incompatible clones demonstrated different fluorescence spectra during the development of the flower (Kovaleva and Roshchina, 1999). The differences in the fluorescence spectra of the self-incompatible clone were already seen, beginning from the flower bud, where pollen of incompatible clone has no expressed maxima, unlike pollen of compatible clone emitted with maxima 475 and 640 nm. Pollen of opening flower from both clones demonstrates one maximum in green 518-520 nm, while in the open flower there were differences. In self-compatible clone we see two maxima at 450-470 and 520 nm and shoulder 640 nm related, perhaps, to azulene or flavonoids rutin or quercetin. Unlike self-compatible clone, in self-

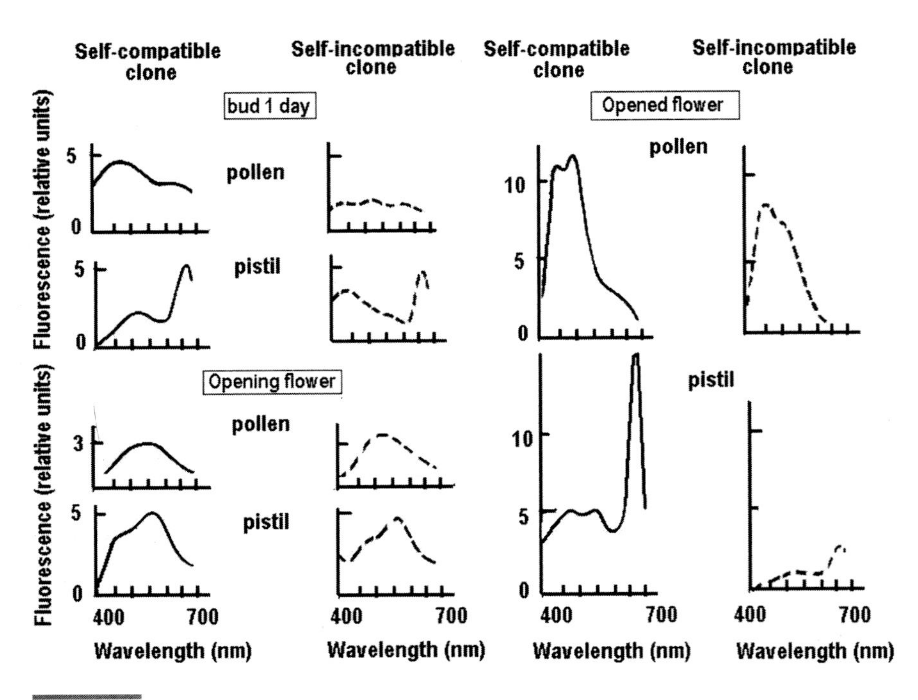

Fig. 6.4. The fluorescence species of generative organs from self-compatible and self-incompatible clones of *Petunia hybrida*

incompatible clone, pollen does not have the same shoulder. Pistil of self-compatible clone had maximum in green at 530 nm (flavins) whereas those from self-incompatible – in green at 440 nm (terpenoids) in the flower bud before flower opening. In the completely open flower, the pistil stigma of the former clone demonstrated maximums 480, 510-520 and 680 nm, but in the latter clone – 480, 510-520 and 695 nm. It appears to be redox reactions on the surface of pollen at the interaction "pollen-pistil" (See Chapter 5, Section 5.1.1). They include reduction-oxidation of the surface components, which differ in self-compatible and self-incompatible clones. The difference in the fluorescence spectra of the clones could also be related to genetic peculiarities, in particular with the activity of the S-gene. Thus, the method of microspectrofluorimetry appears to be used for the earlier diagnostics of self-incompatibility due to their fluorescence spectra of pollen and pistils.

6.3 AUTOFLUORESCENCE IN DIAGNOSTICS OF CELLULAR DAMAGE

Damaging factors, such as ageing, ultra-violet light or high doses of ozone and other factors, influence secretory cells (Roshchina and Roshchina, 1993; 2003; Roshchina et al., 1998 a, b; Roshchina and Karnaukhov, 1999; Roshchina and Melnikova, 2000; 2001). Similar changes in autofluorescence and orange-blue shift were also observed for the pollen treated with ozone (Roshchina and Melnikova, 2001; Roshchina and Roshchina, 2003; Roshchina, 2003). Some examples will be given below.

6.3.1 Diagnostics of Stress or Ageing Effects

Final effects of hazzard factors as ageing, ultra-violet light, high doses of ozone, etc, are often seen as similar cellular damages (Roshchina and Roshchina, 2003; Roshchina, 2003). But in some cases earlier signs of damages differ. Below there are examples of alterations in the cellular fluorescence under above-mentioned damaging conditions.

6.3.1.1 Ageing, γ-irradiation and ultra-violet light

Ageing and ultra-violet light effects are usually seen in the cellular damage (Roshchina et al., 1998e; Roshchina and Roshchina, 2003; Roshchina, 2003). But γ-irradiation (3000 Gy) only decreased the total green fluorescence by 50% in the pollen of *Hippeastrum hybridum* (Roshchina et al., 1998b).

Ageing, seen as the storage after the collection of *Papaver orientale* pollen from 1994 to 1998 years, shows the formation of new fluorescing products with maxima in the more short wavelength region which may be dealt with the ageing (Roshchina et al., 1998e). UV-light (300-350 nm) treatment for 1 h induced the analogous shifts in the pollen fluorescence from orange to green or blue) for *Hemerocallis fulva* and *Dactylis glomerata* (Roshchina et al., 1998e). As seen in Fig. 6.5, the old pollen of *Papaver orientale* (collections of the years 1995-1997) became green-yellow

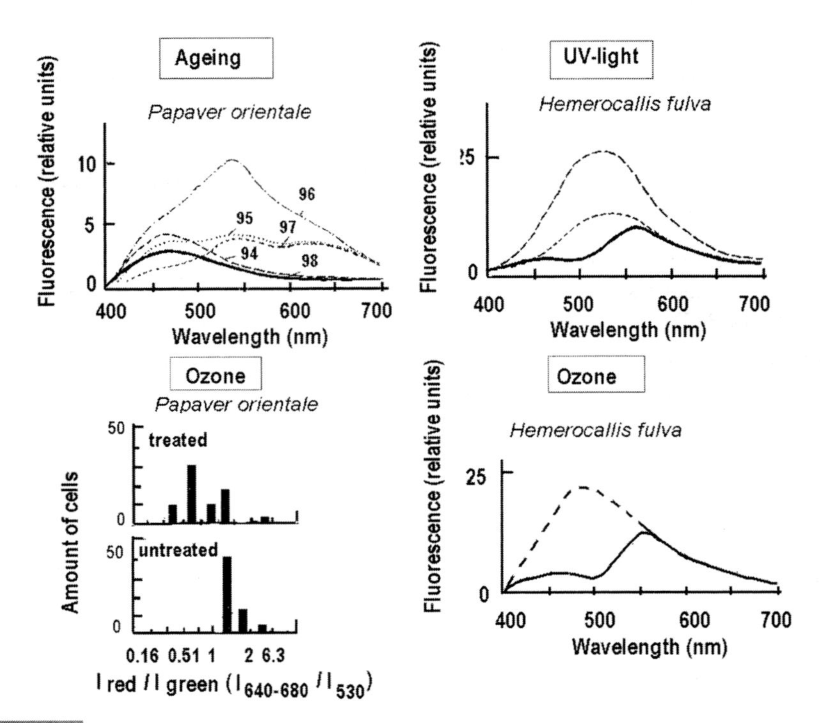

Fig. 6.5 The fluorescence of pollen grains treated with stress factors (ageing, UV-light and ozone). Sources: Roshchina et al., 1998e; Roshchina 2003). The fluorescence spectra are given in upper side (left and right) and in lower side (right) whereas histograms of the fluorescence intensity Ired/Igreen for cells treated or untreated with ozone – in lower side (left). Broken lines shown as 94-97 – storing during 1994-1997 years and unbroken line marked as 98 – fresh collected pollen. In other cases unbroken line – untreated pollen, broken lines - after the treatment with UV-light 300-400 nm (100 W m^{-2} for 1-2 h exposure) or ozone(> 0.1 ppm or 0.1 μl/L). Histograms of the fluorescence intensity Ired/ Igreen are done for cells treated or untreated with ozone (0.1 ppm or 0.1 μl/L).

fluorescing instead blue–fluorescing as in the fresh state. A similar response in a form of the autofluorescence from plant cells treated with UV-light or ozone is arising at the ultraviolet radiation from atomic oxygen (Roshchina and Roshchina, 2003). This occurs due to the factors induced the processes observed at ageing too. Moreover, ageing, UV-light and ozone (see also Section 6.3.1.2) enhanced the fluorescence of the objects 2-3 times in comparison with untreated or fresh pollen. UV-light and ozone also shifted maximum of the fluorescence in the shorter spectral region – instead orange to blue as seen for *Hemerocallis fulva*. Figure 6.5 shows that after treatment with ozone of the pollen grains of *Papaver orientale* the ratio Ired 640-680 nm/I green 530 nm decreases due to the appearance blue-green emission instead green-yellow as in untreated microspores. UV-light effect also includes the ozone contribution (pure effects of ozone can be seen in Section 6.3.1.2). Both UV-irradiation and ozone shifted the autofluorescence maximum from the orange to blue spectral region (Fig. 6.5). The fluorescent products of the reaction are discussed in Section 6.3.2.

6.3.1.2 Ozone and other active oxygen species

Ozone and active oxygen species such as superoxide anion radical and peroxides, which are formed on the secretory cell surface at norm and ozonolysis, can also contribute to fluorescence changes (Table 6.3). It has been demonstrated (Roshchina et al., 1998a,e; 2003; Roshchina and Karnaukhov, 1999; Roshchina and Melnikova, 1998b; 2001) in experiments, in which pollen and pistils from various species treated with ozone, enzyme superoxide dismutase (destroying superoxide radical by dismutation), and peroxidase (cleaving organic peroxides) as well as low-molecular antioxidant ascorbate. The light emission increases when the generator of superoxide anion radical noradrenaline or hydrogen peroxide are added on the surface of secreting cells of the *Hippeastrum* pistil stigma (Roshchina et al., 1998a). A new maximum 600-620 nm is observed if the antioxidants ascorbate and peroxidase are put on the cellular surface.

Ozone. Ozone alters the fluorescence spectra and/or the fluorescence intensity of the many cells possessing a significant autofluorescence (Roshchina and Roshchina, 2003; Roshchina, 2003). Moreover, this compound and other reactive oxygen species (often derived from ozone) induces the ageing of cells. Some examples of the changes are given below. Secretory cells of glands, glandular trichomes and other structures, localized on the surface of the leaf, stem and leaf-derived parts of the flower (petal, sepals, anthers, stamens), interact with ozone. The properties of the individual components are changed (Roshchina and Roshchina, 1993; Roshchina, 1996), showing the changes in the fluorescence of secreting cell (Roshchina and Melnikova, 2000). For instance, the character of the fluorescence spectrum of the *Raphanus sativus* secretory hairs themselves, usually blue light fluorescing, differ significantly after ozone treatment when additional green-yellow emission arises (Roshchina and Roshchina, 2003).

Thus under the influence of ozone the changes inside the hair which occur can be seen as a formation of fluorescing products. Ozone often induces green-yellow fluorescence (Colour Fig. 27, Table 6.3) of the dry vegetative microspores of *Equisetum arvense* (Roshchina and Melnikova, 2000; Roshchina and Roshchina, 2003). This phenomenon seems to belong to fluorescening products of lipid peroxidation occurring under oxidative stress (Roshchina et al., 1998a).

Earlier diagnostics of the ozone damages has been described in the monograph of Roshchina and Roshchina (2003) in detail, here we shall demonstrate new illustrations (Fig. 6.6) for secreting cells of the seedlings sensitive to oxidative stress. Green-yellow fluorescence with maxima at 510-540 nm, which is non-measurable at norm, appeared on the primary leaves of *Plantago major* and *Zea mays* after fumigation by ozone. The fluorescence concentrated in small spots predicts necrosis which developed later in the sites of leaves of *Plantago major*, but in the same sites of *Zea mays* disappeared later which showed the reparation processes.

As seen from Table 6.3, the autofluorescence changes under oxidative treatment with ozone or other active oxygen species so that even the colour of fluorescence may indicate some damages in the cells of microspores chosen as indicators for ozone (Roshchina and Roshchina, 2003). The pollen grains are either enriched in pigments such as carotenoids (*Passiflora, Philadelphus, Hemerocallis*) or phenols (*Hippeastrum*). While vegetative microspores of *Equisetum*, are sensitive to smaller concentrations of ozone, than the pollens chosen.

Secretory cells of vegetative tissues or microspores are suitable objects for O_3 monitoring due to various fluorescent compounds in secretions (Roshchina and Roshchina, 2003). Low doses of ozone < 0.15 μl/l induced green-yellow fluorescence in the leaves of *Plantago major, Zea mays, Raphanus sativus* and *Hippeastrum hybridum* as well as of vegetative microspores of *Equisetum arvense*. The light emission strengthened with the increase of ozone concentration which showed the damage of the

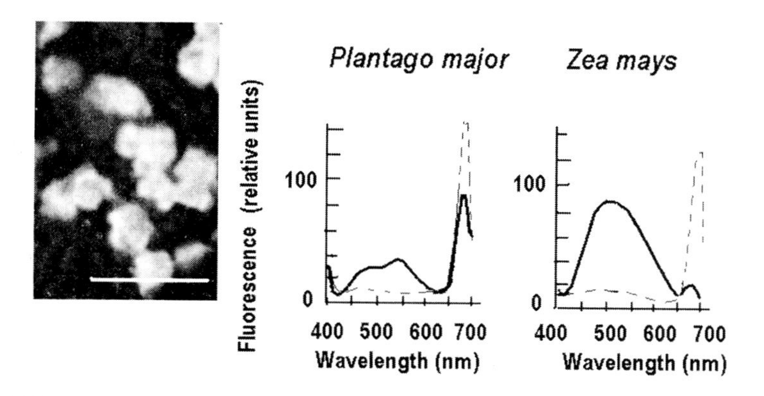

Fig. 6.6 The appearance of the yellow-fluorescing spots on the leaf after the fumigation with 0.05 ppm during 6-12 h of ozone. Photograph of seedlings of *Plantago major* (left, bar 20 μm) and the fluorescence spectra of the species (middle), as well as the fluorescence spectra of *Zea mays* (right). Broken line-untreated cells, unbroken line-treated with ozone.

lightening cells, leading to their necrosis. At higher doses of ozone, blue fluorescence at 420-470 nm was observed in pollen, containing carotenoids such as *Passiflora coerulea, Philadelphus grandiflorus* and *Hemerocallis fulva*, usually lightening in the yellow-orange region 530-560 nm. These pollen grains lost their fertility, and lipofuscin was formed in their cover (Roshchina and Karnaukhov, 1999). Thus, analyzing autofluorescence, we could indicate a damage of the secreting cells induced by the stress factor.

The picture is clearly seen at ozonolysis in generative cells, for instance in pollen (Roshchina and Roshchina, 2003). Pollen, which is a lack of pigments, such as of *Plantago major*, have no significant changes in its fluorescence whereas pigmented pollens, for instance enriched either in carotenoids as *Passiflora coerulea* or phenols as *Hippeastrum hybridum*, had the shifts in the fluorescent spectra. The ratio of secreting microspores such as pollen of *Papaver orientale* may be changed under the treatment with ozone, which alters the value of the Igreen/Ired ratio of fluorescence (Roshchina, 2003). Similar changes in the autofluorescence and the orange-blue shift were also observed for the ozonated pollen from many other plant species (Roshchina and Melnikova, 2001; Roshchina and Roshchina, 2003).

Table 6.3 Indication of cellular damage by oxidant stress based on the microspores autofluorescence (Roshchina and Roshchina, 2003 and unpublished data of the author)

Factor	*Fluorescence*		
	Colour	*Maximum (nm)*	*Intensity (% of control)*
	Pollen of *Passiflora coerulea*		
Control (untreated)	orange	450, 560	100
Ozone 0.6-0.9 ppm (μl/L)	blue	460	100
H_2O_2 10^{-4} M 2-5 min	blue	430	54
24 h	Blue-green	430, 530	120-170
Tert-Butyl peroxide 10^{-4} M 2-5 min	blue	430	30
24 h	orange	445, 550	65
	Pollen of *Philadelphus grandiflorus*		
Control (untreated)	green-yellow	465, 535	100
Ozone 2.4 ppm (μl/L)	blue	475-480	160
H_2O_2 10^{-4}M 2-5 min	blue	435	25
24 h	blue	480, shoulder 585	200-450
Tert-Butyl peroxide 10^{-4} M 2-5 min	blue	430 (470)	30
24 h	blue	490	95-100
	Pollen of *Hemerocallis fulva*		
Control (untreated)	orange	500, 560	100
Ozone 0.6-0.9 ppm (μl/L)	blue	475	78
Ozone 2.4 ppm (μl/L)	green-yellow	510	220
	Pollen of *Hippeastrum hybridum*		
Control (untreated)	blue-green	490-510	100
Ozone 2.4 ppm (μl/L)	yellow-orange	550	320
H_2O_2 10^{-4} M 2-5 min	green-yellow	475, shoulder 550	170-180
24 h	green-yellow	450, 565	235-240
Tert-Butyl peroxide 10^{-4} M 2-5 min	green-yellow	450, 550	10
24 h	yellow	550	120
	Vegetative microspore of *Equisetum arvense*		
Control (untreated)	blue	470, 555	100
Ozone 0.05 ppm (μl/L)	yellow	Shoulder 470, 560	210

Besides prolonged (from 30 min to several d or mon) effects of ozone,active oxygen species such as superoxide anion radical and peroxides formed on the secretory cell surface, quick effects (2-5 min response) can also contribute in the fluorescence changes which was demonstrated in experiments with the treatment of pollen and pistils from various species with antioxidant enzymes – superoxide dismutase, destroying superoxide radical by dismutation, and peroxidase, cleaveging organic peroxides (Roshchina et al., 1998a; Roshchina and Karnaukhov, 1999; Roshchina and Melnikova, 1998b; 2001). The light emission increases when the generator of superoxide anion radical noradrenaline is put on. The reactions are rather fast. If the antioxidants ascorbate and peroxidase are put on the surface, the former induces mainly a new maximum 600-620 nm after 66 s, whereas the latter significantly decreases only the fluorescence intensity, quenching it completely for 110 sec (Roshchina et al., 1998a). The pure water addition slightly reduces the emission intensity (nearly ≈ 10-15%). Total fluorescence of the secreting stigma increases with the addition of 10^{-5} M hydrogen peroxide.

The effects of reactive oxygen species also differ on the self-incompatible and self-compatible clones of *Petunia* (Fig. 6.7). The fluorescence spectra of pollen from both self-compatible and self-incompatible clones are changed differently if they were treated by oxidants such as ozone, hydrogen peroxide and *tert*-butyl peroxide. After the fumigation with ozone (0.5 μl/L) the intensity of the pollen fluorescence

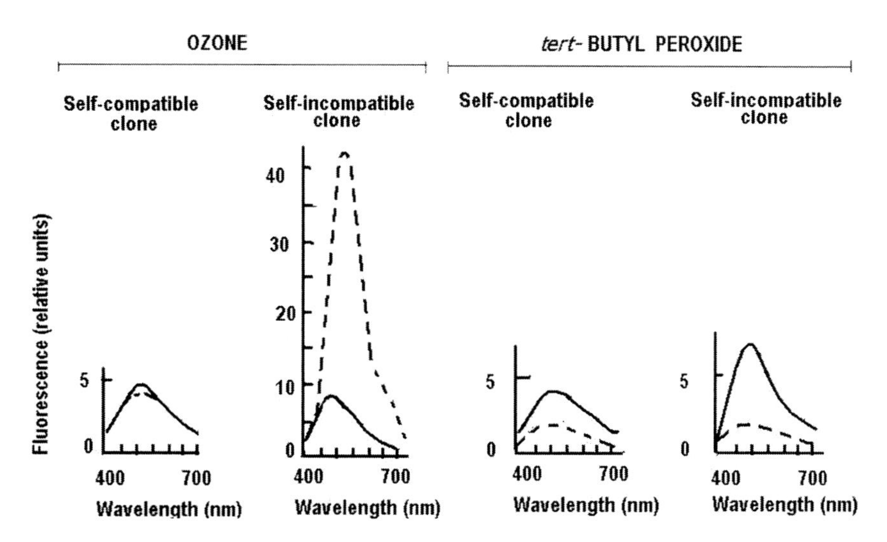

Fig. 6.7 The fluorescence spectra of the pollen from self-incompatible and self-compatible clones of *Petunia hybrida* after the treatment of reactive oxygen species ozone 0.5 ppm or μl/L (left) and *tert*-butyl peroxide 10^{-4} M (right). Unbroken line – untreated pollen, broken line – after the treatment.

belonging to self-incompatible clone increased five times in comparison with the untreated sample while pollen grains from self-compatible clone had no response to the factor. Peroxides, in particular *tert*-butyl peroxide (Fig. 6.7) quenched the fluorescence of both clones.

6.3.2 Fluorescing Products of Damage

Fluorescing products of cellular damage arise due to many stress factors such as ozone and reactive oxygen species or UV-light as well as at ageing. Recently microspectrofluorimetry has been used for the analysis of secondary products formed in leaves treated with herbicides (Hjorth et al., 2006). If the plant cell or tissue is damaged (ageing, UV-light, ozone) arising products of free radical processes, including lipid peroxidation, in living organisms can fluoresce (Haliwell and Gutteridge, 1985; Gutteridge, 1995). The nature of fluorescing spots on leaves and microspores of plants is supposed to connect with the Schiff bases, formed at free radical reactions in a lipid fraction of membranes. Schiff bases arise, when the aldehyde group of the substance react with the amino group of the other compound. Principally, it may be occur with any amino acid or amino acid residue in protein. Aldehydes are formed in many metabolic processes. At stress malondialdehyde often is formed as a result of the lipid peroxidation which leads to the formation of the fluorescent products with emission in blue (420-460 nm). The scheme is presented below on Fig. 6.8. Deeper changes result in the arising of fluorescing pigment lipofuscin, which usually find out in growing old tissues of plants as a final product of lipid peroxidation (Merzlyak, 1988). The contribution to the autofluorescence at 410-450 nm may be also done by pigment lipofuscin.

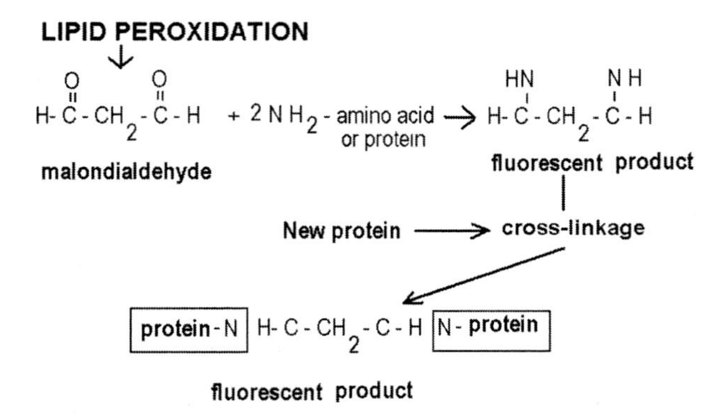

Fig. 6.8 The scheme of the fluorescent products formation at stress and ageing

The fluorescence could be connected with the group of ageing pigments know as as a whole lipofuscin which form in animals from terpenoid compounds (Karnaukhov, 1973). Chloroform soluble products fluoresced with maximum at 440 nm and similar with lipofuscin of animals was also found by Maguire and Haard (1975) in ripening fruits of banana (*Musa cavendishii*, var. Valery) and pear (*Pyrus communis*, var. Barlett), mainly in peel and pulp tissue.

This pigment may be also derived from photosynthesizing cells (Merzlyak et al., 1984; Merzlyak, 1988). In many cases, for example in corn samples, lipofuscin - like pigments, fluorescing at 420-450 nm are formed under air pollution by ozone and nitrogen dioxide (Brooks and Csallany, 1978). Evidence of the participation of lipofuscin in the orange-blue shift in the fluorescent spectra of intact pollen of *Philadelphus grandiflorus* under the ozone influence was noted by Roshchina and Karnaukhov (1999).

Ageing also induces the lipid peroxidation and formation of fluorescent products (Roshchina, 2003). For instance, the products of fresh and stored pollen differ as can be seen from their fluorescence spectra on the chloroform-ethanol extracts from pollen *Hippeastrum hybridum* and vegetative microspores of *Equisetum arvense* (Fig. 6.9). Changes in their composition are seen from the fluorescence spectra of intact microspores (pollen and vegetative microspores) and their chloroform-ethanol (2:1) extract. The chloroform-ethanol extracts from pollen also demonstrated the presence of lipophilic compounds fluoresce in yellow-orange (carotenoids), orange-red (anthocyanins or azulenes and proazulenes) as described earlier (Roshchina et al., 1995; Roshchina et al., 1998a, e;

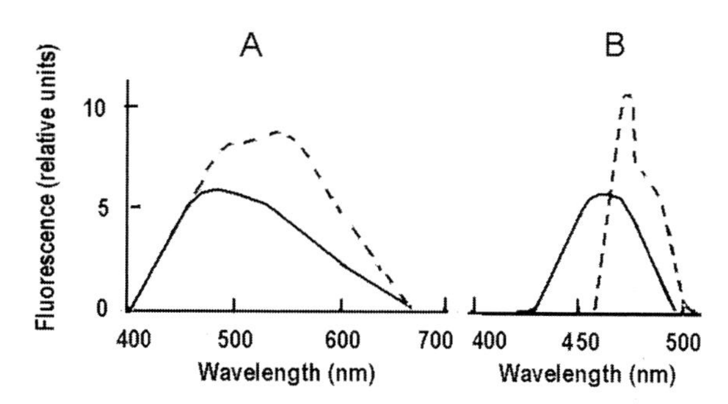

Fig. 6.9 The fluorescence spectra (Left) of chloroform-ethanol extracts (1:1 v/v) from pollen of *Hippeastrum hybridum* (A) and from vegetative microspores of *Equisetum arvense* (B). (Source: Roshchina, 2003). Fluorescence excited by light 380 nm. Unbroken line – fresh material; broken line – after the 1 y of storage.

Roshchina, 2003; 2004a). New fluorescent products in the more wavelength spectral region are observed if it is compared with fresh microspores. Thus, analysis of the autofluorescence could indicate a damage of the secreting cells induced by a stress factor earlier than visible alterations.

6.4 DIAGNOSTICS OF CELLULAR PROCESSES ON THE FLUORESCING CELLULAR MODELS

The search for the most suitable model-object is important for cellular biology. Thus plant microspores, which serve for plant breeding (Pryer et al., 2001) could be recommended as their development and the sensitivity to various external factors are clearly seen (Roshchina et al., 1998f; 2003; Roshchina and Roshchina, 2003). The main advantage of the object is their autofluorescence excited by UV-light. There are vegetative (in spore-breeding plants such as horsetails, mosses and ferns) and generative (male gametophyte called pollen, which is peculiar to Gymnosperms, mainly conifers, and the blossomed species of Angiosperms) microspores. They are single cells covered by the multilayer cellulose wall which protects against unfavourable external factors of an environment. The microspores were found to fluoresce under a luminescence microscope. Their emission may be first registered by microspectrofluorimeters during the development of the microspores (Roshchina et al., 2002). Confocal imaging of secreting plant cells was studied for grass pollen analysis (Salih et al., 1997) and vegetative microspores of horsetail (Roshchina et al., 2004). Special studies devoted to the autofluorescence of plant microspores as cellular models for purposes of cellular biology concerning the problem of chemosignaling (Roshchina, 2005c, 2007a). In a similar unicellular system changes may be seen in the fluorescence during development and at the intercellular interactions (see Chapters 4 and 5). This is important for the study of the allelopathy mechanisms (Roshchina, 2004a, b; 2005a, b) and the relationships between contractile proteins and chemosignaling of any cell (Roshchina, 2004; 2005a, b; 2006b).

6.5 FLUORESCENCE OF CELLS AS BIOSENSORS AND BIOINDICATOR REACTIONS

Fluorescence of intact plant secretory cells are sensitive to various factors as seen from the above-mentioned sections of the chapter. This may be useful in the search of biosensors and bioindicators (Roshchina, 2003).

Bioindicator reactions may often be peculiar to special sensitive cells, which respond on the external factors, in particular pollution ionizing radiation, etc. Autofluorescence of such cells could be recommended for this aim (Roshchina, 2003). The phenomenon is useful for the ecological monitoring in the same degree as the chlorophyll emission (Karnaukhov, 2001; Roshchina, 2006a). The interaction of the cells with secretions and their individual components is registered by luminescent microscopy and is used as fluorescent markers and dyes in cell biology (See also Chapter 7). Biosensors are usually chosen or constructed, based on bioindicator reactions (Thompson ed., 2006).

Biosensors are the analytical systems, which contain sensitive biological elements and detectors. Plant intact cells as possible biosensors have a natural structure which determines their high activity and stability. The criteria in the screening of the plant cells as biosensors should be as follows: 1. Reaction is fast which based on the time of the response; 2. Reaction is sensitive to small doses of the analyzed compounds or their mixtures; 3. Method of detection such as biochemical, histochemical, biophysical (in particular, spectral changes of absorbance or fluorescence) are easy in the laboratory and in the field. Searching for biosensors, which are suitable in the analysis of the mechanisms of effects for biologically active substances or external factors of the environment among plant species, is a real problem.

Generative (pollen, male gametophyte) or vegetative microspores, pistil stigma in the flower and some secretory cells of multicellular plant structures are considered as potential biosensors for active oxygen species such as ozone, free radicals and peroxides (Roshchina and Roshchina, 2003) or for allelochemicals (Roshchina 1999; 2001a, b; 2004a; Roshchina, 2005c) and pharmaceuticals (Roshchina, 2006a, b). The first possibility is described in the monograph of Roshchina and Roshchina (2003) for ozone-sensitive objects, in particular for microspores of *Equisetum arvense*, pollen of *Philadelphus grandiflorus*, secretory leaf hairs of *Raphanus sativus* and idioblasts of root cellular culture of *Ruta graveolens*. Vegetative microspores of *Equisetum arvense* and secretory hairs of *Raphanus sativus* became yellow-fluorescing after fumigation of ozone in concentrations > 0.008 ppm (μl/L), while normally they fluoresce in the blue and red spectral regions, relatively (as an example see colour Fig. 27 in Appendix 2). On the contrary, pollen of *Philadelphus grandiflorus* and idioblasts of root cellular culture of *Ruta graveolens* missed normal yellow-orange emission in UV-light and fluoresced in the blue spectral region.

Changes in the fluorescence of unicellular microspores of plants (generative microspores or pollen of *Hippeastrum hybridum* and vegetative microspores of *Equisetum arvense*) as well as some unicellular and multicellular secretory structures of various plants organs were considered as possible biosensor reactions for the study of allelopathic mechanisms (Roshchina, 2005a, b) and for analysis of the ozone concentrations and damages in the environment (Roshchina and Roshchina, 2003). Fluorescence of the cells is recommended as a sensitive test-process for the analysis of the excretions (Roshchina, 2004a; 2005c). Fluorescing allelochemicals, isolated from the species enriched in tannins, furanocoumarins, alkaloids and sesquiterpene-containing plants were studied (Roshchina, 2005b, c).

Other perspectives originating from the experiments with biologically active compounds, which can be not only allelochemicals in biocenosis, but serve as medicinal drugs too (Roshchina, 2006a). The mechanisms of the sensitivity in chemosignal processes with participation of biologically active compounds – neurotransmitters (acetylcholine, dopamine, noradrenaline, serotonin, histamine) and natural oxidants (ozone, free radicals and peroxides) as well as antioxidants are discussed on the basis of cellular fluorescence changes after the treatment with the compounds (see Sections 5.1. and 6.3). Plant microspores may be used in laboratory practice as the objects of study and in the testing for the determination of sensory systems. Moreover, they could serve as the biosensors for medicinal drugs known as agonists and antagonists of neurotransmitters, instead animals with a necessity of vivisection for analysis. The redox reactions at the secretions' binding with the cell-acceptor appear to take place which reflects the fluorescence changes too. If the compound penetrates the cell and fluoresces within the cell, selectively staining some compartment, similar biosensors are suitable for determining the characteristics of the affecting drug or allelochemical (Roshchina, 2004a, b; 2005a, b, c; 2006a, b). More details of this aspect will be described in Chapter 7.

6.6 CYTODIAGNOSTICS IN BOTANICAL INVESTIGATION AND EDUCATION

The information presented in the above-mentioned chapters may be also used for a wide range of botanical investigation and education. For example, luminescence microscope and its new modifications, including confocal microscopy and microspectrofluorimetry, could be applied in

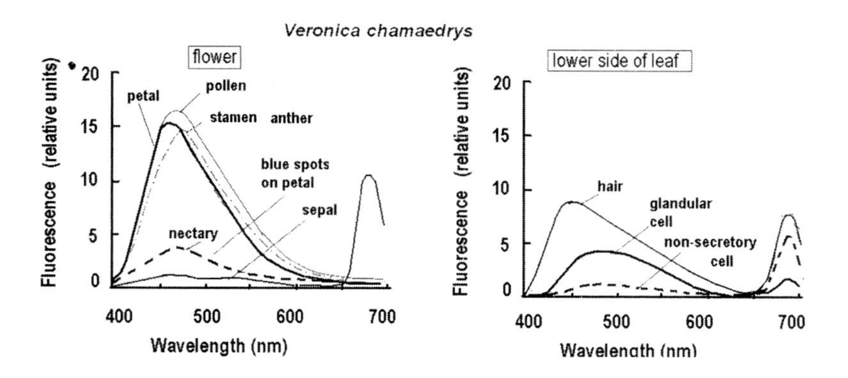

Fig. 6.10 The fluorescence spectra of different parts of *Veronica chamaedrys*, including secretory cells

many laboratories and universities. By their non-invasive methods, fluorescent cells and their separate parts can be seen. Moreover, anyone may be able to distinguish secretory cells in the complex tissues without histochemical dyes. Many observations are possible during cellular and plant development (see Chapter 4). Some examples of similar plant analysis are given in Figs. 6.10 and 6.11, where fluorescent parts of various organs are observed. Blue fluorescence (440-460 nm) is peculiar to various secretory cells of the *Veronica* species, while in *Hibiscus* and *Geum* species there are differences between secretory cells emitted in blue (440-460 nm)

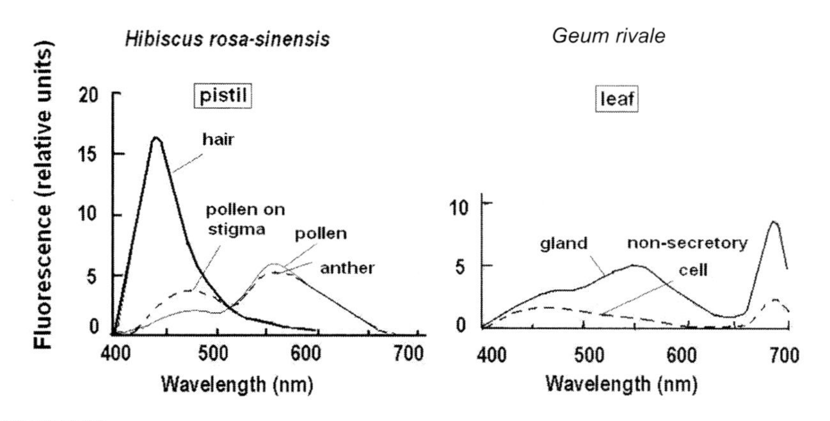

Fig. 6.11 The fluorescence spectra of different parts of *Hibiscus rosa-sinensis* and *Geum rivale*

and in yellow-orange (550-560 nm). Pollen lying on pistil stigma fluoresces in the shorter spectral region, than pollen of the female gametophyte.

Conclusion

Fluorescent analysis of plant secretory cells has the potential for cellular diagnostics both in fundamental studies and in practice. This is because of the possibility of getting precise information in intact tissues and single cells without procedures of tissue homogenization and long biochemical manipulations. One of the non-invasive applications could be the luminescence observation of plant materials in order to know if is it ready or not for pharmacy as the main valuable drugs are concentrated in secretory structures. The second possibility deals with the estimation of cell viability which is especially required for normal pollination and fertilization in genetic analysis for selection in agriculture and biotechnology. The third potential for the use of the fluorescence changes is analysis of cell damage which is necessary for ecological monitoring. Earlier diagnostics of the disturbances induced by the oxidative stress, ageing or pest invasion may be based on the fluorescence of secretory products.

CHAPTER 7 Individual Components of Secretions as Fluorescent Dyes and Probes

Fluorescence of living secretory cells excited by ultraviolet light is used in various diagnostic procedures as seen in Chapter 6. The same released secretion may include fluorescent components, which interact with other cells as the chemical signal. Some fluorescent substances from plant secretions such as phenols, alkaloids, terpenoids can be used for histochemical staining in the fundamental and applied studies of cell biology, signalling, allelopathy, etc, as well as in university courses and manuals of biophysics, physiology, botany and ecology (Roshchina, 2002, 2003, 2004a, b, c). When the pure fluorescent substances (10^{-6}-10^{-5} M) were added to the cell-acceptor, the changes in their fluorescence could be seen (which may be used in a modelling of allelopathic interactions (Roshchina, 2004a, b, c; 2005a, b). Certain compounds from secretory cells of weeds, medicinal or poisonous plant species applied to pharmacology (Ebadi, 2002) can act on known targets in cells (Balandrin et al., 1985; Duke et al., 2000). If they fluoresce, their ability to light emission under a luminescence microscope becomes useful for histochemical staining or analysis of inter- and intracellular binding.

In this chapter, using fluorescent components of secretions as various natural dyes and probes for analysis of mechanisms of the cell-cell interactions and for histochemical studies will be discussed.

7.1 INTERACTION OF CELLS WITH FLUORESCENT COMPONENTS OF THE SECRETIONS

Fluorescent substances contained in plant secretions may either penetrate or not penetrate into the cell, after binding with the certain cellular component the changes in the fluorescence of cellular compartment or the fluorescence of the accepted compound can be seen. The targets of binding will be discussed below.

7.1.1 Binding with the Cell Wall

The cell wall binds phenols or sesquiterpene lactones partly (they may also penetrate into the cell) which leads to changes in the bright blue-green fluorescence and even to arising of red emission on cellular models as plant microspores. For example (Fig. 7.1), tannin chamaeriol alters the fluorescence spectra of *Hippeastrum hybridum* pollen. Pure chamaeriol fluoresces with maximum 440-450 nm, while pollen without any additions, moistened with water, has no significant fluorescence. But added tannin binds with the cell wall, which emits in the same spectral region with maxima 440-450 nm. The emission concentrates on the cell surface. Usually, water-soluble tannins react with proteins (in particular, enzymes esterases, hydrolases and phosphatases) located in the cell wall.

Sesquiterpene lactones such as azulene and grosshemine soluble in ethanol or in the ethanol-water mixtures are also bound with cellulose, and, as seen in Fig. 7.1, the cell wall fluoresce in the orange-red region of spectrum with maximum 620-640 nm (Roshchina et al., 1995; 2004, 2005a). The example of the sesquiterpene lactone testing on the cellular model as pollen of *Hippeastrum hybridum* is shown in Fig. 7.1 (Left). The fluorescence of the cell wall differs from other cellular compartments. Sesquiterpene lactones, tauremizine and artemisinine induced blue fluorescence of the cell wall (Fig. 7.1, Right) for *Equisetum arvense* vegetative microspores. Moreover, artemisinine stimulated (up to 5 times) the blue emission with maximum 455 nm in elaters, which are the derivatives of the cell wall.

Tannins may interact with proteins (Goodwin and Mercer, 1983) whereas sesquiterpene lactones – with cellulose and hemicellulose (Roshchina et al., 1995) in the cell wall. Among the possible targets are DNAase, ATPase, RNAse and various oxidases.

7.1.2 Binding with Receptors

Hydrophilic components of the plant secretions slowly penetrate into a cell-acceptor, concentrating on the cell surface. They may be binding on the cell surface with special receptors or sensors (usually enzymes) and show the location of the structures. Among them are many substances known as

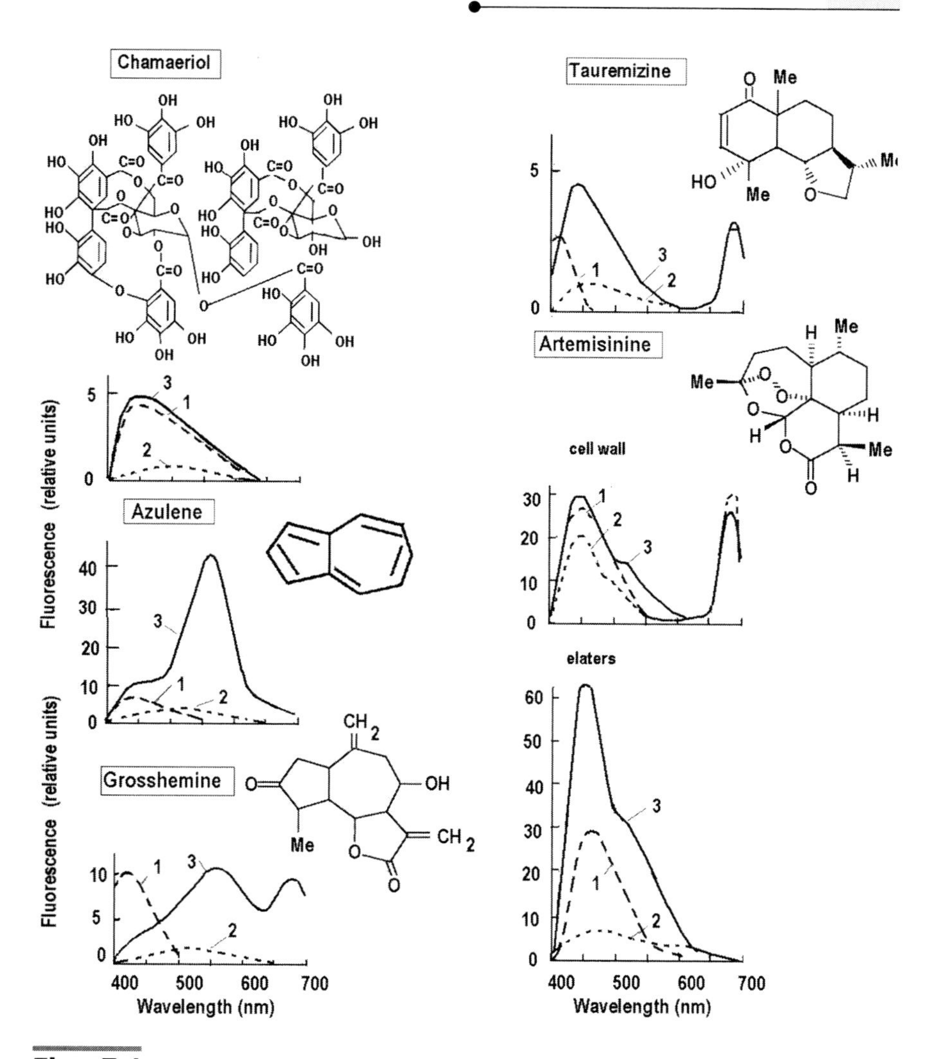

Fig. 7.1 The effects of tannin chamaeriol and sesquiterpene lactones on the fluorescence spectra of *Hippeastrum hybridum* pollen (Left) and of *Equisetum arvense* vegetative microspores. Source: Roshchina (2004a; 2005b, c). 1. pure compound chamaeriol 10^{-4} M or sesquiterpene lactones 10^{-5} M in ethanol; 2. Cell without treatment; 3. Cell treated with compound studied.

neurotransmitters and antineurotransmitter agents – alkaloids (Roshchina, 2001a, b). Neurotransmitters acetylcholine and biogenic amines themselves has no fluorescence in the visible spectral region, but fluorescent artificial label in the ligand usually links to the main molecule (Roshchina et al.,

2003b). Among natural components of the plant secretions are agonists and antagonists of the neurotransmitters, which fluoresce, when bound with the cellular surface, and may be used as fluorescent markers and dyes (Roshchina, 2004a, 2005a, b).

7.1.2.1 Agonists and antagonists of neurotransmitters

In non-synaptic systems, biogenic amines (catecholamines and serotonin) may regulate the development of animal and plant cells (Buznikov, 1987; 1990; Buznikov et al., 1996; Roshchina, 1991; 2001a). The mechanisms of action of these compounds have not been studied enough, which is partly due to insufficient development of adequate experimental methods. One of these methods is fluorescent analysis, which requires the use of corresponding fluorescent probes. Recently the possibility of using fluorescent agonists and antagonists of some neurotransmitters for the analysis of the location of binding of neurotransmitters by living animal and plant cells has been analyzed. For this purpose, synthetic fluorescent lipophilic analogues of dopamine, serotonin and acetylcholine or natural agonists and antagonists are applied (Roshchina et al., 2003c, d; 2005c; Bezuglov et al., 2004). First steps in a similar direction have been done with synthetic fluorescent analogues of neurotransmitters (Roshchina et al., 2003c, d) and with their natural agonists and antagonists (Roshchina, 2004a; 2005a, b, c).

Lipophilic analogues of neurotransmitters as the algorithm for fluorescent analysis. Lipophilic derivatives of dopamine (BODIPY-DA), serotonin (BODIPY-5HT), and acetylcholine (BODIPY-ACH) containing the 4,4-difluoro-5,7-dimethyl-4-bora-3a,4a-diaza-2s-indacene-3-dodecanoic acid residue, were used as fluorescent probes (Roshchina et al., 2003d; 2005c; Bezuglov et al., 2004). The probes were synthesized from the corresponding amines and fluorescent dodecanoic acid by the method of mixed anhydrides, as described in the paper of Bezuglov et al., (2004). The excitation wavelengths were 360-380, 410-440, 460, and 480 nm. Upon excitation by actinic light, these compounds fluoresce in the green part of the spectrum (maximum at 518 nm). The cells were treated both with the fluorescent probes and non-fluorescent dopamine and serotonin at concentrations 10^{-7}-10^{-5} M (Fig. 7.2). After the treatment, fluorescence of vegetative microspores of horsetail *Equisetum arvense* and the mouse embryo cells were analyzed (Fig. 7.2). Dry plant microspores had weak self-fluorescence; while in the case of mouse embryos, any self-fluorescence in the visible part of the spectrum was absent at all developmental stages. Histochemical staining of both (animal and plant) types of cells with

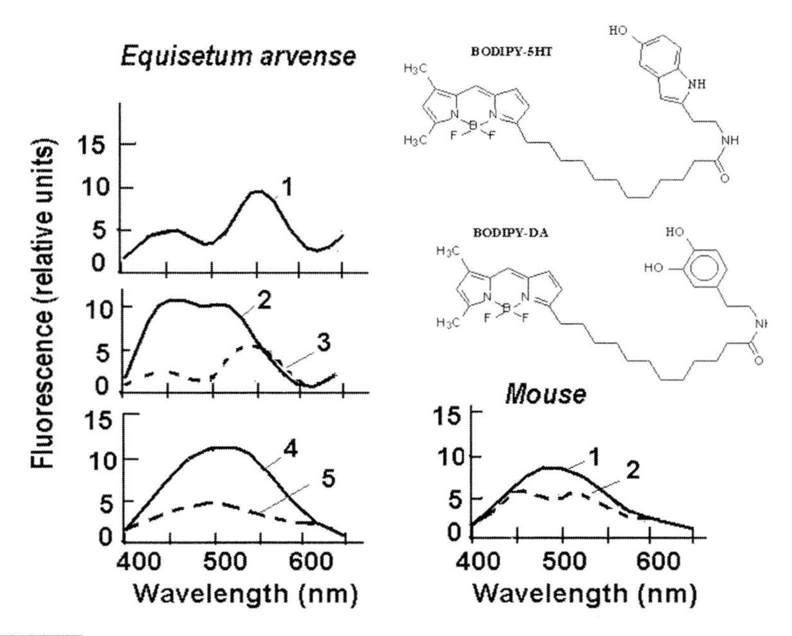

Fig. 7.2 Fluorescence spectra of vegetative microspores *Equisetum arvense* (left) and cells of the mouse embryo (right) after treatment with fluorescent neurotransmitters (excitation wavelength, 360-380 nm). Adopted from Roshchina et al., (2003c, d). Left - (1) Untreated cells (control); (2) the treatment with 10^{-7} M BODIPY-DA; (3) the treatment first with unlabelled 10^{-7} M dopamine and then with 10^{-7} M BODIPY-DA; (4) the treatment with 10^{-7} M BODIPY-5HT; and (5) the treatment first with unlabelled 10^{-7} M serotonin and then with 10^{-7} M BODIPY-5HT. Right.- the mouse embryo at the stage of two blastomers after treatment with 10^{-6} M fluorescent neurotransmitter BODIPY-DA : (1) the treated cell; (2) the reduction body (outside the cell).

BODIPY-DA and BODIPY-5HT resulted in the appearance or enhancement of fluorescence intensity in the green part of the spectrum at 500-530 nm (Fig. 7.2 curves 2 and 4). But the fluorescence intensity at 500-530 nm increased slightly if the cells were preliminarily treated with non-fluorescent dopamine and serotonin (Fig. 7.2, curves 3 and 5). This leads to the assumption that neurotransmitters and their fluorescent analogues bind mostly to the same sites in the cell. Green fluorescence was especially strong in peripheral cell areas, where the plasma membrane (and the cellulose wall in plant cells) is located (See colour Figs. 28 and 29 in Appendix 2). Laser-scanning confocal microscopy allowed one to see the concentration of the green fluorescence at the pollen surface of *Plantago lanceolata* and sea urchin eggs *Paracentrotus lividus* (Roshchina et al., 2005) as also seen in colour Fig. 28 in Appendix 2. The location of fluorescent transmitters predominantly outside the cell and a significant decrease in fluorescence intensity after the pretreatment of cells with dopamine or serotonin suggest that fluorescent derivatives are bound on the cell surface by the structures that bind natural neurotransmitters (e.g., receptors or transporters). This binding (possibly fairly specific) prevents a large part of lipophilic BODIPY derivatives from penetrating into the cell. Due to the presence of a lipophilic fluorophore, these fluorescent probes should penetrate across the plasma membrane; however, they concentrate outside the cell. Thus, similar to non-fluorescent dopamine and serotonin, BODIPY-neurotransmitters are bound on the cell surface (possibly, by the corresponding plasma-membrane receptors). However, it cannot be ruled out that these compounds may be bound inside the cell, because fluorescent probes contain a lipophilic moiety and, therefore, should penetrate into the cell and bind with organelles.

Intracellular binding of neurotransmitters located in secretory vesicles within the cell is possible too, when the compounds are released in the interior. If neurotransmitters are released from secretory vesicles within the cell they can also be bound with the organelles (Buznikov, 1987; 1990; Buznikov et al., 1996; Roshchina, 1991; 2001a). Non-fluorescent dopamine and serotonin do not penetrate into the cells, but stimulate the synthesis of RNA in isolated nuclei by binding to nuclear proteins (Arkhipova et al., 1988; Tretyak and Arkhipova, 1992). Experiments with isolated nuclei from the white petals of *Philadelphus grandiflorus* showed that both artificial and natural agonists and antagonists of some neurotransmitters may fluoresce on the surface of the intact nuclei, as it was seen for the plasmalemma of various living cells (Roshchina et al., 2005). The example of binding of BODIPY-dopamine is demonstrated in Colour Figs. 28 and 30 in Appendix 2. After histochemical staining of cells with BODIPY-neurotransmitters, there was green fluorescence of the reduction body, located on the surface of the mouse embryo. The reduction body consists of a large nucleus and a very narrow strip of cytoplasm, i.e., it is "enriched" in DNA (Fig. 7.2). The experiments were in accordance with the data (Roshchina et al., 2003c, d) received on the model systems analyzing the fluorescence of individual cell components (albumin, DNA, RNA, and β-carotene, which often binds to chromatin). As shown in Fig. 7.3, the fluorescence of water-wetted preparations (except for β-carotene) was weak or absent. The addition to these preparations of 10^{-6}-10^{-7} M aqueous

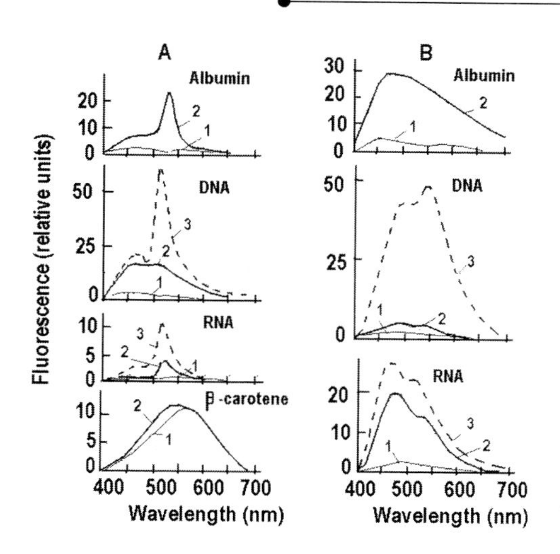

Fig. 7.3 The fluorescence spectra of individual cell components (albumin, DNA, RNA, and β-carotene) after the treatment with 10^{-7} M fluorescent neurotransmitters (A) BODIPY-DA or (B) BODIPY-5HT (excitation wavelength, 360-380 nm). Adopted from Roshchina et al., (2005 c, d) (1) Fluorescence of untreated component (moistening with water, 2 mg/0.02 ml; the control); (2) treatment with BODIPY-neurotransmitter; (3) fluorescence of the nucleic acid—protein mixture (ratio 1 : 1) after the treatment with BODIPY-neurotransmitter.

solution of a fluorescent probe always enhanced fluorescence in the green part of the spectrum. The effects of BODIPY-DA and BODIPY-5HT differed. The fluorescence spectrum of BODIPY-DA-treated albumin has a narrow maximum at 517-530 nm and a not very pronounced maximum at 465 nm, whereas a pronounced maximum at 465 nm and a not very pronounced maximum at 500-600 nm are present in the spectrum of BODIPY-5HT-treated albumin. Two maxima at 517-530 and 465 nm were detected in the fluorescence spectra of DNA and RNA in the experiments with both neurotransmitters. The value of the maximum at 518 nm sharply increased in the presence of albumin. In the experiments with BODIPY-5HT, the value of the maximum at 465 nm also increased. In the fluorescence spectrum of β-carotene, the maximum was shifted from 585 to 520-530 nm. Non-fluorescent serotonin and dopamine at concentrations of 10^{-6}-10^{-7} M did not cause similar changes in the fluorescence spectra. It can be assumed that the most significant changes in the fluorescence of the compounds studied were due to the presence of albumin in the medium. Taking into account that all dopamine and serotonin receptors are of the same nature as protein, the experiments with albumin lead to the assumption that the compounds studied probably bind to cell receptors. The determination of the fluorescence intensities of individual cell sites after histochemical treatment showed that as much as 70-90% of fluorescent neurotransmitters bind on the cell surface, mainly with receptors or within cells, for example with nuclei. Currently, the use of such fluorescent probes may be recommended for histochemical studies on the regulation of the cell system development in plants and animals by biogenic amines.

Natural agonists and antagonists of neurotransmitters. Chemoreception of neurotransmitters as chemosignals studied with their agonists and antagonists. The first response to neurotransmitters was alterations in cellular fluorescence excited by UV (360-380 nm) light. Examples of our experiments are shown in Figs. 7.4 and 7.5. The fluorescence of the cells was estimated by spectrofluorimetry and microspectrofluorimetry after the

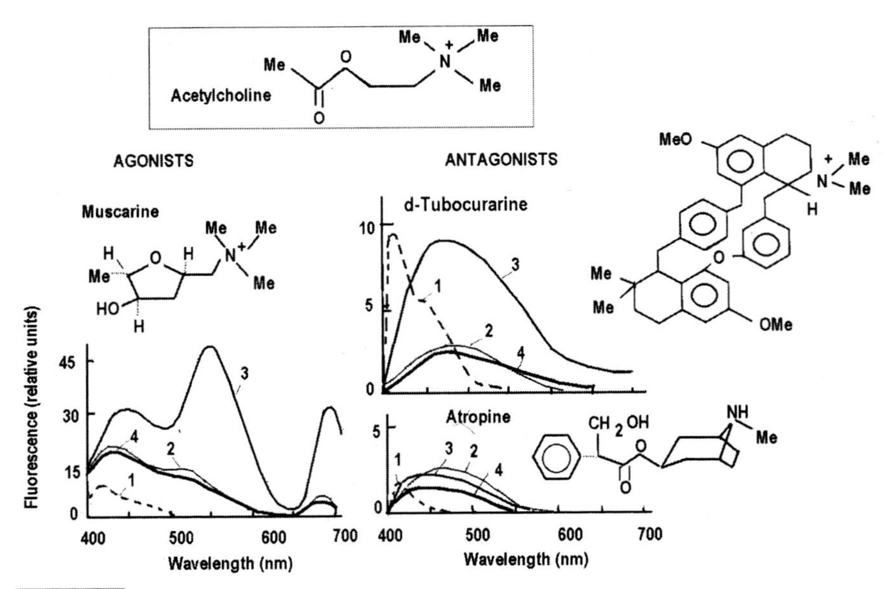

Fig. 7.4 Fluorescence spectra of vegetative microspores of *Equisetum arvense* (left) and pollen of *Hippeastrum hybridum* (right) registered with spectrofluorimeter or microspectrofluorimeter. Sources: Roshchina et al., 2003e; Roshchina, 2004a; 2005c). 1. solutions of the acetylcholine agonist muscarine 10^{-3} M or antagonists d-tubocurarine and atropine 10^{-5} M in 1 cm cuvette; 2. untreated microspore; 3. microspore treated with the agonist 10^{-6} M or the antagonists 10^{-6} M; 4. pretreatment of microspore with acetylcholine 10^{-6} M, then additions of the agonist or antagonist 10^{-6} M.

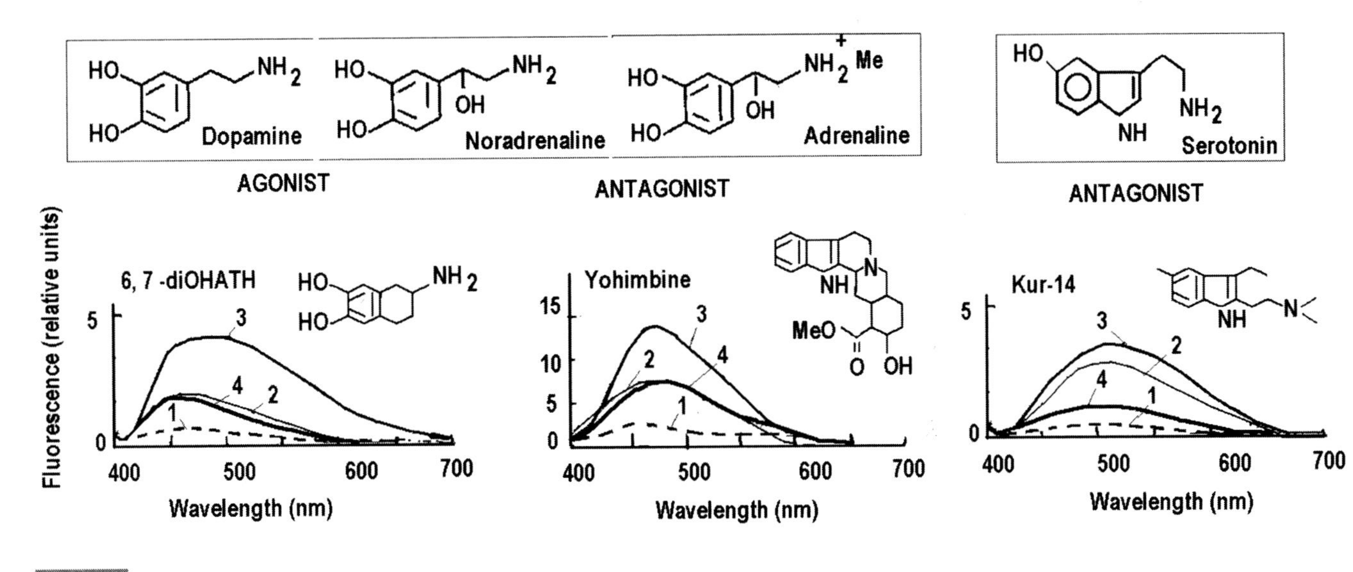

Fig. 7.5 Fluorescence spectra of pollen of *Hippeastrum hybridum* registered with spectrofluorimeter or microspectrofluorimeter. Adopted from Roshchina et al., (2003e). 1. solutions of agonist and antagonist of biogenic amines 10^{-5} M in 1 cm cuvette; 2. untreated microspore; 3. microspore treated with the agonist 10^{-6} M or the antagonists 10^{-6} M; 4. pretreatment of microspore with neurotransmitter 10^{-6} M, then additions of the agonist or antagonist 10^{-6} M.

treatment with agonist of acetylcholine muscarine, agonists of dopamine 6,7-diOHATN (2-amino-6,7-dihydroxy-1,2,3,4-tetrahydronaph-thalene hydrobromide) and BODIPY-DA, agonist of serotonin BODIPY-5HT, and antagonists of acetylcholine atropine, tubocurarine and quinuclidinyl benzylate, antagonists of noradrenaline propranolol, yohimbine, labetalol, prazosin, antagonist of serotonin Kur-14 (5-methyl- 2-α-dimethyl-aminoethyl-3-β-ethylindole) and antagonist of histamine tavegyl (clemastin) (Roshchina, 2003e; 2004a; 2005b, c).

Self-fluorescence of studied pure compounds added on the subject glass was small, but the emission increased 5-10 times after their interactions with the microspores. After the cell treatment with muscarine, d-tubocurarine, atropine, diOHATN, mesaton, yohimbine, labetalol, prazosin and tavegyl microspores emitted, mainly, in blue (max. 450-470 nm) whereas BODIPY-DA, BODIPY-5HT and Kur-14 - in green (max. 518-520 nm). The green or blue rings of the studied agonists or antagonists were seen under the luminescent microscope on the cellular surface where they were bound with proposed receptors. Among studied agents, BODIPY-derivatives of neurotransmitters, muscarine, 6-7- diOHATH, d-tubocurarine, yohimbine and Kur-14 may be recommended as possible fluorescent probes for the study of cellular location of cholinoreceptors and receptors of biogenic amines. Besides, blue-fluorescent alkaloids capsaicin from *Capsicum annuum* L., prototype of vanilloid receptor (see Chapter 3), and arecoline from *Areca catechu* L., agonist of acetylcholine, have been, perhaps, perspective as fluorescent dyes. Anthraquinone hypericin, bounded on the cell surface (see colour Fig. 26 in Appendix 2), inhibits protein kinases (Haugland, 2000) and has an affinity to glutamate N-methyl-D-aspartate receptor (Kubin et al., 2005). This compound related to surface receptors also may be fluorescent dye that needs special investigations in a future.

7.1.3 Binding with the Surface Enzymes (ATP ase and Cholinesterase)

Cellular surface of many species includes enzymes as possible sensors of the secretory products released by other cells. Among the enzymes are Na^+/ K^+- ATPase inhibited by ouabain (G-strophanthin which is glycoside of unsaturated steroid lactone) and cholinesterase inhibited by alkaloids berberine and glaucine belonging to plant species from the family

Papaveraceae, predominantly *Berberis vulgaris* L. and *Glaucium flavum* Crantz (Roshchina, 2005b; Budantsev and Roshchina, 2005; 2007). These fluorescing agents may bind appropriate enzymes on the cell surface, as shown in Fig. 7.6. The Fluorescence spectra of the cellular surface of vegetative microspore from *Equisetum arvense* demonstrate the shift of the maximum to longer wavelengths for ouabain – from 435 and shoulder 450 nm to maxima 480 nm and 560 nm whereas the fluorescence intensity of the cells increased 3-5 times in comparison with untreated microspores. However the chlorophyll fluorescence was quenched in this case. Berberine stimulated upto 5 times the fluorescence at 540-550 nm without a large shift of the maximum. Unlike this alkaloid, glaucine (pure solution) emitted in blue with maximum 430 nm, and, when being added to the cell studied, induced not only the enhanced emission at the same spectral region, but also a new maximum in green-yellow (530 nm) arose. Ouabain, berberine and glaucine may be recommended as fluorescent dyes.

7.1.4 Binding with Sensors Participating in Intracellular Regulation of Cyclic AMP Concentration

Fluorescent secretory products such as forskolin from *Coleus forskohlii*, theophylline from the tea plant *Camellia sinensis* and *Coffea arabica* and derivative of cyclic AMP dibutyryl-cAMP are able to increase the intracellular amount of cAMP (Mashkovskii, 2005). Figure 7.7 shows how the interaction with the plant model cells – microspores changes the light emission. Mechanism of the interactions consists as follows. Forskolin stimulates the activity of adenylate cyclase, enzyme catalyzed the synthesis of cyclic AMP, while theophylline inhibits cAMP phosphodiesterase which destroys cAMP. Perhaps, the compounds bind with to the enzymatic systems of cyclic AMP.

7.1.5 Binding with DNA-containing Organelles

Sesquiterpene lactones (azulenes and proazulenes) and alkaloids may bind with DNA-containing structures in a cell (Roshchina, 2002; 2004a; 2005 b, c). Sesquiterpene lactones can fluoresce in solutions, mainly in blue (at 420-460 nm). This fluorescence is related to the 7-membered ring(s) as the chromophore (Roshchina et al., 2004a). Emission of alkaloids is in all visible spectrum. This is useful in the analysis of the cellular interactions. This phenomenon is useful in the analysis of the cellular

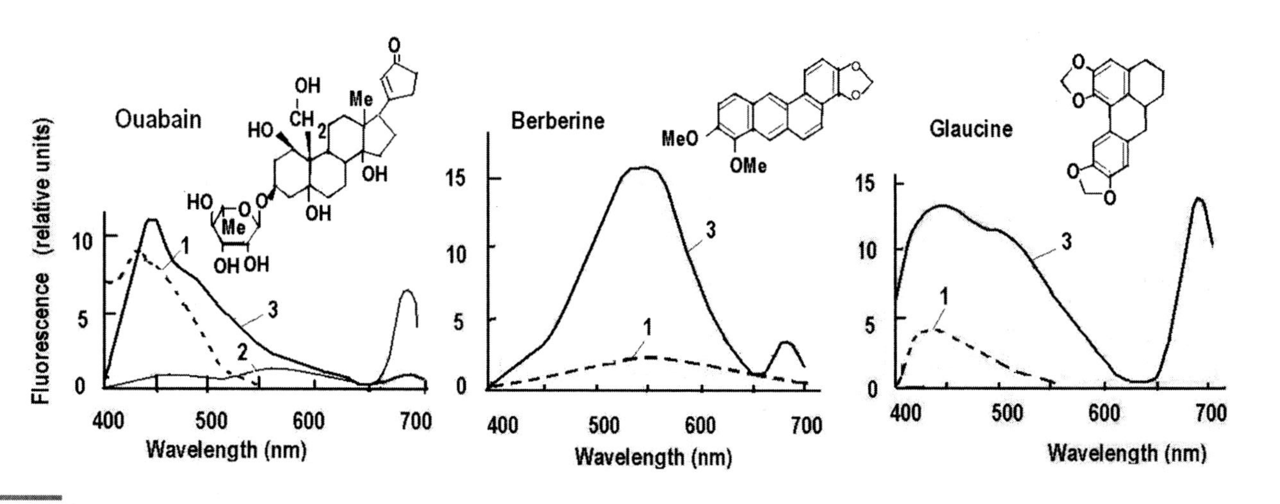

Fig. 7.6 Fluorescence spectra of vegetative microspore from *Equisetum arvense*. Source: Roshchina (2005b) and unpublished data. 1. solutions of agent studied 10^{-5} M (water solutions) in 1 cm cuvette; 2. untreated microspore; 3. microspore treated with the agent 10^{-5} M.

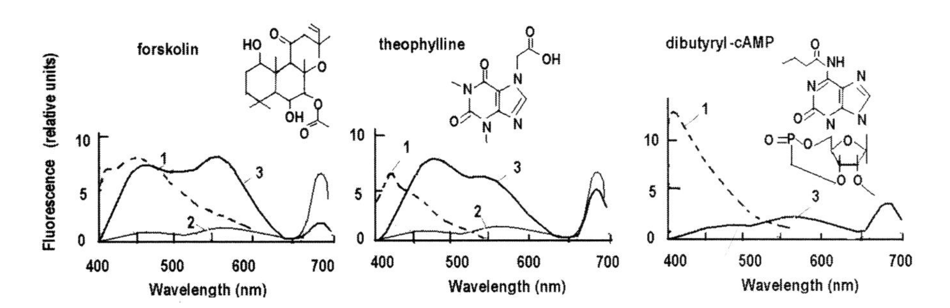

Fig. 7.7 The fluorescence spectra of 10^{-5} M substances, which increase the intracellular amount of cAMP, and of the vegetative microspores from *Equisetum arvense* treated with the compounds. 1. pure compound in 1 cm^2-cuvette; 2. and 3. cells before and after the interactions with the compound studied, relatively.

interactions. The structures are stained by the compounds as histochemical dyes.

Sesquiterpene lactones. Lipophilic sesquiterpene lactones are able to pass through the plasmic membrane inside the cell. As shown in the paper (Roshchina, 2004a), azulene from *Achillea* and *Artemisia* genera and proazulenes such as gaillardine from secretory cells of *Gaillardia pulchella* and artemisinine from plants of *Artemisia* genera could be used as fluorescent dyes on the nucleus, in particular at the staining of pollen and vegetative microspores. These sesquiterpene lactones stain nuclei and chloroplasts in unicellular models: pollen of *Hippeastrum hybridum* and vegetative microspores from *Equisetum arvense* (Fig. 7.8). Sesquiterpene lactones themselves fluoresce in the blue (400-430 nm) spectral region whereas pollen cells (without any treatment) -at 490-510 nm. The staining with the sesquiterpene lactones induced the increased blue fluorescence of the both generative and vegetative microspores. In particular, the light emission is clearly seen for nucleus and chloroplasts of microspores from *Equisetum arvense* or nuclei in pollen of *Hippeastrum hybridum*. The study of the pollen fluorescence in the presence of sesquiterpene lactones showed significant changes – instead of blue, the colour of fluorescence was green (500-530 nm). Azulenes and proazulenes from *Gaillardia*, *Artemisia* and *Achillea* genera may also contribute in the earlier observed luminescence of the allelopathically active plants (See Chapter 5).

The study of the pollen fluorescence in the presence of azulene on other models such as onion *Allium cepa* or zygota of the mouse showed significant changes (Fig. 7.8). In the first object, besides the blue colour (maximum 465-470 nm) of fluorescence, there is maximum 530-535 nm in green-yellow, while the treated second object (untreated has no fluorescence) emitted only in blue (maximum 455-465 nm). In all cases, the nuclei fluorescence is clearly seen. Similar effects were observed for gaillardine. The effects of azulene and proazulenes studied on the fluorescence of individual cellular components such as nucleic acids DNA, RNA, and albumin (as a model of any protein) were varied (Roshchina,

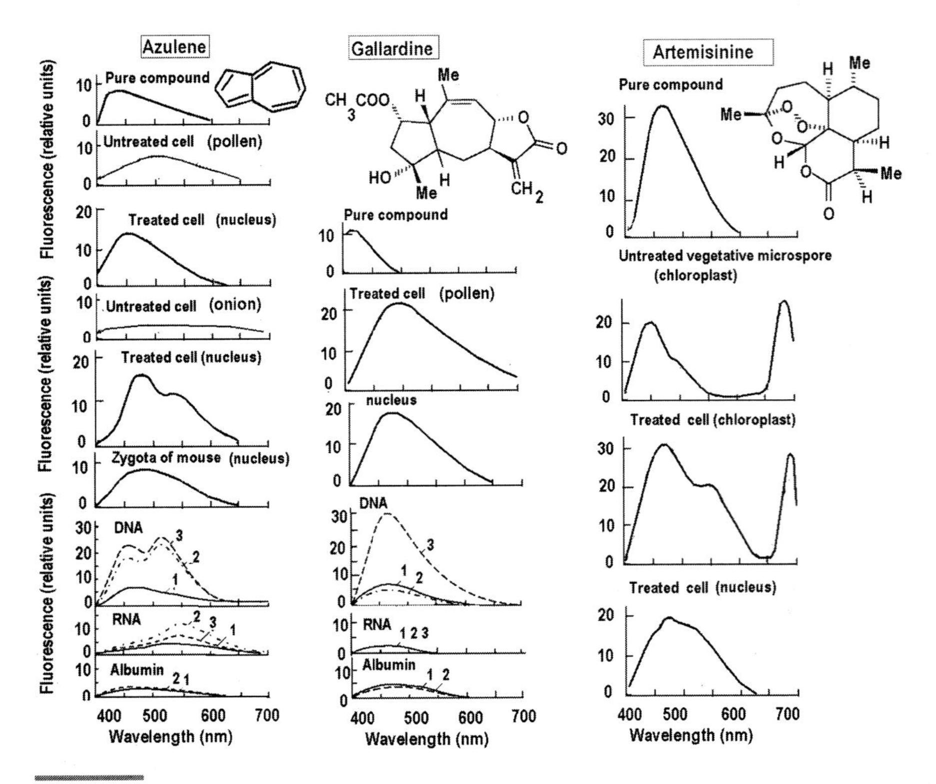

Fig. 7.8 The fluorescent spectra of cells and cellular individual constituents (DNA, RNA and bovine serum albumin) treated with 10^{-5} M sesquiterpene lactones, which may interact with DNA-containing organelles. Sources: Roshchina (2004a; 2005b) and unpublished data. Pollen-pollen of *Hippeastrum hybridum*; onion-cells of bulb from *Allium cepa*; zygota of the mouse treated (There is no emission in untreated samples), vegetative microspore–vegetative microspore of *Equisetum arvense*. DNA, RNA and Albumin - the fluorescence spectra of DNA, RNA and albumin as the nucleus components. DNA, RNA and albumin - 2mg/ml. Mixtures of nucleic acid: albumin = 1:1 v/v. 1. individual compounds DNA or RNA or albumin; 2. 1 + sesquiterpene lactone; 3. DNA or RNA + albumin + sesquiterpene lactone.

2004a). (It should be noted that self-fluorescence of nucleic acids and albumin was observed only for their dried preparations, and samples treated with water had no light emission). Most changes were observed for DNA samples in the preparations treated with azulene, grosshemine and gaillardine, while the effects for RNA and albumin were neglected. There was a change in the fluorescence colour of the nucleic acid from blue to green. Unlike azulene, the blue fluorescence of DNA alone slightly increased in the presence of proazulenes grosshemine and gaillardine, however, this stimulation strengthened in the mixture DNA-albumin. Unlike azulene, grosshemine and gaillardine, proazulene austricine had a weak effect on DNA, and, mainly, stimulated the albumin blue fluorescence. Thus, azulene bounded predominantly with the DNA (both separately and in the mixture with the protein), whereas grosshemine and gallardine – with a protein in a mixture with DNA. Austricine appears to have an effect only on the nuclear protein.

Major mechanisms of their action consist in the interaction with nucleic acid and in the influence on the protein synthesis. The compounds predominantly may bind with DNA directly (azulene) or with protein of nucleus (gaillardine) as shown in our paper (Roshchina, 2004a). Artemisinine also induced the fluorescence of nuclei and chloroplasts in the vegetative microspores of *Equisetum arvense* (Roshchina, 2005b). As seen in Fig. 7.8, enhanced blue fluorescence arose in nuclei and chloroplasts. Sesquiterpene lactones can inhibit the number of enzymes and other essential macromolecules (Schmidt, 1999), mediate cellular death by triggering apoptosis (Dirsch et al., 2001) or inhibit the transcription factor NF-χB (Rüngeler et al., 1999), as summarized in papers (Roshchina, 2004a; 2005b, c). The effects of azulene and proazulenes may be due to the changes in the redox state. Azulene and artemisinine form alkoxy radicals in the reactions catalyzed with (by) Fe^{2+} (Posner et al., 1995; Roshchina, 2003) that may influence the fluorescence. Azulenes and proazulenes are widely used in medicine and cosmetic preparations, but may be applied to cell biology for analysis of the nuclei state.

Alkaloids. The binding of alkaloids such as rutacridone from roots of *Ruta graveolens*, glaucine from the plant *Glaucium flavum* and casuarine from the Australian plant *Casuarina equisetifolia* (fam. Casuarinaceae) or *Eugenia jambolana* (family Myrtaceae) leads to the fluorescence of DNA-containing organelles within a cell (Roshchina, 2002; 2003; 2005a). Figure 7.9 demonstrates the fluorescence spectra of the compounds after

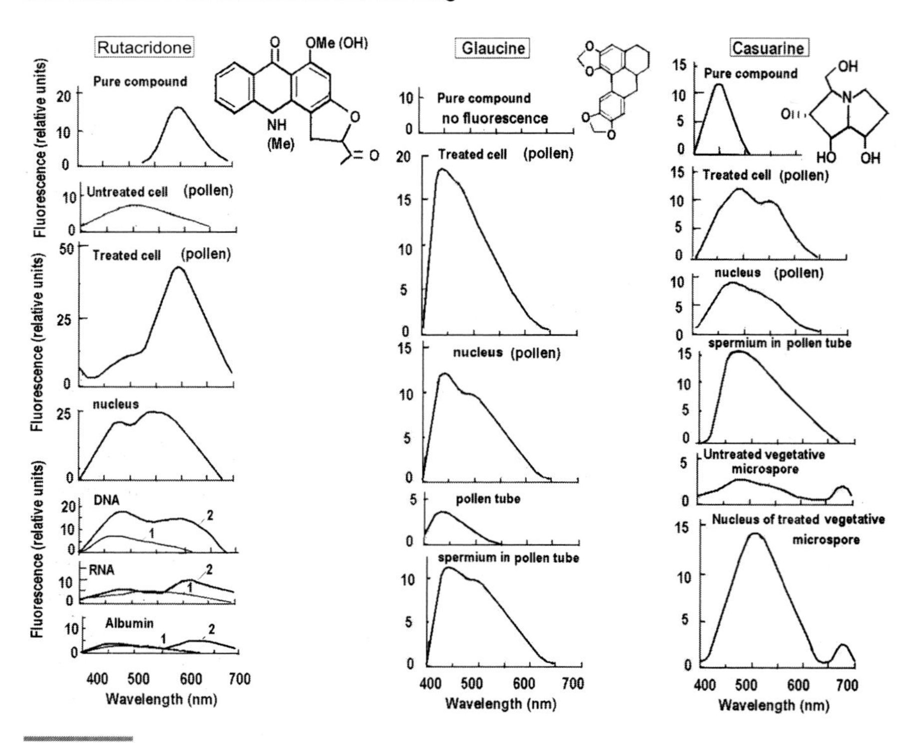

Fig. 7.9 The fluorescence spectra of cells (pollen of *Hippeastrum hybridum* and vegetative microspore of *Equisetum arvense*) and nuclear components (DNA, RNA and albumin) after the treatment with 10^{-5} M alkaloids. Sources: Roshchina (2002; 2005b, c) and unpublished data of the author. Solutions of pure compound were follows: rutacridone in DMSO, others–in water or ethanol. Concentrations of DNA, RNA and bovine serum albumin – 2 mg/ml.

interaction with cellular models – pollen of *Hippeastrum hybridum* and vegetative microspores of *Equisetum arvense*. Orange-fluorescing rutacridone (maximum 590-595 nm) penetrates the cells, when being dissolved in dimethylsulfoxide or chloroform. As a result, the nuclei in all cells and chloroplasts (if the cells are able to photosynthesis) became green fluorescent (with maxima 470 and 530-540 nm) whereas other cellular compartments were stained, emitting in orange (maximum 590-595 nm). Only DNA showed similar green fluorescence at the model interactions with nucleic acids (Roshchina, 2002).

Water-soluble glaucine and ethanol-soluble casuarine induced blue-blue-green fluorescence (maxima 450-465 nm) in cellular models used (Fig.

7.9). The nuclei and chloroplasts were stained and had clear fluorescence differing from surrounded organelles.

The staining of DNA-containing organelles with the alkaloids may be of special interest because the compounds mentioned are known as medicinal drugs: rutacridone – as an antiviral agent (Baumert et al., 1982; Southon and Buckingham, 1989), glaucine as a broncholitic compound (Golovkin et al., 2001; Mashkovskii, 2005) and casuarine as beneficial in treating AIDS patients (Bell et al., 1997; Denmark and Hurd, 2000).

7.1.6 Binding with Contractile Proteins

One of the targets for intracellular binding of fluorescent components of plant secretions may be contractile proteins. Usually weak autofluorescence related to actin filaments may be seen only at wounding of plant cells (Goodbody and Lloyd, 1990). Known molecular probes of natural origin which inhibit cellular motility are cytochalasins from the fungi *Helminthosporium dematioideum* and colchicines from plants belonging to the genera *Colchicium, Merendra, Ramond* and *Gloriosa* and others. The compounds are secretory products and may interact with sensitive living cells. Among them are cytochalasin B, which blocks the actin polymerization, or the antitubulin reagents colchicine and vinblastine which inhibit a polymerization of tubulin. These substances, when they have been released by the cell-donor, can interact with the cell-acceptor as allelochemicals and demonstrate fluorescence (Roshchina, 2004b; 2005a, b, c). The compounds are used as pharmaceuticals and components of cosmetic preparations according to their bactericidal and anti-inflammatory characteristics. Moreover, colchicine may depress germination of seeds, pollen and vegetative microspores (Vaughan and Vaughan, 1988; Roshchina, 2004b; 2005a). Recently, colchicine and cytochalasin B have been shown to penetrate the cell-acceptor and to induce fluorescence of some cellular compartments (Roshchina, 2005a; b). This is clearly seen for colchicine.

It is known that colchicine solutions, if excited by UV-light, fluoresce and the light emission increases at contact with protein tubulin (Arai and Okuyama, 1973; Bhattacharyya and Wolff, 1974; Croteau and Leblanc, 1977). The ability of colchicine to fluoresce in solutions with purified tubulin, mainly in the monomer form, has been described earlier by (Bhattacharyya et al., 1986). Its own emission of pure compound was ~ 300 times lesss. So this tubulin-binding agent, which blocks its polymerization in microtubules, may bind with monomer forms of the contractile protein which are in the environment of the microtubules. This emission shows the sites of the tubulin accumulation. According to Bhattacharyya et al., (1986), the promotion of the drug fluorescence at colchicine-tubulin interaction appears to be related to the B ring of colchicine, a determinant recognized by the binding site on tubulin. The N-acyl group of the B-ring may participate.

As seen in Fig. 7.10, the ethanol (10^{-5} M) solutions of colchicine fluoresce with maximum 460 nm. The dilution of the stock solution with water medium up to concentration 10^{-5} M shifts the maximum to 430-440 nm. But the diluted solutions (10^{-5}–10^{-7} M) had small emission on the object glasses, if any, as will be seen below. These solutions were added to living cells (cellular models) of vegetative microspores from horsetail *Equisetum arvense* and generative microspores (pollen) from knight's star *Hippeastrum hybridum* (Roshchina, 2004a; 2005b, c). Pollen is suitable for studies as the paper of Rowley (1967) described the first microtubules in pollen grains as object for microscopic observation. The pollen tube is considered a cellular model for the study of the non-muscle motility because it is possibile to see the movement of organelles, in particular nuclei (Chen et al., 2001; Vidali et al., 2001). The fluorescence spectra of the microspores excited with ultra-violet 360-380 nm and observed under a luminescence microscope, showed maxima in the blue-green spectral region (Fig. 7.10). In the spectra of horsetail vegetative microspores treated with colchicine there are maxima 500 nm and 680 nm (chlorophyll fluorescence), while in pollen and pollen tube – maximum 520 nm. This fluorescence was 2-5 times higher, than a weak autofluorescence of untreated cells and the solutions of the anticontractile agents themselves, which were applied to the object glasses. Perhaps, the fluorescent sites are microtubules, where soluble monomers of tubulin are accumulated. In particular, nuclei were shown to emit intensively due to fluorescing complex "colchicine-tubulin" of the microtubules, which are located around these contractile structures (Roshchina, 2005a).

Microspores stained with colchicines were also analyzed by Laser Scanning Confocal Microscope as seen in colour Figs. 31 and 32 in Appendix 2 (Roshchina et al., 2006; 2007a). On LSCM-images, there is green fluorescence of the vegetative microspores. But especially bright lightening was seen in elaters of vegetative microspores from *Equisetum arvense* (colour Fig. 31). Elaters are motile, contractile derivatives of the spore exine (the cover layer), which serve for the spore anchoring to

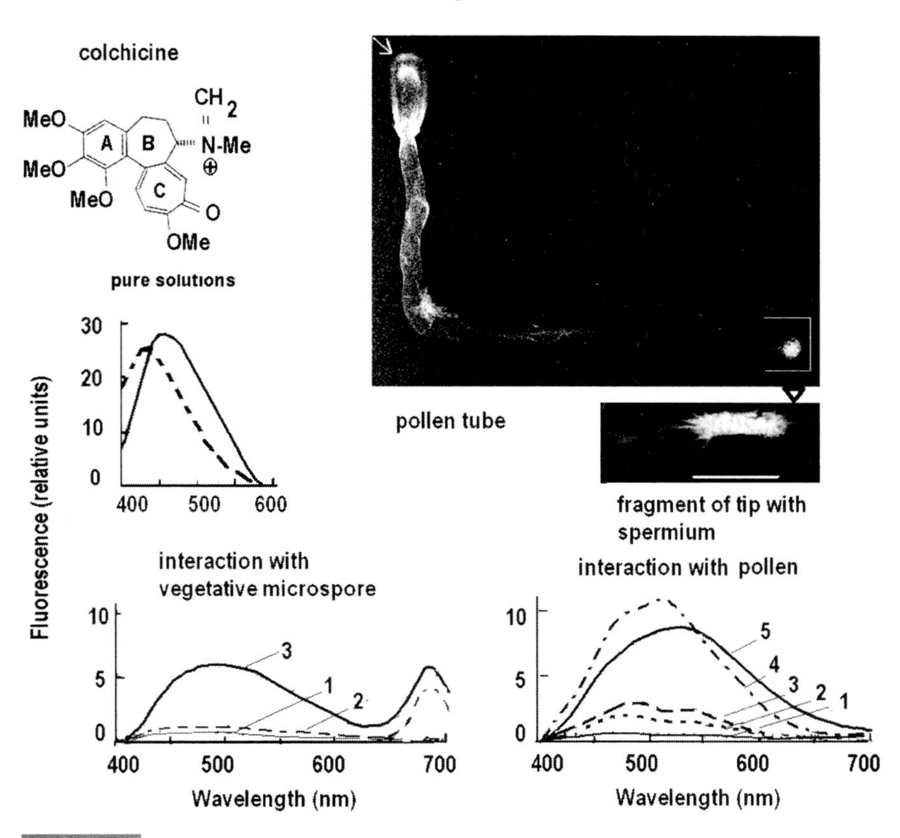

Fig. 7.10 The fluorescent spectra of pure 10^{-5} M colchicine (unbroken line - ethanol and broken line - ethanol–water 1:1000 solutions) and after ther interaction of their 10^{-6}M solutions with vegetative microspores of *Equisetum arvense* or pollen *Hippeastrum hybridum*. Sources: Roshchina (2005a, c) and unpublished data. The excitement was with wavelength 360-380 nm. 1. and 2. solution of anticontractile agent on the object glass and the untreated cell studied, relatively for both cells types. 3. the emission for *E. arvense* after the treatment with colchicine; 3, 4, 5- the emission of cell fluorescing strands, nucleus of pollen grain and spermium in pollen tube, relatively for *H. hybridum* after the treatment with colchicines. On photographs bright emission is observed in pollen grain (arrow shows dividing nucleus and spindle of contractile elements in anaphase) and in the some parts (tubulin-related bundles) of the pollen tube. Fluorescening spermium is seen at the tip of the pollen tube (more clear fluorescent strands around spermium in the magnified fragment). Bar = 20 μm

substrate, in particular to a soil. Thus, microtubules are not only near the cellular surface of the spore, but in the elaters too. Bright green emission is observed in the pollen grain and in some parts of the pollen tube (colour Fig. 32). There are bright green-lightening strands in interceptions of the pollen tube.

Thus, colchicines may be recommended as a fluorescent marker of tubulin for various investigations of cells.

7.2 POSSIBLE MECHANISMS OF THE EMISSION OF THE SECRETORY PRODUCTS AT THE INTERACTION WITH LIVING CELLS

Depending on whether a secretory product is hydrophilic or hydrophobic, it is possible for different mechanisms of the selective emission of the cellular compartments to be stained. The targets in the cellular compartments for the secreted products which induced the fluorescence may be proteins, lipids, nucleic acids, high molecular and low molecular secondary metabolites.

7.2.1 Proteins

The first barrier for the secretory product acting on the cell-acceptor may be a cell wall (for bacteria and plants) and the plasmalemma (for every cell), which include proteins – from water-soluble to those bonded in the structures.

Hydrophilic, slowly penetrated compounds, such as neurotransmitters and antineurotransmitter agents (alkaloids) act on the protein state of the receptors or surface sensors (enzymes). Most reactive are double bonds of the substances affected, in this case, the attachment occurs to amino acid groups of protein. But the chemical composition of the compound acted is also significant. Mechanism of the light emission may be related to the formation of Schiff bases in reactions between amino acid or amino acid residues of proteins and aldehyde groups of the secretory component or their derivatives, in particular malondialdehyde formed in lipid peroxidation.

The structure of protein plays a certain role in the arising of the emission. This process may be similar with those which occurred in fluorescent proteins of sea animals, which have a glass-like (goblet-like) structure that is responsible for the visible emission (Tsien, 1998; Labas et al., 2001).

Enzymes and other proteins of cellular wall also react via amino groups. Redox reactions between oxygen of the proteins and the secretory products are also real, when the secretory product comes in contact with the cellular compartment. There is possible oxidation of the protein or oxidation/hydrogenation (protonation) of the fluorescent secretory component itself as well as interactions with their transformed derivatives.

The mechanism of the emission of the anticontractile agents such as cytochalasins and colchicines appears to connect with the cellular contractile protein binding, rather than with a transformation of the anticontractile agents themselves. The visible maxima of the treated cells shifted to longer wavelengths in comparison with the fluorescence of the solutions of the compounds. Colchicine and its derivatives are known as toxins-mitotic agents which induced polyploidy, but in many cases they act as herbicides and can retard a germination of seeds, vegetative microspores and pollens (Vaughan and Vaughan, 1988; Roshchina, 2005a, b). Up to now, the location of contractile proteins in cells was studied with fluorescent antibodies against actin, myosin and tubulin, or with special fluorescent dyes, in particular Texas-red phalloidin (Haugland, 2000; Chen et al., 2001; Vidali et al., 2001). This problem is real for many types of living cells, especially in a field of non-muscle motility and chemosignaling (Roshchina, 2004a, 2005a). In practice, some anticontractile agents found among secretory products may be used as fluorescent probes in cellular biology, in particular colchicine for the study of cytoskeleton and cell division.

7.2.2 Lipids

The mechanism of the cell stained with the fluorescent secretory product may be also connected with the interactions of the lipid part of the cell wall, plasmalemma and cellular organelles. The chain processes of lipid peroxidation induced by a secretory product may participate in the changes of the fluorescence. It occurs via the formation of Schiff bases between amino acid residues and malondialdehyde as a final metabolite of lipid peroxidation.

Sesquiterpene lactones may react as lipophilic compounds which react with the cell wall structures, part of the molecules also enter cells and interact with organelles. Compounds containing 7-chain rings have the ability to fluorescence and at contact with cellulose fluoresce in the orange-red spectral region as azulenes and some proazulenes (Roshchina et al., 1995; Roshchina, 2003; 2004a, b, 2005a, b). They also bind with DNA-containing organelles that can be seen in their blue or blue-green fluorescence within a cell (Roshchina, 2004a; 2005a, b, c). Lipophilic alkaloids, in particular rutacridone (Roshchina, 2002), demonstrated a similar ability to stain these structures and induce green fluorescence.

7.2.3 Nucleic Acids

Sesquiterpene lactones and some alkaloids had the ability to make structures containing nucleic acids visible due to the arising fluorescence as seen from Section 7.2.2. The first group of the compounds (azulene and some proazulenes) appears to be able to bind either with DNA directly or with proteins bound with the nucleic acid (Roshchina, 2004a). A similar tendency is also there for alkaloids. For example, alkaloid rutacridone gave orange fluorescence in cytoplasm and green emission for nuclei and chloroplasts which may include direct interaction with DNA or, in a lesser degree, with RNA (Roshchina, 2002; 2004a, 2005b).

7.2.4 High Molecular and Low Molecular Secondary Metabolites

Polysaccharides and phenol cellular polymers such as lignin and suberin may be targets for some secretory products which either enhance the fluorescence intensity or alter the colour of the emission, in particular at pH changes, redox reactions, etc (Wolfbeis, 1985; Rost, 1995; 2000). The mechanism of the processes is only proposed, but needs the experimental confirmation in model systems. As for the interaction of low-molecular compounds, we have very little information (Roshchina, 1999a,b; 2003) of reactions of secretory products with other secondary metabolites such as phenols (esculetin, quercetin, kaempferol and rutin) and terpenoids (menthol, geraniol, linalool, etc). Perhaps, it is also important for the fluorescence phenomenon.

Conclusion

The fluorescence of released components of secretions may be seen in cell-acceptor at intercellular interactions that is perspective for the microscopic analysis. When the pure fluorescent substance is added to the cell-acceptor, changes in their fluorescence are observed. The changes allows one to see the fluorescent cellular compartment which make the secretory product suitable for histochemical staining. The compounds may be used in the studies of the mechanisms of the intercellular and intracellular interactions as well as in any laboratory practice as fluorescent dyes and probes.

Conclusion

Autofluorescence excited by ultraviolet, violet or blue light is peculiar to many plant secretions that show the phenomenon in intact secretory cells. Any modification of a luminescence microscope – from a simple technique to microspectrofluorimetry and confocal microscopy allows one to see it. The main fluorescence of secretory cells is related to secondary metabolites concentrated in the extracellular space, vacuoles, secretory vesicles as well as some amount of the fluorescent secretions released out. Non-invasive observation of the secretory cells, based on their autofluorescence, allows the analysis of the vital state of the structures.

Observation of autofluorescence of plant secretory structures opens new possibilities in fluorescent analysis. On the one hand, a source of new discoveries may be in every secretory cell, when comparing their autofluorescence and the chemical composition of the secretions. On the other hand, although our fundamental knowledge of autofluorescence is thought to be insufficient, this phenomenon is of interest in the global problem of living cell luminescence. We know little about mechanisms of the emission arising within a cell as well as its quenching. The emission is often quenched or stimulated by other components of the secretions Therefore, specialists in biophysics, biochemistry and physiology have the widest field for investigations as pioneers.

Up to now the possibility of plant vegetative tissue to fluoresce under ultra – violet or violet irradiation of a luminescence microscope is supposed to be associated with chlorophyll. Meanwhile secretory cells contain fluorescing substances such as phenols, flavines, alkaloids and others that make possible fluorimetric analysis of the intact cell *in vivo*. Various groups of substances have significant differences in their spectral features.

Observed fluorescence of intact cells is a sum of emissions of several different groups of compounds both excreted out or accumulated within the cell and linked on the cellular surface. Biochemistry of components containing secreting structures has not been studied enough yet. However, compounds occurring in the secretions possess characteristic maxima in the absorbance and fluorescence spectra which is suitable for the identification of the compounds in mixtures.

The fluorescence spectra of intact secretory cells, containing various substances, differ in the maxima position and the intensity. This allows a preliminary discrimination of dominating components in the secretory cell that is tested. As shown in Chapters 1 and 2, microspectrofluorimetry is especially useful for such studies. Recently, the *Leica* company advertised the possibility of registering the fluorescence spectra of the microscopic objects by Laser Scanning Confocal Microscopy (Horizons of Modern Microscopy, 2007) which also may have potential for plant secretory cells.

The visible fluorescence of the secreting cell depends on chemotaxonomy and metabolism of secreting products, stored within the secretory structures as well as evacuated on the surface of the secreting cell (see Chapters 1 and 2). Scheme shows the fluorescence position in the visible part of the spectrum for the main groups of secondary metabolites occurring in secretory structures. Most of them fluoresce in blue, and only a few – in green, yellow and orange-red. Their emission differs from chlorophyll, but is similar with reduced pyridine nucleotides and lipofuscins which also lighten in the blue and blue-green spectral regions, relatively.

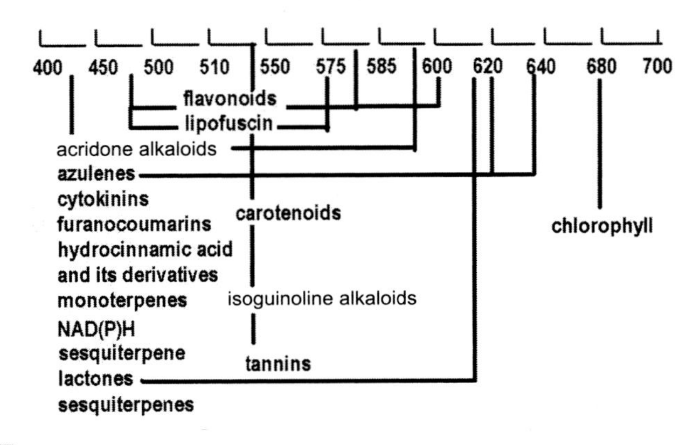

Scheme Spectral regions, where secretory products fluoresce

Among the factors influencing light emission, is the physiological state of the fluorescing cell, organ, and whole organism. Autofluorescence of the intact secretory cells changes during the cell development and under various damaging factors as seen from Chapters 4 and 6. The filling of the cells with a secretion and its removal are easily observed in these conditions which can serve as an indicator of the cell state *in vivo*. This reflects the alterations in a composition or/and redox state of the secretory products accumulated. The phenomenon could be an indicator of the structures' formation, based on the appearance of their secretions at the earliest stages. The light emission of secreting structures may be useful for the analysis of pollen maturing and ageing or self-incompatibility in pollen-pistil relations. Moreover, autofluorescence could be a sensitive parameter for the analysis of cell-cell interactions as have been described in Chapter 5.

A special focus may be on autofluorescence as an approach to the study of intercellular contacts and signalling (see Chapter 5). One cell releases fluorescent secretions and another cell receives this chemosignal. Depending on the composition of the secretion, changes in the fluorescence of the cell-acceptor take place, such as a shift in the colour and maxima of the emission or/and alterations in the intensity of lightening. Moreover, if the fluorescent compound penetrates into a cell it may induce some cellular compartments to fluoresce too under actinic light. Thus, the investigator is able to analyze a mechanism of the intercellular interactions. The base of the method is a feature of most biologically active compounds present in secreting cells, especially with double bonds, to fluoresce under the irradiation of ultra-violet light (UV-light). Some secretory products, such as sesquiterpene lactones and alkaloids may selectively stain the cellular compartments that make them potentials for natural fluorescent dyes as shown in Chapter 7.

Some of the problems in the autofluorescence study of secretory cells are the effects of various factors – temperature, the light intensity, pH of medium, composition of medium, the ability of the external chemical to oxidize or reduce the fluorescent substance and others as shown in Chapter 3. Some data concerning pH influence is available only for narrow groups of secondary metabolites – flavonoids. The information about effects of redox factors is also not sufficient.

An unresolved problem is also the analysis of stained multicomponent secretions released or stored in secretory structures *in vivo*. One of the approaches may be receiving values of quantum yields of the fluorescence for individual components of the secretions, mainly secondary metabolites. Another possibility is the analysis of the fluorescence spectra of secretions and secretory cells in Gaussian curves or by modelling of natural secretions in composition of various individual compounds. All these possibilities requires knowledge of the nature of secretory products that also may be received in the future.

Besides the advantages in the above-mentioned possibilities in the application of autofluorescence, there are some difficulties depending on the methods of the registration of the phenomenon. Except photosynthesis and the technique used for its analysis, there are no procedures of quick resolution of the fluorescence applied to secretory cells yet. Kinetics of the emission changes are not so far measurable. Quantum yields of many plant secondary products are not known, and we cannot use these characteristics. Moreover, we have only a few possibilities to obtain the fluorescence spectra by microspectrofluorimetry or the confocal microscopy technique, but microspectrophotometers are not used widely for the absorbance analysis of plant secretory structures. This makes it difficult to estimate the quantum yields of the secretions within intact cells.

Today, we have better perspectives on the application of the autofluorescence phenomenon of secretory cells in practice. They are as follows:

1. Spectral characteristics of individual substances in secretory cells could be measured *in vivo* which is significant for their identification. The observation of fluorescence allows one to identify secretory cells among non-secretory ones in all plant organs and to observe the accumulation of secretions on the surfaces of herbaceous and woody plants. Non-invasive fluorescent analysis of secreting cells and tissues may become one of the normal methods in botanical studies, education, criminology and pharmacology because there is no necessity to use artificial dyes for the examination of the plant material;
2. The phenomenon of secretory cells' autofluorescence may be used for the analysis of changes in the surrounded medium because the cells are sensitive to environmental shifts. Autofluorescence may serve as a biosensor and bioindicator reaction. It may be applied in pharmacology for the analysis of a plant material without expensive biochemical procedures, based on the scales of the individual secretory products accumulated in similar cells.

3. Autofluorescence may be used in the analysis of secretory structures for photographs of their internal and external images without histochemical staining and fixation. The images of cellular components appear as botanical visual aids for education and handbooks. Examples of similar galleries for secretory cells are given in colour Figs. 2, 6 and 30 in Appendix 2. Moreover, the design of such images may have artistic value in polygraphy and for the promotion of biological knowledge.
4. The ability to fluorescence under UV- or violet light of luminescence microscopy is also useful in the search of new natural fluorescent dyes among fluorescent components of plant secretions.

Notwithstanding the little information about the nature of secretory products and their location within secretory structures, we believe, that the problem will be studied by specialists of fundamental sciences. In addition, this new line in the study of the autofluorescence of plant secretory cells already has many applications in practice. Fluorescent analysis of plant secretory cells has the potential for cellular diagnostics due to a possibility to receive express- information in intact tissues and single cells without procedures of tissues homogenization and long biochemical manipulations (see Chapter 6). One of the non-invasive applications could be the luminescence observation of plant materials in order to know if it is ready or not for pharmacy as the main valuable drugs are concentrated in secretory structures. The estimation of cell viability is also especially required for normal pollination and fertilization in genetic analysis for selection in agriculture and biotechnology. Moreover, there are some possibilities to use the fluorescence changes in analysis of cell damage which is necessary for ecological monitoring. Earlier diagnostics of the disturbances induced by the oxidative stress, ageing or pest invasion may be based on the fluorescence of secretory products.

Among natural dyes from plants only a few are used as fluorescent probes (Haugland, 2000). Their design coupled with the vast array of commercially available primary and secondary antibodies have provided the biologist with a powerful arsenal through which to probe the minute structural details of living organisms with this technique.

Fluorescent secondary metabolites found in secretory cells may also be applied to cell biology for histochemical staining. Apart from this, the compounds may be used in the study of the mechanisms of intercellular and intracellular interactions as well as in any laboratory practices as fluorescent dyes and probes.

Thus, the magic ray of actinic light in a luminescence microscope may open a new world – a fluorescing world of plant secreting cells and their secretions. The time is ripe for a modern Leeuwenhoeks and Hookes.

APPENDIX

1 Glossary of Biological Terms

Emergencies are surface plant projections which arise from sub-epidermal plant cells. They are often filled with secretions and are functionally similar to trichomes, that have a more complex structure.

Glands – specialized multicellular structures on the surface of leaves, stems and flowers (filled, mainly, with terpenoid- or phenol- containing secretory products).

Hairs (trichomes) are unicellular or multicellular (mainly glandular) secretory structures that contain various secretions.

Hydathodes represent modified cells of stomata (cells, regulating aperture for gases) on overgrown parts of the plant. They release water and water-soluble substances out of the plant.

Idioblasts are single specialized cells which are found among non-secretory cells of vegetative and reproductive organs and differ substantially from other cells of the same tissue in their form, structure and content. In many cases idioblasts lack organelles, and the secretion spreads within the interior of the cells which means a secretory function is present.

Laticifers are living cells with a latex (included alkaloids and terpenoids) secreting in vacuoles. They are found in many plants, especially in the family Euphorbiaceae.

Microspores – secreting unicellular structures for breeding. Vegetative microspores (non-sexual cells) are known as spore-breeding plants (horsetails, ferns, mosses) whereas generative male microspores, which are called pollen, pollen grains or male gametophyte, are peculiar to plants, belonging to Gymnosperms (mainly, the conifer species) and Angiosperms (species with flowers).

Nectaries – secretory structures filled with nectar and located in the flower (floral nectaries) or on the leaf and stem (extrafloral nectaries).

Pollen means a male gametophyte (see microspores). When the cell is added (artificially or, as in nature, by wind or insect-pollinators) to the surface of the pistil stigma in a flower or on the surface of female Gymnosperm microstrobils, it can germinate: forming an amoebe-like emergence called a pollen tube (Stanley and Linskens, 1974). The division of the pollen nucleus leads to the appearance of spermia which move along the growing pollen tube to the egg cell. So fertilization occurs.

Resin ducts and reservoirs are secretory structures that contain resin (complex terpenoid composition). The structures are especially abundant in conifer plants.

Trichomes (also see hairs) are unicellular or multicellular secretory structures which originate from epidermal cells on plant surfaces.

Vegetative microspores – non-sexual cells serve for vegetative breeding of spore-breeding plants (see microspores). During their development the unicellular organism converts to multicellular thallus, which later forms sexual organs.

Viable or non-viable microspores – cells, which are able to germinate in nature or in the artificial nutrient medium.

Main examples of secretory structures

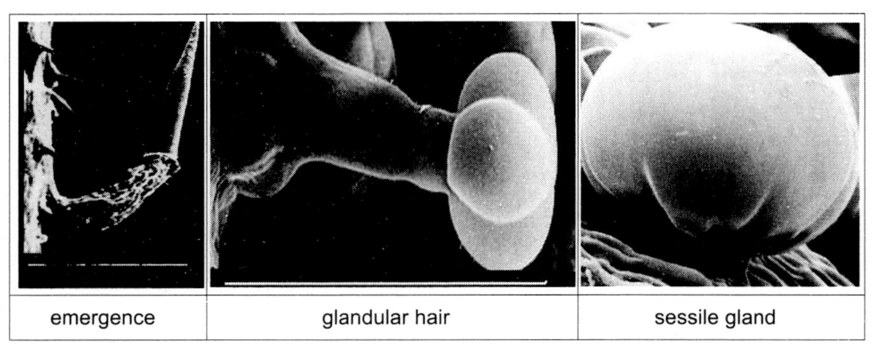

emergence | glandular hair | sessile gland

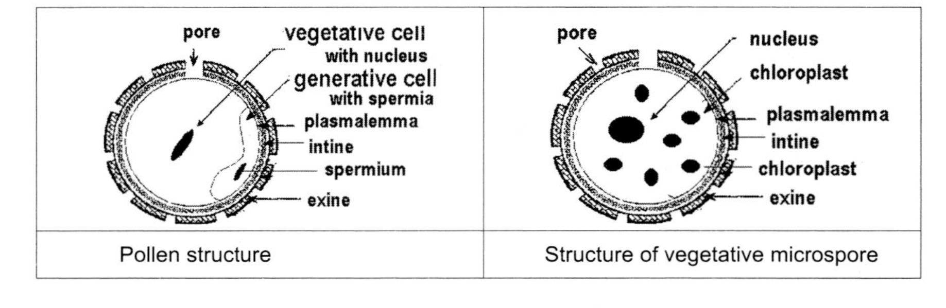

Pollen structure | Structure of vegetative microspore

APPENDIX

2 Colour Photographs of Secreting Cells

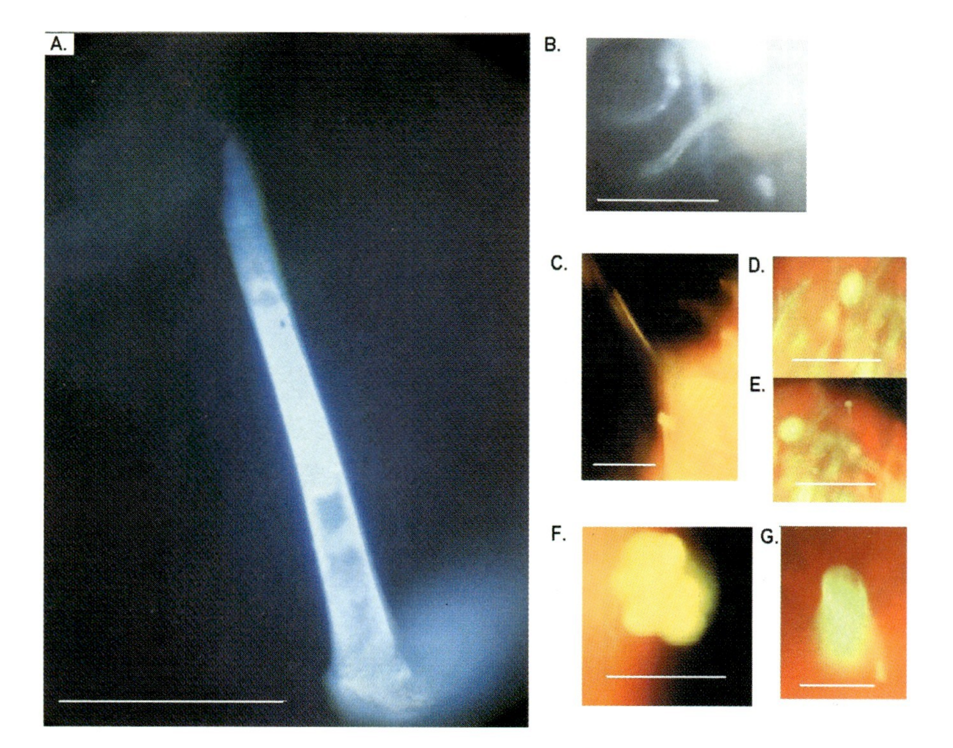

Fig. 1 The fluorescence colour images of plant secretory structures under a luminescence microscope. A - Stinging hair (emergence) on stem of *Urtica dioica*; B - Secretory trichome on leaf of *Achillea millefolium*; C-E - Secretory hairs on leaf of *Lycopersicon esculentum*; F - Gland on leaf of *Mentha piperita*; G - glandular hair on leaf of *Artemisia absinthinum*. Bar = 100 μm.

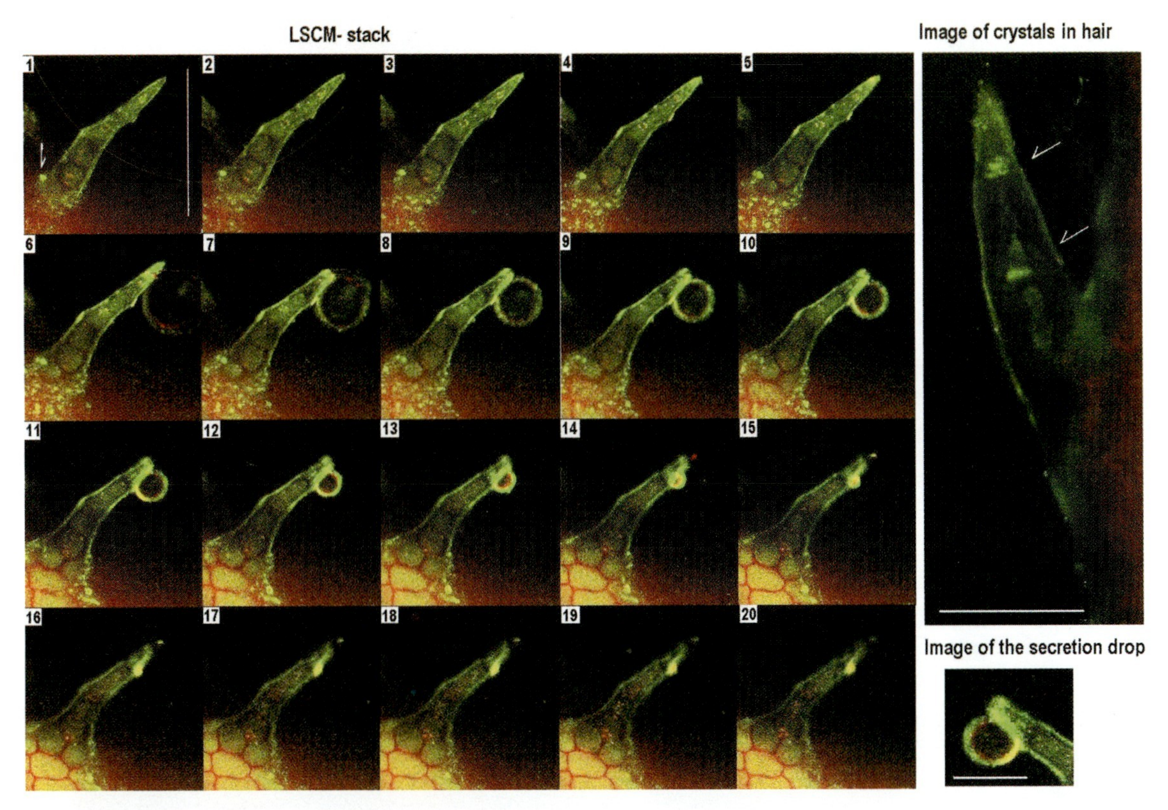

Fig. 2 The LCSM images of the secretory leaf hair of *Solidago virgaurea* L. The stack of optical slices cut through 1 μm (excitation by laser beam in track 1-400 nm and in track 2 – 488 nm, emission in channel 1 400-465 nm and in channel 2 560-700 nm). Bars on section 1 = 150 μm, for separate hair = 10 μm, for drop of secretion = 50 μm.

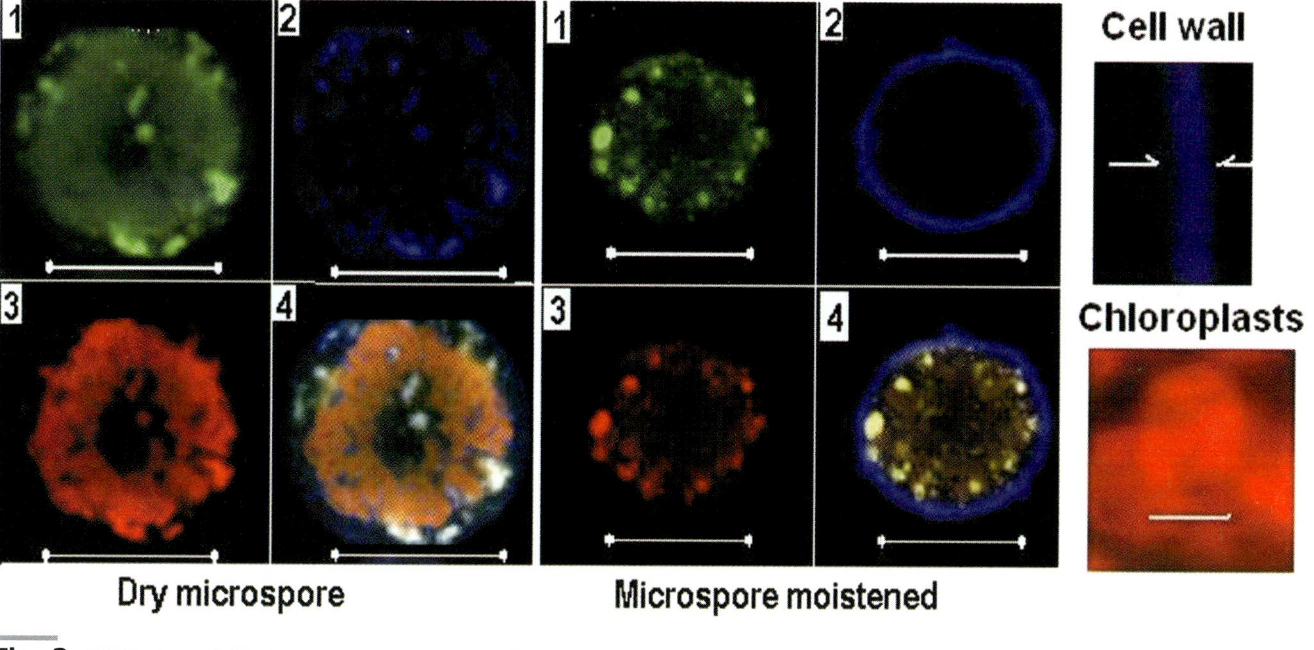

Fig. 3 LSCM view of dry and moistened (15 min) fluorescent vegetative microspore of *Equisetum arvense* under three laser excitation. 1 – channel 488 nm; 2 – channel 533 nm; 3 – channel 633 nm; 4 – summed image with mixed (in a superposition pseudocolors. 1 bar = 20 μm.

A. Cell wall of root

B. Cell wall of stomata

C. Isolated vacuole

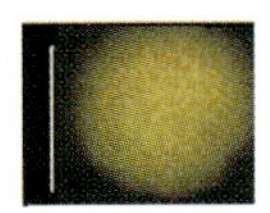

D. Chloroplasts in guard cells of stomata

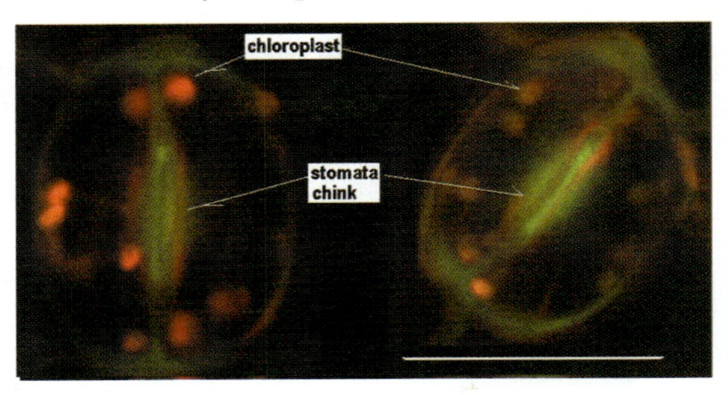

Fig. 4 The fluorescence images of various cellular organelles from plants photographed under a luminescence microscope (A-C) and LSCM view (D). A. root of *Pisum sativum*, bar = 20 μm; B. leaf of *Solidago virgaurea*, bar = 50 μm; C. vacuole from the bulb scale of *Allium cepa*, bar = 50 μm; D. leaf of *Solidago canadensis*, bar = 50 μm.

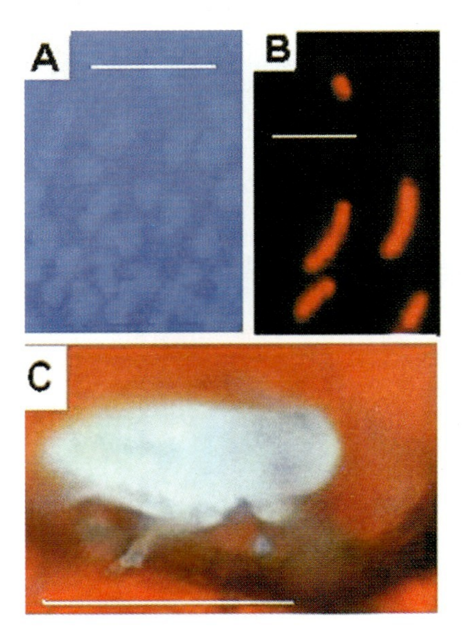

Fig. 5 The fluorescence images of bacteria *Pseudomonas aeruginosa* (A), *Synechococcus* sp. (B) and tick on the leaf of stinging nettle *Urtica dioica* (C) photographed under a luminescence UV light microscope. Bars for A, B, and C = 10, 10 and 200 μm, relatively. *Pseudomonas aeruginosa* breaks the fluorogenic compound to release the fluorogen which shows blue fluorescence whereas *Synechococcus* sp. demonstrates red fluorescence due to the presence of bacteriochlorophyll. Tick, which is on red-fluorescing leaf, fluoresces in blue.

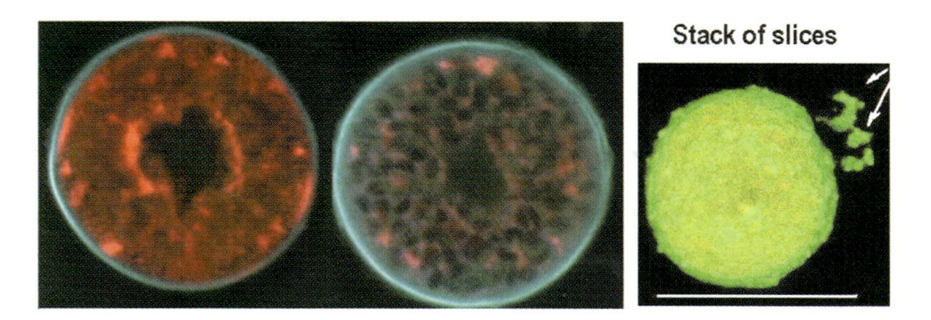

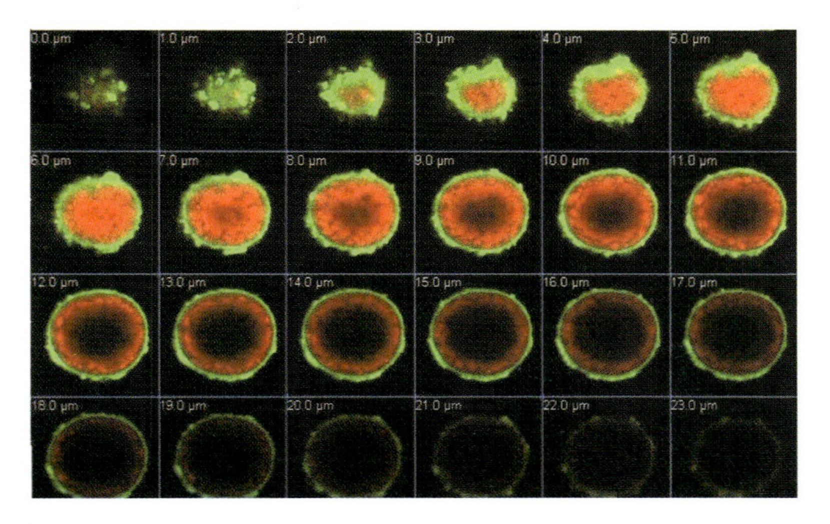

Fig. 6 LSCM views as optical slices of fluorescent vegetative microspore of *Equisetum arvense* under three laser beam excitation. Upper left – two views of slices (1 channel 458 nm; 2 channel 533 nm; 3 channel 633 nm). Upper right - stack of slices (1 channel 488 nm; 2 channel 533 nm; 3 channel 633 nm). 1 bar = 20 μm. On the stack of optical slices elaters are shown with arrows. Lower view demonstrated gallery of optical slices (through 1 μm). Excitement with laser beams 488 nm (emission > 520 nm, green pseudocolor), 543 and 633 nm (fluorescence is observed at 650-750 nm, red pseudocolor).

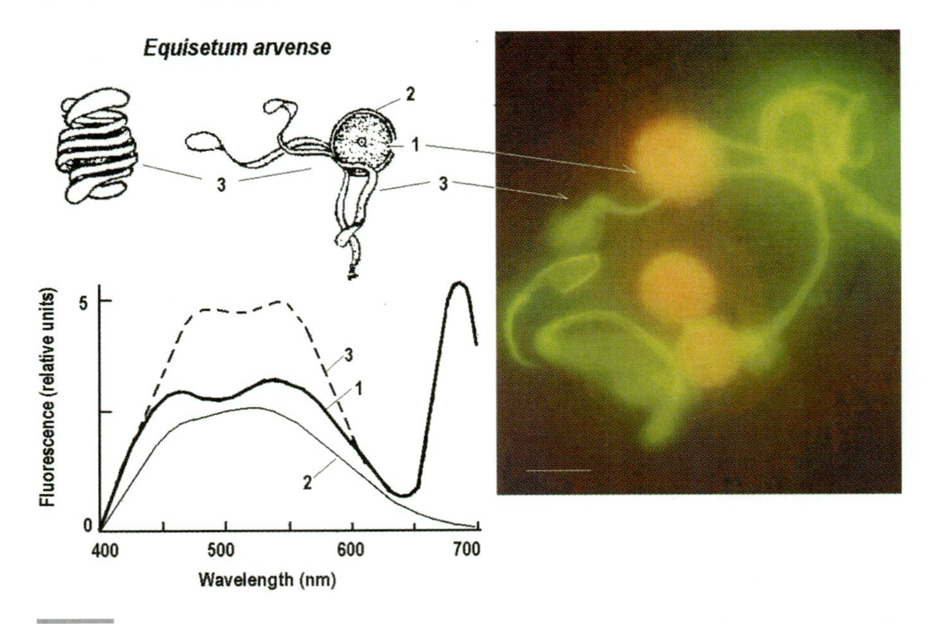

Fig. 7 The fluorescence spectra (left) and fluorescence image (right) under luminescent microscope of microspore from spore-bearing plant horsetail *Equisetum arvense.* 1. Middle of the spore; 2- cover (exine); 3- elaters. Bar = 20 μm.

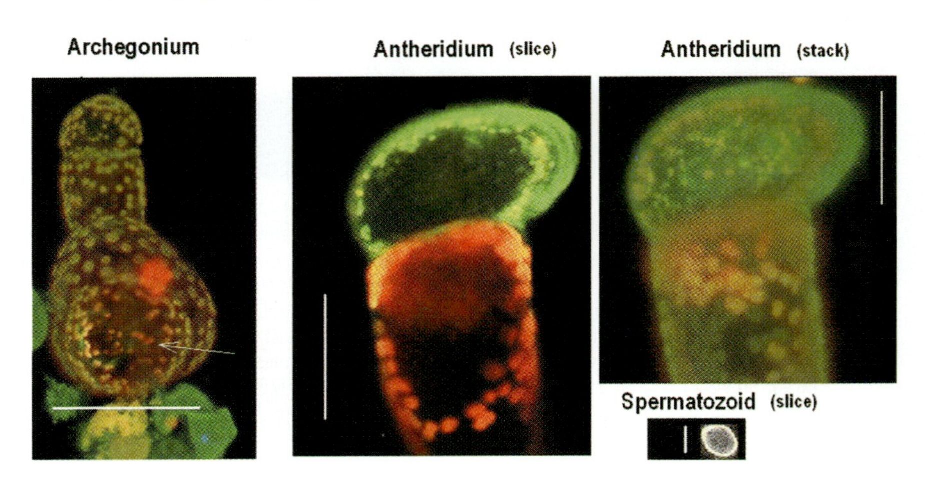

Fig. 8 The LSCM image of sexual organs on the gametophytes of *Equisetum arvense.* Bars (left to right) = 100, 150, 100 and 10 (spermatozoid) mm, relatively for every image. Laser excitation 488 nm. Arrow shows the egg cell location in archegonium.

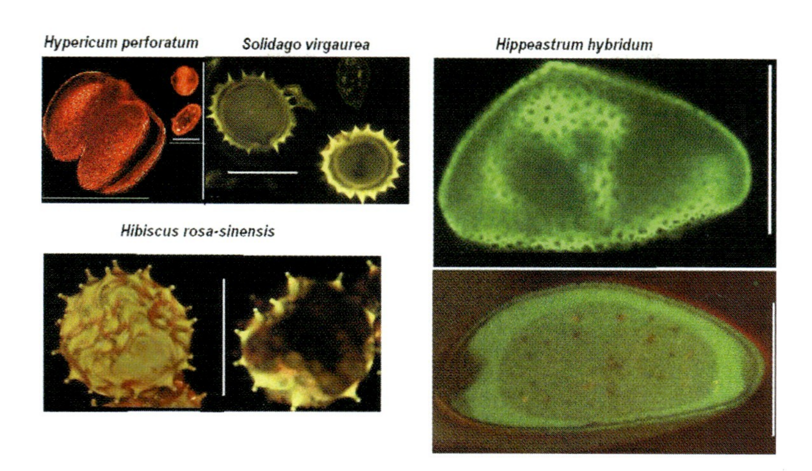

Fig. 9 The LSCM –images of the anther and pollens. Laser excitation 488 nm. Source: Roshchina *et al.,* 2007b; Yashina *et al.*, 2007). Anther (left, bar = 500μm) and pollen slices (right, bar = 20μm) from *Hypericum perforatum.* 1 bar = 100 μm for *Solidago virgaurea* and *Hibiscus rosa-sinensis,* and =50 μm for *Hippeastrum hybridum.*

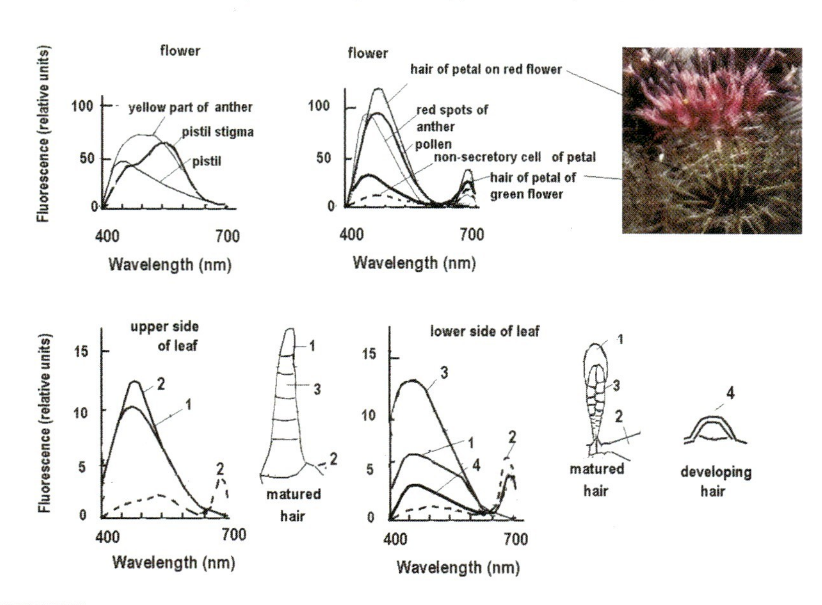

Fig. 10 The fluorescence spectra of flower and leaf secretory cells of Angiosperm plant *Arctium tomentosum.* Most significant fluorescence is for flower surface. Unbroken lines – secretory cells, broken line – non-secretory cells. 1,2,3,4 in lower spectra show the analysed parts of developing and matured secretory hairs.

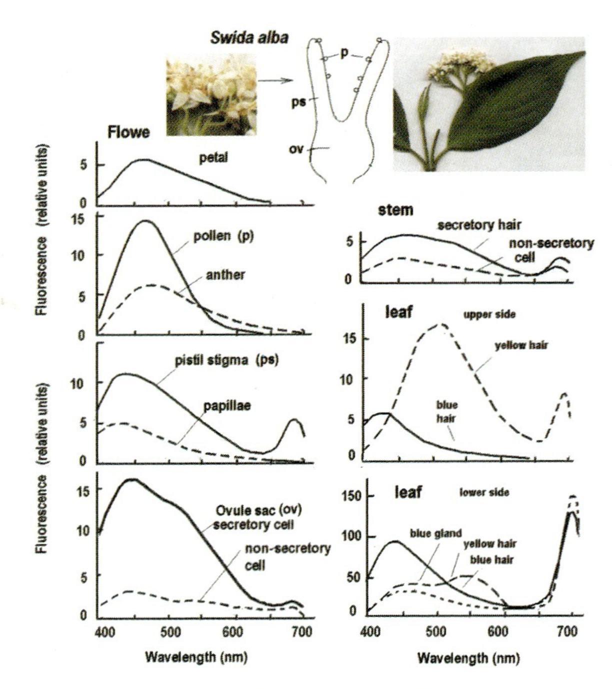

Fig. 11 The common view of the blossom and the fluorescence spectra of secretory cells of *Swida alba.* ps-pistil stigma, p-pollen, ov-ovule sac.

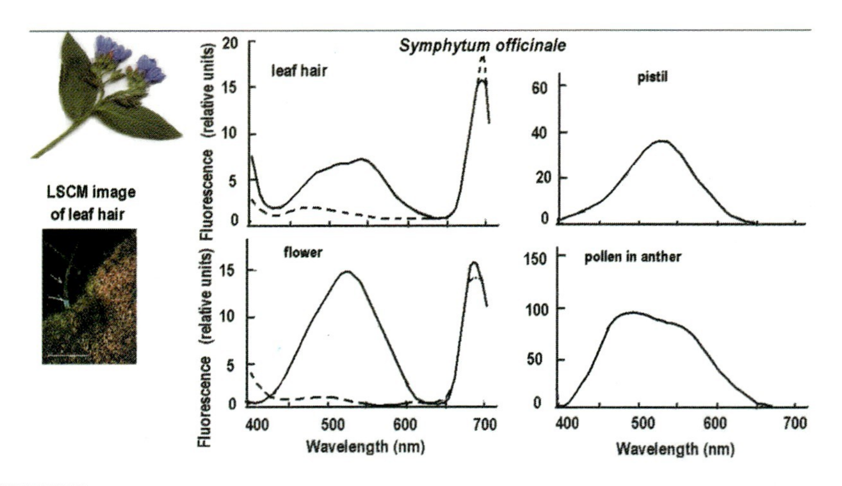

Fig. 12 The fluorescence spectra of secretory cells from various organs of *Symphytum officinale,* which contain alkaloids. Arrows on LSCM image show the fluorescing part of secretory hair (Bottom of the hair fluoresces in blue). Bar = 150 μm.

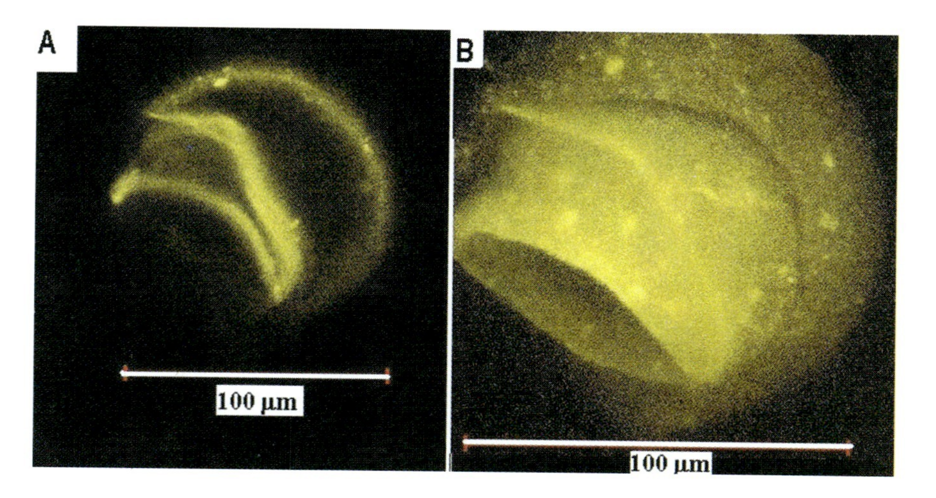

Fig. 13 The fluorescence image of *Chenopodium album* salt gland seen in laser-scanning confocal microscope. The excitation wavelength 488 nm. A- image of the gland slice (big fluorescent crystals are seen); B- stack of the slices (through 1 µm) of the gland. Small crystals are observed.

Urtica dioica

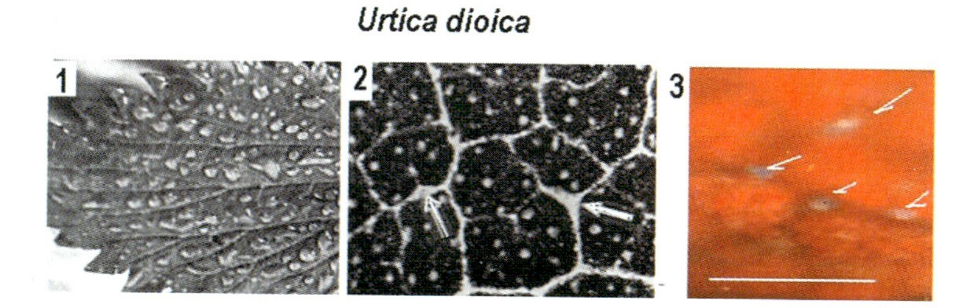

Fig. 14 The common views (1- drops on the surface, seen under transparent white light of a microscope; 2) and the fluorescence image (3) of the hydathodes (blue-fluorescing spots shown with arrows) and surrounded cells on the leaf of *Urtica dioica*. Bar = 100 µm.

Solidago virgaurea

Fig. 15 The fluorescence image of the hydathode as stomata on the leaf of *Solidago virgaurea*. Bar = 100µm. The guttation water has no measurable emission.

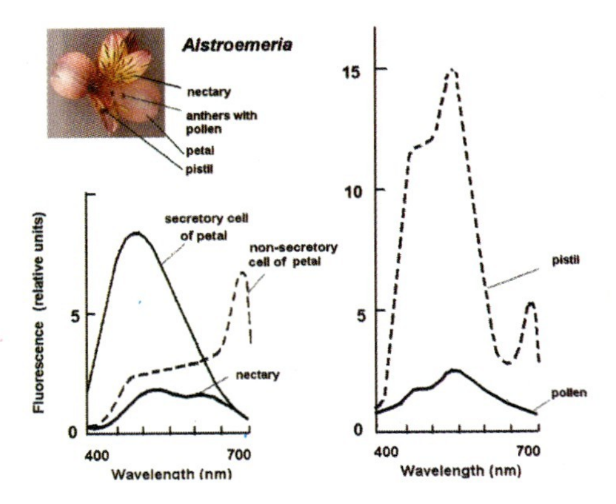

Fig. 16 The fluorescence spectra of the flower secretory cells of Angiosperm plant *Alstroemeria*. Sources: Roshchina and Melnikova (1999) and unpublished data of Roshchina.

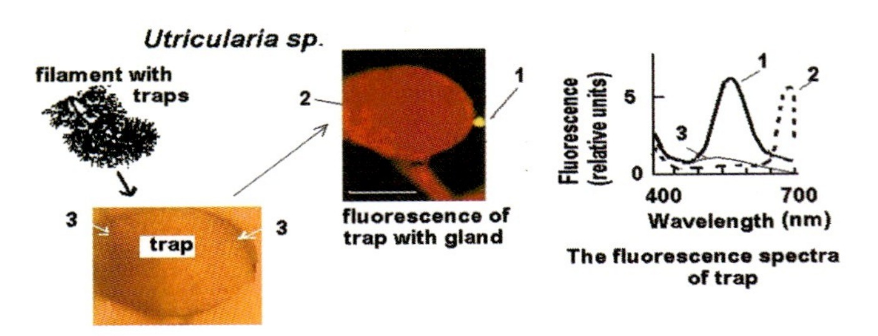

Fig. 17 The scheme and image of trap stained with red analogue of Ellman reagent (left), fluorescence image (middle) and the fluorescence spectra (right) of trap secretory cells in carnivorous plant *Utricularia vulgaris* L. Adopted from Roshchina et al., (2000a). Bar = 0.5 mm. 1- trap gland, 2- the trap surface (non-secretory cells), 3- slime. The excitement of the autofluorescence was by UV-light 360-380 nm.

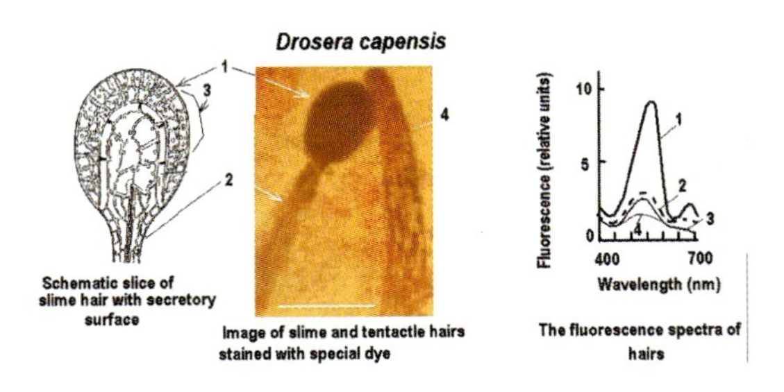

Fig. 18 The scheme (left), image of cells stained with red analogue of Ellman reagent (middle) and the fluorescence spectra (right) of secretory hairs in the carnivorous plant *Drosera capensis* L. 1- slime hair, 2- stalk part of slime hair, 3- slime on the surface of the slime hair, 4- tentacle hair. Bar = 3 mm. The excitement of the autofluorescence was by UV-light 360-380 nm.

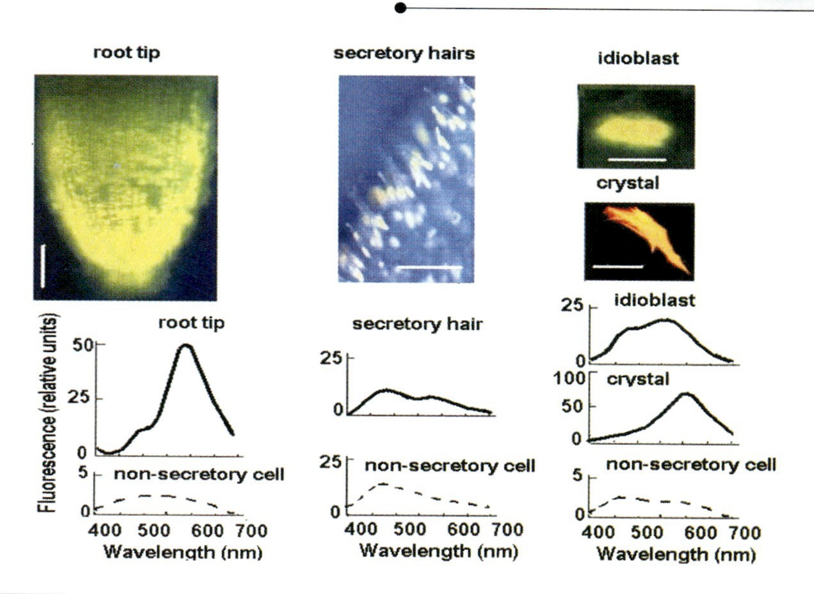

Fig. 19 The images seen under a luminescence microscope (upper side) and fluorescence spectra (below) of root secretory cells from *Ruta graveolens*. Adopted from Roshchina (2005b) and unpublished data of the author. Bar (From left to right) = 200, 100 and 20 μm, relatively.

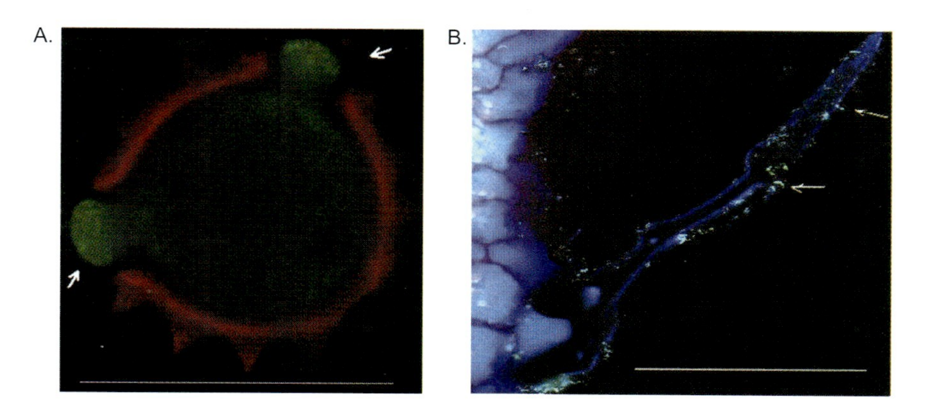

Fig. 20 LSCM- image comparison of pollen and leaf hair of *Solidago canadensis*. Excitation – 458nm, emission – 488-550 nm and 650 nm.The similarity in the green fluorescence between the excretions in apertures (shown by arrows) of pollen grain (A) and crystals of the excretions (shown by arrows) within leaf secretory hairs (B). A – bar = 20 μm; B – bar = 50 μm.

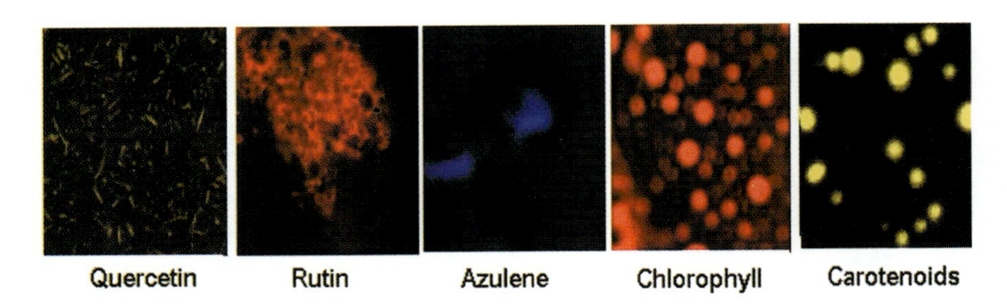

Fig. 21 The fluorescence image of individual substances under Laser Scanning Confocal Microscope (LSCM). Excitation by wavelengths 458, 543 nm and 633 nm. Emission > 488 nm. Chlorophyll or carotenoids are a sum of chlorophylls *a* and *b* or a sum of carotenoids from pea leaves purified on silicagel by thin layer chromatography.

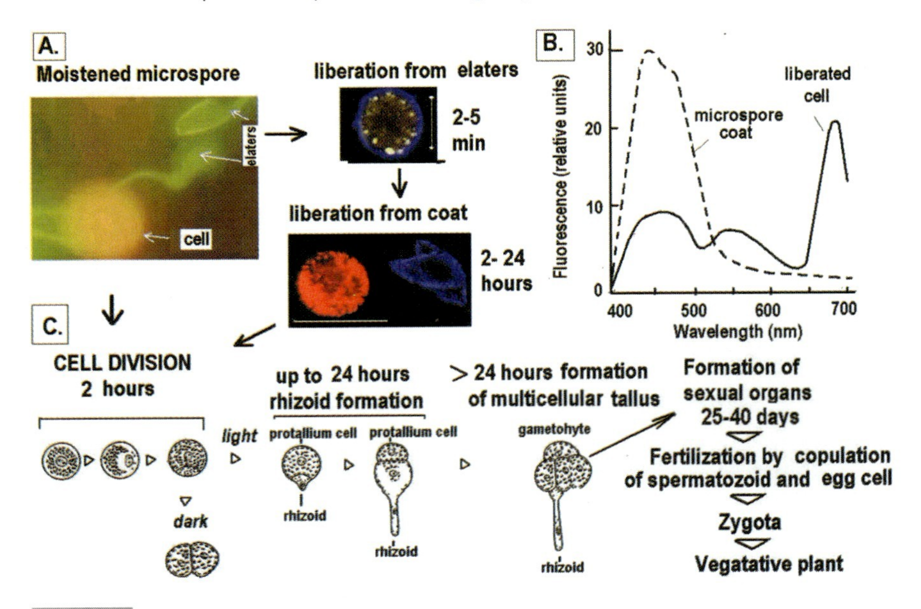

Fig. 22 The development and fluorescence of vegetative microspores of horsetail *Equisetum arvense.* Adopted from Roshchina (2006a). A. The fluorescence images of microspores from the moment of moistening to the liberation from the coat (rigid cover). Bar = 30 μm. B. The fluorescence spectrum of the microspore that has put off the coat. C. Scheme of the development of microspore in dark and in day light.

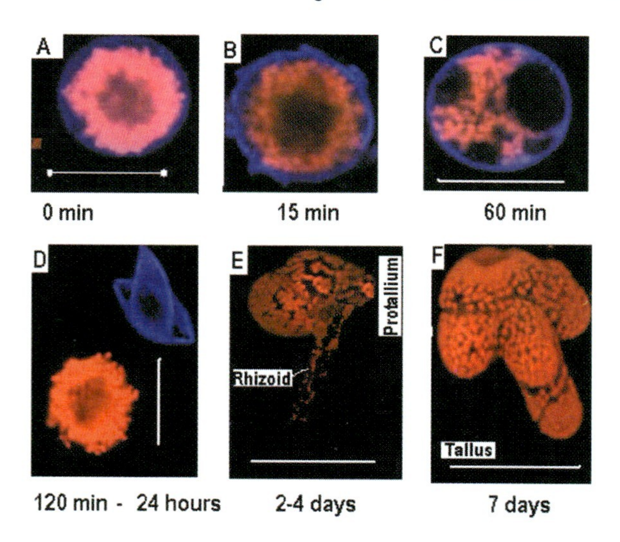

Fig. 23 The changes in the LSCM images of fluorescing microspores of *Equisetum arvense* during their development. The excitation by lasers with wavelengths 458 and 633 nm, the registration of the autofluorescence was at a superposition 505-630 and 650-750 nm, relatively. A. dry microspore; B. swollen microspore after 15 min of moistening; C. division of chloroplasts in the microspore; D. developed liberated red-fluorescing cell without cover and blue-fluorescing rigid cover; E. formation of protallium with rhizoid; F. Formed thallus. 1 bar for A and B = 20 μm, for C and D = 50 μm; for E and F = 100 μm.

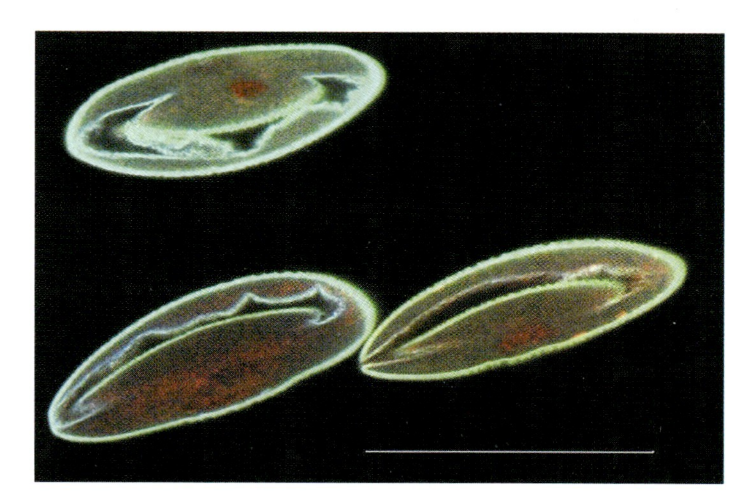

Fig. 24 The LSCM images of mature and immature pollen of *Hippeastrum hybridum*. Laser excitation 458 nm. Emission at wavelength > 483 nm. Immature pollen contains chlorophyll fluoresced in red.

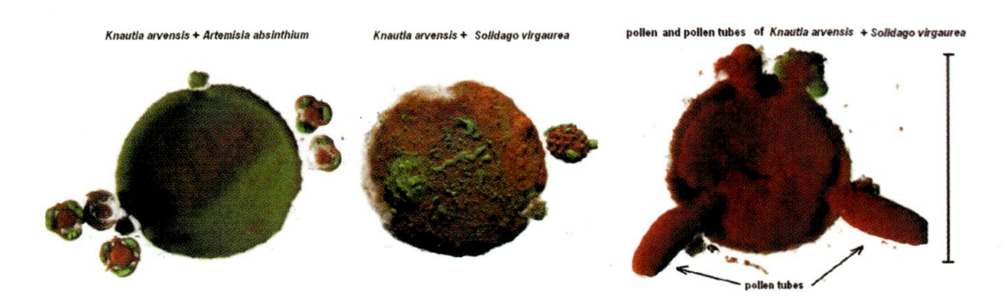

Fig. 25 The LSCM images at the interaction of large pollen grains from *Knautia arvensis* with smaller pollen grains from *Artemisia absinthium* (left and middle) and *Solidago virgaurea* (right). Left and middle – relatively, frontal slices and surface of non-germinated pollen from *Knautia*, Right – the stack of germinated pollen of *Knautia* with pollen tubes and pollen from *Solidago.* Bar = 100 μm. Wavelength of excitation - 488 nm, and emission – at 500-550 and 650 nm.

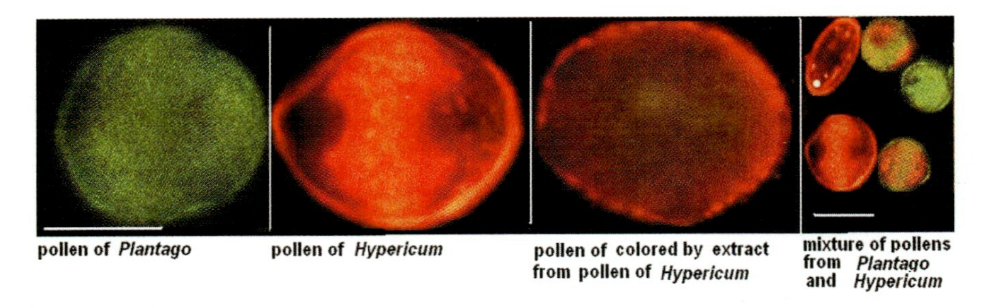

Fig. 26 Modelling of the interactions between pollens of *Plantago major* and *Hypericum perforatum.* Wavelength of excitation – 1 channel 458 nm and 2 channel -545 nm. Emission – at 500-550 and 650-700 nm. Right three LSCM-images with bar = 20 μm, while fouth picture 1 bar = 20 μm. Water extract from pollen of *Hypericum perforatum* (2 mg of pollen grains + 0.5 ml of water, exposure 30 min). Red-fluorescing excretions or water extract from *Hypericum perforatum* interact with green-fluorescing pollen grains of *Plantago major* that induced the staining of the surface, which became red-fluorescing.

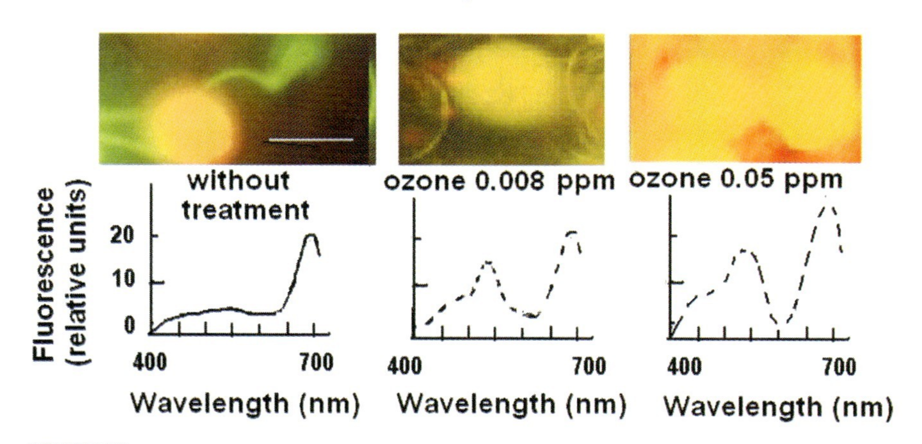

Fig. 27 The images under luminescence microscope at excitation by light 360-380 nm (Upper) and the fluorescence spectra (Below) of dry vegetative microspores of *Equisetum arvense* untreated and treated with ozone. Adopted from Roshchina (2005d).

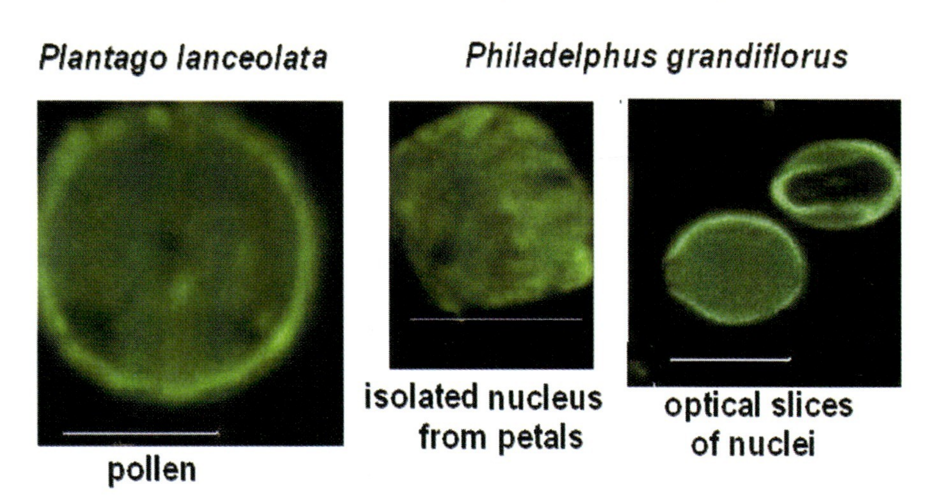

Fig. 28 LSCM images of pollen *Plantago lanceolata* and isolated nuclei from petals of *Philadelphus grandiflorus* treated with BODIPY-DA (fluorescent analogue of dopamine). Excitation by laser 488 nm. Emission > 500 nm. Adopted from Roshchina et al., (2005).

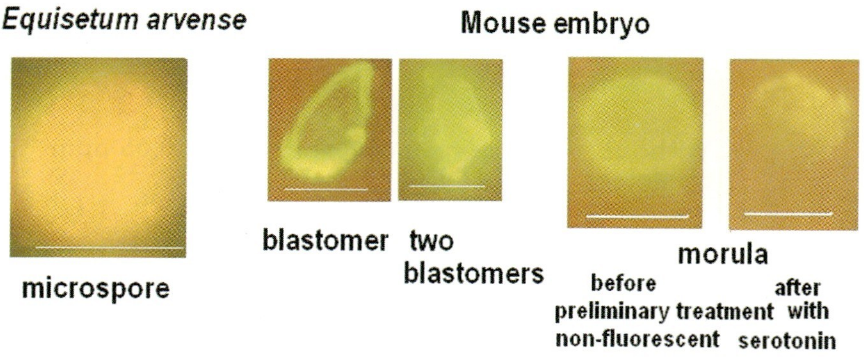

Equisetum arvense
Mouse embryo
blastomer
two blastomers
morula
microspore
before
after
preliminary treatment with
non-fluorescent serotonin

Fig. 29 The images under a luminescence microscope at excitation by light 360-380 nm of cells of dry vegetative microspores of *Equisetum arvense* and of a mouse embryo on the different phases of development staining with Bodipy-5 HT (fluorescent analogue of serotonin). Green fluorescence is concentrated on the cellular surface. 1 bar = 20 μm.

A.
B.
1
2
3
4
5
6
7
8
9
10
11
12

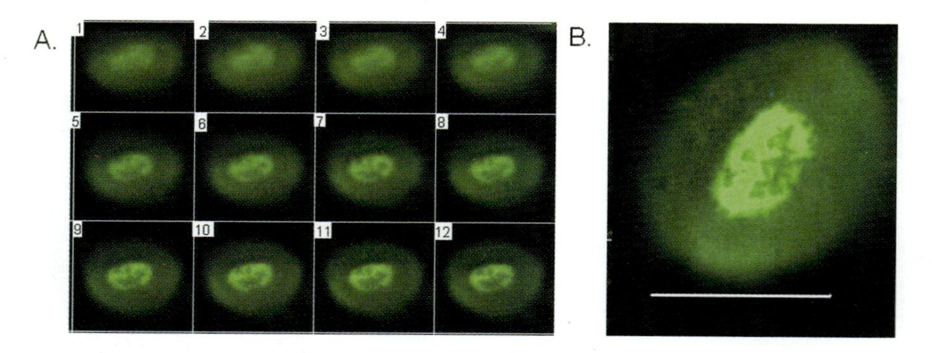

Fig. 30 The LSCM images of damaged nucleus (in drop of cytoplasm) isolated from white petals of *Matricaria chamomilla* (according to Roshchina, 2006b) and stained by Bodipy-dopamine 10^{-7} M. Laser excitation 488 nm. Bar = 10 μm. A. – galery of optical slices through 1 μm; B. one from optical slices.

Optical Slice

Front projection

50 μm

50 μm

Three-dimensional reconstruction

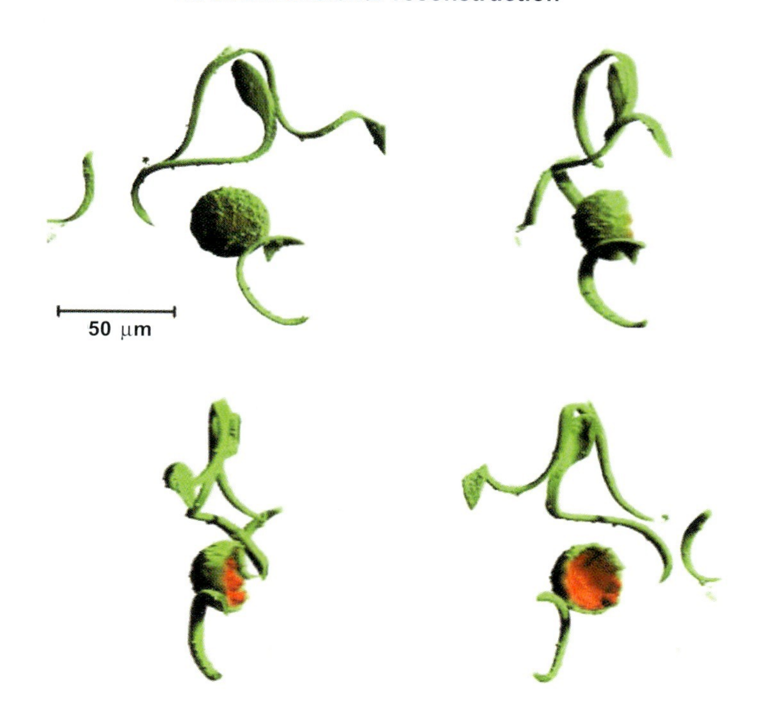

Fig. 31 The LSCM images of vegetative microspores of *Equisetum arvense* stained with colchicine 10^{-7} M. Laser excitation wavelength 458 nm. Emission > 550 nm. The green colour of the emission is seen as elaters (threads served for spore anchoring to a substrate such as soils) and cytoplasmic periphery of microspore. Interior of the spore looks red due to the presence of chloroplasts. Excitation by laser 488 nm. Source: Roshchina et al., (2006).

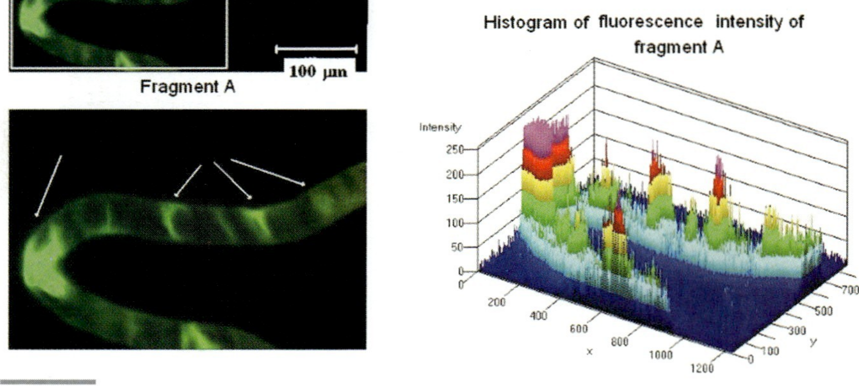

Fig. 32 The LSCM images of *Hippeastrum hybridum* pollen tube stained with colchicine 10^{-7} M. Laser excitation wavelength - 458 nm. The bright green emission is observed in pollen grain and in the some parts of the pollen tube. Source: Roshchina et al., (2006).

Bibliography

Acharya, A.S. and Manning, J.M. (1983). Reaction of Glycolaldehyde with Proteins: Latent Crosslinking Potential of alpha-hydroxyaldehydes. *Proceedings of the National Academy of Sciences of the United States of America* 80 (n 12, Part 1): 3590-3594.

Agaltzov, A.M., Gorelik, V.S. and Rakhmatulaev, I.A. (1997). Spectral, energetic and time characteristics of two-photon excited fluorescence of ZnSe crystal at the blue region of spectra. *Physics and Technique of Semi-conductors* 31(12): 1422-1424

Ahmad, A., Barry, J.P., and Nelson, D.C. (1999). Phylogenetic affinity of a wide, vacuolate, nitrate-accumulating *Beggiatoa* sp. from Monterey Canyon, California, with Thioploca spp. *Applied and Environmental Microbiology* 65(1): 270-277.

Ahmed, A.A., Jakupovic, J.P. and Mabry, T.J. (1993). Sesquiterpene glycosides from *Calendula arvensis*. *Journal of Natural Products* 56 (10): 1821-1824.

Alberti, S., Park, D.R., and Herzenberg, L.A. (1987). A single method for substraction of cell autofluorescence in flow cytometry. *Cytometry* 8 (2): 114-119.

Alexandrov, V.Ya., and Sveshnikova, I.N. (1956). The application of fluorescent microscopy in paleobotany. *Botanical Journal* (USSR) 41 (2) 206-212.

Aliotta, G. and Cafiero, G. (1999). Biological properties of rue (*Ruta graveolens* L.). Potential use in sustainable agricultural systems. In: *Principles and Practices in Plant Ecology. Allelochemical Interactions*. S. Inderjit, K.M.M., Dakshini, and C.L. Foy (eds). pp. 551-563. CRC Press. Boca Raton, Fl, USA.

Ames, R.N., Ingham, E.R., and Reid C.P.P. (1982). Ultraviolet-induced autofluorescence of arbuscular mycorrhyzal root infections: an alternative to clearing and staining methods for assessing infections. *Canadian Journal Microbiology* 28: 351-355.

Anaya, A.L., Hernandez-Bautista, B.E., Jimenez-Estrada, M., and Velasco-Ibarra, L. (1992). Phenylacetic acid as a phytotoxic compound of corn pollen. *J. Chem Ecol.* 18: 897–905.

Andon, T.H. and Denisova, G.A. (1974). The location of coumarin compounds in the secretory receptacles of *Ruta graveolens. Plant Resources (Rastytelye Resyrsi)* 10 (4): 526-540.

Andreoni, A., Colasanti, A., Colasanti, P., Mastrocinque, M., Riccio, P., and Roberti, G. (1994). Laser photosensitization of cells by hypericin. *Photochemistry and Photobiology* 59 (5): 529-533

Anet, E., Lythogoe, B., Silk, M.H., and Trippett, S. (1952). Oenanthotoxin and cicutotoxin. *Chemical Industry* 31: 757.

Anet, E., Lythogoe, B., Silk, M.H., and Trippett, S. (1953). Oenanthotoxin and cicutotoxin. Isolation and structure. *J. Chem. Soc.* 309-322.

Angell, P., Arrage, A.A., Mittelman, M.W., and White, D.C. (1993). On line, non-destructive biomass determination of bacterial biofilms by fluorometry. *J. Microbiological Methods* 18: 317-327.

Anyos, T. and Steelink, C. (1960). Fluorescent petal constituents of *Chrysanthemum coronarium* L. *Arch. Biochem. Biophys.* 90: 53-67.

Arai, T. and Okuyama, A. (1973). Purification and properties of colchicine-binding protein from the bovine brain. *Seikagaku* 45: 19-29.

Arkhipova, L.E., Tretyak, T.M. and Ozolin, O.N. (1988). The influence of catecholamines and serotonin on RNA-synthesing capacity of isolated nuclei and chromatin of brain and rat liver. *Biochemistry* (USSR) 53(7): 1078-1081.

Asbeck, F. (1955). Fluorescezieren der Blutenstaub. *Naturwissenschaften* 42 (5): 632.

Ascensao, L. and Pais, M.S. (1987). Granular trichomes of *Artemisia campestris* (ssp.*maritima*): ontogeny and histochemistry of the secretory product. *Bot Gaz* 148: 221-227.

Ashman T.L., Jim Swetz, J., and Shivitz, S. (2000). Understanding the basis of pollinator selectivity in sexually dimorphic *Fragaria virginiana. OIKOS* 90: 347–356.

Aslan , K., Gryszynski, I., Malicka, J., Matveeva, E., Lakowicz., J.R., and Geddes, C.D. (2005). Metal-enhanced fluorescence: an emerging tool in biotechnology. *Current Opinion in Biotechnology* 16 (1): 55-62.

Atkinson, L.R. (1938). Cytology. In: *Manual of Pteridology.* Fr. Verdoorn (ed) pp. 196-232. Martinus Nijhoff. The Hague.

Aubin, J.E. (1979). Autofluorescence of viable cultured mammalian cells. *J. Histochemistry and Cytochemistry* 27 (1): 36-43.

Aucoin, R.R., Scneider, E., and Arnason, J.T. (1992). Evaluating the phytotoxicity and photogenotoxicity of plant secondary compounds. In: *Plant Toxin Analysis.* H.F. Linskens and J.F. Jackson (eds). pp. 75-86. Springer-Verlag. Berlin, Heidelberg, Germany

Bachmann, R., Burda, C., Gerson, F., Scholz, M., and Hansen, H.J. (1994). Radical anions of polyalkylazulenes: an ESR and ENDOR study. *Helvatica Chimica Acta* 77: 1458-1465.

Baeyens, W.R.G. (1985). Fluorescence and phosphorescence of pharmaceuticals. In: *Molecular Luminescence Spectroscopy. Methods and Applications.* S.G. N. Y. Schulman (ed). Pt1. pp. 29-166. Wiley and Sons. Chichester, UK.

Bailey, P.S. (1958). The reactions of ozone with organic compounds. *Chem. Rev.* 58: 926-1110.

Bailey, P.S. (1973). *Ozonation in organic chemistry.* Vol. 1. Olefinic compounds: Acad. Press. New York, USA.

Bailey, P.S. (1978). *Ozonation in organic chemistry.* Vol. 1: Olefinic compounds. Acad. Press.New York, San Francisco, London.

Baird, C. (1995). *Environmental Chemistry.* Freeman and Co. New York, USA.

Baker, H.G. and Baker, I. (1975). Nectar constitution and pollinator-plant evolution. In: *Animal and Plant Coevolution.* (Eds. L.E. Gilbert and P.H. Raven). Pp 100-140. Austin: University of Texas.

Baker, H.G. and Baker, I. (1983). A brief historical review of the chemistry of floral nectar. In: The Biology of Nectaries. (Eds. B. Bentley and E.T. Elias). Pp 126-152. New York: Columbia University Press.

Baker, H.G., Baker, I., and Opler, P.A. (1973). Stigmatic dates and pollination. In: *Pollination and Dispersal.* (Eds. N.B.M. Brantjes and H.F. Linskens).pp. 47-60. Nijmegen: University of Nijmegen.

Balandrin, M.F., Klocke, J.A., Wurtele, E.S., and Bollinger, W.H. (1985). Natural plant chemicals: Sources of industrial and medicinal materials. *Science* 228: 1154-60.

Balcar, H., Sedláěk, J. Zedník, J., Blechta, V., Kubát, P., and Vohlídal, J. (2001). Polymerization of isomeric N-(4-substituted benzylidene)-4-ethynylanilines and 4-substituted N-(4-ethynylbenzylidene) anilines by transition metal catalysts. Preparation and characterization of new substituted Polyacetylenes with Aromatic Schiff Base Type. *Polymer* **42** (16): 6709-6721.

Barenboim, G.M., Domanskii A.N., and Turoverov, K.K. (eds) (1969). *Luminescence of Biopolymers and Cells.* Plenum Press.New York, USA.

Barriero, A.T., Sanchez, I.F., Astarejos, I., and Zafra, M.I. (1992). Homoditerpenes from the essential oil of *Tanacetum annuum. Phytochemistry* 31 (5): 1727-1730.

Baumert, A., Kuzovkina, I.N., Krauss, G., Hieke, M., and Grogewer, D. (1982). Biosynthesis of rutacridone in tissue culture of *Ruta graveolens* L. *Plant Cell Reports.* 1: 168-171.

Beer, R. (1909). The development of the spores of *Equisetum. New Phytol.* 8(7): 261-266.

Bell, A., Pickering, L., Watson, A.A., Nash, R.J., Pan, Y.T., Elbein, A.D., and Fleet, G.W. (1997). Synthesis of casuarines [Pentahydroxylated pyrrolidizines] by sodium hydrogen telluride-induced cyclisations of azidodimesylates. *Tetrahedron Letters.* 38: 5869-5872.

Beneš, M., Hudeček, J., Anzenbacher, P., and Hof, M. (2001). Coumarin 6, hypericin, resorufins, and flavins: suitable chromophores for fluorescence correlation spectroscopy of biological molecules. *Collection of Czechoslovak Chemical Communications* 66(6): 855-869.

Benson, R.C., Meyer, R.A., Zaruba, M.E., and McHann, G.M. (1979). Cellular autofluorescence – is it due to flavins? *Journal of Histochemistry and Cytochemistry* 27(1): 44-48.

Berger, F. (1934). Das Verhalten der heufieber-erregenden Pollen in filtrien ultravioleten Licht. *Beitrage der Biologische Pflanzen.* 22 (1) 1-12.

Berlman, I.B. (1971). *Handbook of Fluorescence Spectra of Aromatic Molecules.* Second Edition. New York, London: Academic Press, 474 pp.

Bezuglov, V.V., Gretskaya, N.M., Esipov, S.S., Polyakov, N.B., Nikitina, L.A., Buznikov, G.A., and Lauder, J. (2004). Fluorescent lipophilic analogs of serotonin, dopamine and acetylcholine: synthesis, mass-spectrometry and biological activity. *Bioorganic Chemistry* (Russia) 30(5): 512-519.

Bhattacharyya, B. and Wolff, J. (1974). Promotion of fluorescence upon binding of colchicine to tubulin. *Proc. Nat. Acad. Sci, USA* 71 (7): 2627-2631.

Bhattacharyya, B., Howard, R, Maity, S.N., Brossi, A, Sharma, P.N., and Wolff, J. (1986). B ring regulation of colchicine binding kinetics and fluorescence. *Proc Natl Acad Sci USA.* 83 (7): 2052–2055.

Bicchi, C., Frattini, C., and Sacco T. (1985). Essential oils of three asiatic *Artemisia* species. *Phytochemistry* 24 (10): 2440-2442.

Biophysical and Biochemical Aspects of Fluorescence Spectroscopy. (1991). T.G. Dewey (ed) Plenum Press.London. New York.

Bittner, K. (1905). Über Chlorophyllbildung im Finstern bei Kryptogamen. *Österr. Bot.Ztschr.* 302-313.

Blehova, A., Erdelsky, K., Repcak, M., and Garcar, J. (1995). Production and accumulation of 7-methyljuglone in callus and organ culture of *Drosera spathulata Labill. Biologia* (Bratislava) 50: 397-401.

Bloch, R. (1965). Polarity and gradients in plants: A survey. In:*Handbuch der Pflanzenphysiologie. (Encyclopedia of Plant Physiology),* Bd.15, teil1, ss. 234-274 Springer-Verlag.Berlin, Heidelberg, New York.

Bloemberg, G.V., Wijfjes, A.H.M., Lamers, G.E.M., Stuurman, N., and Lugtenberg B.J.J. (2000). Technical Advance Simultaneous Imaging of *Pseudomonas fluorescens* WCS365 Populations Expressing Three Different Autofluorescent Proteins in the Rhizosphere: New Perspectives for Studying Microbial Communities. *MPMI* 13 (11): 1170–1176.

Bondar, O.P., Pivovarenko, V.G., Rowe, E.S. (1998). Flavonols – new fluorescent membrane probes for studing the interdigitation of lipid bilayers. *Biochimica et Biophysica Acta* 1369 (1): 119-130.

Bopp, M. (1965). Entwicklungphysiologie der Moose. *Handbuch der Pflanzenphysiologie. (Encyclopedia of PlantPhysiology)*, Bd. 15, teil 1, pp. 802-843. Springer-Verlag.Berlin, Heidelberg, New York.

Bors, W. (2005). Radikalfangende Wirkung von Pflanzeninhaltsstoffen. (Radical capturing effects of plant constituents). *GSF-Forschungszentrum für Umwelt und Gesundheit GmbH in der Helmholtz- Gemeinschaft Annual Reports.* ss.55-60.

Bourett, T.M., and Howard, R.J. (1992). Actin in penetration pegs of the fungal rice blast pathogen *Magnaporthe grisea*. *Protoplasma* 168: 20-26.

Branscheidt, P. (1930). Zur Phisiologie der Pollenkeimung und ihrer experimentellen Beeinflussung. *Planta* 11: 368-456.

Brooks, R.I. and Csallany, A.S. (1978). Effects of air,ozone, and nitrogen dioxide exposure on the oxidation of corn and soybean lipids. *J. Agricult. and Food Chem.* 28 (5): 1203-1209.

Brumberg, E.M. (1955). Fluorescence microscopes. *J. General Biology* (USSR) 16 (3): 222-237.

Brumberg, E.M. (1956.) Ultra-violet fluorescence microscopy. *J. General Biology* (USSR) 17 (6): 401-412.

Brumberg, E.M. (1959). Fluorescent microscopy of biological objects using light from above. *Biophysics* 4 (1): 97-104.

Buchachenko, A.A., Khloplyankina, M.S., and Dobryakov, S.N. (1967). Electron-excited state quenching by organic radicals. *Optics and Spectroscopy* (USSR), 22: (4): 554-559.

Budantsev, L.E. (ed) (1996). *Plant Resources of Russia and Surrounded Countries.* Mir Sem'ya. Sankt-Peterburg, Russia.

Budantsev, A. Yu. and Roshchina, V.V. (2005). Testing of alkaloids inhibitory activity to acetylcholinesterase. *Plant Resourses (Russia)* 41: 131-138.

Budantsev, A. Yu., and Roshchina, V.V. (2007). Cholinesterase activity as a biosensor reaction for natural allelochemicals: Pesticides and pharmaceuticals. In: *Cell Diagnostics in Allelopathy*. S.S. Narwal and V.V. Roshchina (eds). pp. 127-146. Science Publisher. Enfield, Plymouth.

Buschmann, C. and Lichtenthaler, H.K. (1998). Principles and characteristics of multi-colour fluorescence imaging of plants. *J. Plant Physiol.* 152: 297-314.

Buschmann, C., Langdorf, G., and Lichtenthaler, H.K. (2000). Imaging of the blue, green and red fluorescence emission of plants: An overview. *Photosynthetica* 38 (4): 483-491.

Butkhuzi, T., Kuchukashvili, Z., Sharvashidze, M., Natsvlishvili, G. and Gurabanidze V. (2002). Cytodiagnostics on intact pollen using photoluminescence. *J. Biological Physics and Chemistry* 1 (1/2): 53-55.

Buvat, R. (1989). *Ontogeny, cell differentiation and structure of vascular plants.* Springer-Verlag. Berlin, Heidelberg, New York.

Buznikov, G.A. (1987). *Neurotransmitters in Embryogenesis.* Nauka, Moscow, Russia.

Buznikov, G.A. (1990). Neurotransmitters in embryogenesis. Chur: Harwood Academic Press. 526 p.

Buznikov, G.A., Shmukler Yu. B, Lauder J.M. (1996). From oocyte to neuron: do neurotransmitters function in the same way throughout development? *Cell. Molec. Neurobiol.,* 16 (5): 532-559.

Cai, G., Romagnoli, S., Moscatelli, A., Ovidi, E., Gambellini, G., Tiezzi, A., and Cresti, M. (2000). Identification and Characterization of a Novel Microtubule-Based Motor Associated with Membranous Organelles in Tobacco Pollen Tubes. *Plant Cell* 12 (9): 1719–1736.

Candy, H. (1985). Photomultiplier characteristics and practice relevant to photon counting. *Rev. Sci. Instrum.* 56: 183-193.

Cantiello, H.F. (1997). Role of actin filaments organization in cell volume and ion channel regulation. *Journal of Expt. Zool.* 279 (2): 425-435.

Castle, H. (1953). Notes of the development of the gametophyte of *Equsetum avense* in sterile media. *Botanical Gazette* 114 (3): 323-328.

Ceska, O., Chaundhary, S., Warrington, P., Poulton, G., and Ashwood-Smith, M. (1986). Naturally-occuring crystals of photocarcinogenic furanocoumarins on the surface of parsnip roots sold as food. *Experientia* 42: 1302-1304.

Chalchat, J.C., Garry, R.P., Michet, A., and Remery, A. (1985). The essential oils of two chemotypes of *Pinus sylvestris*. *Phytochemistry* 24 (10): 2443-2444.

Chance, B. and Thorell, Bo. (1959). Localization and kinetics of reduced piridine nucleotide in living cells by microfluorimetry. *J. of Biological Chemistry* 234 (11): 3044-3050.

Chappelle, E.W., Wood, P.M., McMuitrey, J.E., and Newcourt, W.W. (1985). Laser-induced fluorescence of green plants. 3: LIF spectral signatures of five major plant types. *Appl. Optics* 24: 74-80.

Chappele, E.W., McMurtrey, J.E., and Kim, M.S. (1990). Laser induced blue fluorescence in vegetation. In: *Remote Sensing Science for the Nineties IGARSS'90*: R. Mill(ed) Vol 3, pp. 199-1922 10th Annu. Int. Geosci. and Remote Sensing Symp.Washington, D.C., 1990.May 20-24, College Park Univ. Maryland, The Institute of Electrical and Electronic Engineers, INC. Maryland, USA.

Char, M.B.S. (1977). Pollen Allelopathy. *Naturwissenschaften* 64: 489-490.

Chen, C.Y., Wong, E.I., Vidali, L., Estavillo, A., Hepler, P.K., Wu, Hen-ming., and Cheung, A.Y. (2001). The Regulation of Actin Organization by Actin-Depolymerizing Factor in Elongating Pollen Tubes. *The Plant Cell.* 14: 2175-2190.

Chen, P.K. and Leather, G.R. (1990). Plant growth regulatory activities of artemisinin and its related compounds. J. Chem Ecol 16: 1867-1876.

Cheng, P.C. and Summers, R.G. (1990). Image contrast in confocal light microscopy: In: *Handbook of Biological Confocal Microscopy*. J.B Pawley (ed). pp. 179-195. Plenum. New York, USA.

Christensen, L.P. and Lam, J. (1990). Acetylenes and related compounds in Cynareae. *Phytochemistry* 29(9): 2753-2785.

Cresti, M., Blackmore, S., and van Went, J.L. (1992). *Atlas of Sexual Reproduction in Flowering Plants*. Springer-Verlag. Berlin, Heidelberg.

Croateau, R. and Leblanc, R.M. (1977). Colchicine fluorescence measured with a laser spectrofluorimeter. *Journal of Luminescence* 15: 353-356.

Cruden, R.W. (1972). Pollination biology of *Nemophilla menziesii* (Hydrophyllaceae) with comments on the evolution of oligolectic bees. *Evolution* 26 (3): 373-389.

Cruz-Ortega, R. and Anaya, A.L. (2007). Biochemical approach to study oxidative damage in plants exposed to allelochemical stress: A case study. In: *Plant Cell Diagnostics* (Eds. V.V. Roshchina and S.S. Narwal). Pp. 117-126. Science Publisher: Enfield, Jersey (USA), Plymouth.

Curtis, L.C. (1943). Deleterious effects of guttated fluids on foliage. *Amer. J. Bot.* 130 (4): 778-781.

Curtis, J.D. and Lersten, N.R. (1990). Internal secretory structures in *Hypericum* (Clusiaceae) *H. perforatum* L. and *H. balearicum* L. *New Phytologist* 114 (3): 571-580.

Dahse, I., Bernstein, M., Müller, E., and Petzold, U. (1989). On possible functions of electron transport in the plasmalemma of plant cells. *Biochem. Physiol. Pflanzen.* 185 (3/4): 145-180.

Davies, K.L., Stricżynska, M., and Gregg, A. (2005). Nectar-secreting Floral Stomata in *Maxillaria anceps* Ames & C. Schweinf. (Orchidaceae). Annuals of Botany 96(2): 217-227.

Demchenko, A.P. (1991). Fluorescence and dynamics of proteins. In: *Topics in Fluorescence Spectroscopy*. J.R. Lakowicz (ed) Vol. 3, pp. 65-111. Plenum Press. New York and London. Denmark, S.E. and Hurd A.R. (2000). Synthesis of (+)-casuarine. *J. Org. Chem.* 65: 2875-2886.

Demchenko, A.P. (2005a). Optimization of fluorescence response in the design of molecular biosensors. *Anal. Biochemistry* 343(1): 1-22.

Demchenko, A.P. (2005b). The future of fluorescent sensor arrays. *Trends in Biotechnology* 23(9): 456-460.

Denmark, S.E. and Hurd, A.R. (2000). Synthesis of (+)-casuarine. *J. Org. Chem.* 65: 2875-2886.

Digman, M.A., Brown, C.M., Sengupta, P., Wiseman, P.W., Horwitz, A.R., and Gratton, E. (2005). Measuring fast dynamics in solutions and cell with a laser scanning microscope. *Biophisical Journal* 89 (2): 1317-1327.

Dirsch, VM, Stuppner, H., and Vollmar, A.M. (2001). Cytotoxic sesquiterpene lactones mediate their death inducing effect in leukemia T cells by triggering apoptosis. *Planta Medica* 67: 557-559.

Dodd, N.J.F. and Ebert, M. (1971). Demonstration of surface free radicals on spore coats by ESR techniques. In: *Sporopollenin*: Proc. Symp. at Geology Department,

Imperial College, London, 1970, 23-25 September. J. Brooks, P.R. Grant, M. Muir, and P.R. Gijzel (eds) pp. 408-421. Acad. Press. London, New York.

Dougal, D.T. (1903). The influence of light and darkness upon growth and development. Mem. of the New York *Bot. Garden* 11: 1-319.

Drabent, R., Pliszka, B., and Olszewska, T. (1999). Fluorescence properties of plant anthocyanin pigments. I. Fluorescence of anthocyanins in *Brassica oleracea* L. extracts. *Journal of Photochemistry and Photobiology*, B, 50 (1): 53-58.

Driessen., M.N.B.M., Willemse, M.T.M., and van Luijn, J.A.G. (1989). Grass pollen grain determination by light- and ultra-violet microscopy. *Grana* 28 (2):115-122.

Duke, S.O. Dayan, F.E., Romagni, J.G., and Rimando, A.M. (2000). Natural products as sources of herbicides: current status and future trends. *Weed Research* 40: 99-111.

Dumas, C., Bowman, R.B., Gaude, T., Gully, C.M., Heizman, Ph., Roeckel, P. and Rougier, M. (1988). Stigma and stigmatic secretion reexamined. *Phyton* 28: 193-200.

Dunant, Y. (1994). Hormones and neurotransmitters release: four mechanisms of secretion. *Cell Biology Internacional* 18 (5): 327-336.

Dirsch, V.M., Stuppner, H., and Vollmar, A.M. (2001). Cytotoxic sesquiterpene lactones mediate their death inducing effect in leukemia T cells by triggering apoptosis. *Planta Medica* 67: 557-559.

Dzhabiev, T.S., Moiseev, D.N., and Shilov, A.E. (2005). Six-electron water oxidation into ozone in natural photosynthesis. *Doklady of Russian Academy of Sciences* 402 (4): 555-553.

Ebadi, M. (2002). *Pharmacodynamic Basis of Herbal Medicine*. CRC Press. Boca Raton, Fl, USA.

Eftink, M.R., and Ghiron, C.A. (1981). Fluorescence quenching studies with proteins. *Analytical Biochemistry* 114(2): 199-227.

Eftink, M.R. (1991). Fluorescence quenching: Theory and applicatins. In: *Topics in Fluorescence Spectroscopy*. J.R. Lakowicz (ed), Vol. 2, pp. 53-126. Plenum Press. New York and London.

Egner, A. and Hell, S.W. (2005). Fluorescence microscopy with super-resolved optical sections. *Trends in Cell Biology* 15 (4): 207-215.

Egorov, I.A. and Egofarova, R.Kh. (1971). The study of essential oils of blossom pollen of grape. *Doklady AN SSSR* 199: 1439-1442.

Eilert, U., Wolter, B. and Constabel, F. (1986). Ultrastructure of acridone alkaloid idioblasts in root and cellular culture of *Ruta graveolens*. *Can. J. Bot.* 64: 1089-1096.

Eisner, T., Eisner, M., Hyypio, P.A., Aneshansley, D., and Silberglied, R.E. (1973). Plant taxonomy: Ultraviolet patterns of flowers as fluorescent patterns in pressed herbarium specimens. *Science* 179(4072): 486-487.

Eldred, G.E. and Katz M.L. (1988). Possible mechanism for lipofuscinogenesis in the retinal pigment epithelium and other tissues. In: *Lipofuscin - 1987; State of the Art.*/Zs-Nagy I. (ed) pp. 185-211 Akademiai Kiado, Budapest, Amsterdam.

Elenevskii, A.G., Solov'eva, M.P., and Tikhomirov, V.N. (2000). *Botany of Higher or Overground Plants.* Academia. Moscow, Russia.

Ellinger, P. (1940). Fluorescence microscopy in biology. *Biol. Rev.* 15: 323-350.

El-Seedi, H.R., Ringbom, T., Torssell, K., and Bohlin Lars. (2003). Constituents *oi Hypericum laricifolium* and Their Cyclooxygenase (COX) enzyme activities. *Chem. Pharm. Butt.* 51(12): 1439-1440.

Elston, D.M. (2001). Fluorescence of fungi in superficial and deep fungal infections. *BMC Microbiol.* 1: 21.

Ermolaev, V.A., Bodunov, E.N., Sveshnikova, E.B., and Shakhverdov, T.A. (1977). *Non-radiating (emitted) transfer of energy from electron excitation.* Nauka. Leningrad, Russia.

Erokhova, L.A., Brazhe, N.A., Maksimov, G.V., and Rubin, A.B. (2005). Researching conformational alteration of carotenoids in neurons under exposure to neorotransmitters. *Doklady of Russian Academy of Sciences* 402 (4): 548- 550.

Essau, K. (1965). *Plant Anatomy.* Wiley. New York, USA.

Essau, K. (1977). *Anatomy of Seed Plants.* Vol. 1-2. Wiley. New York, USA.

Evans, L.S. and Ting, J.P. (1974). Ozone - sensitivity of leaves: Relationship to leafwater content, gas transfer resistance and anatomical characteristic. Amer. J. Bot. 61: 592-597.

Executive Summary. Special of Atmospheric Science Panel J. Photochemistry and Photobiology. (1998) 46 (1-3): 1-4.

Fahn, A. (1979). Secretory tissues in plants. London: Academic Press.

Fahn, A.(1988). Secretory tissue in vascular plants. *New Phytologist* 108: 229-257.

Feder, W.A. and Manning, W.J. (1979). Living plants as indicators and monitors. In:. *Handbook of Methodology for the Assessment of Air Pollution Effects on Vegetation.* W.W. Heck, S.V. Krupa, and S.N. Linzon (eds) pp. 9-1, 9-14. Air Pollution Control Association. Pittsburgh, USA.

Federico, R. and Angelini, R. (1986). Occurence of diamine oxidase in the apoplast of pea epicotyls. *Planta.* 167: 300-302.

Fedorin, G.F. and Georgievskii, V.P. (1974a). Influence of the substitutes on the fluorescence spectra of simple coumarins. *Journal of Applied Spectroscopy* (USSR) 20(1): 153-154.

Fedorin, G.F. and Georgievskii, V.P. (1974b). Influence of solvents on the fluorescence spectra of coumarins. *Journal of Applied Spectroscopy* (USSR) 21 (1): 164-167.

Feofilov, P.P. (1944). About the connection of organic dyes with its chemical structures. *Dokl. USSR Acad. Sci.* 45 (9): 387-389.

Ferreira da Silva, P., Lima, J. C., Quina, F. H., and Maçanita A. L. (2004). Excited-State Electron Transfer in Anthocyanins and Related Flavylium Salts. *J. Phys. Chem. A*, 108 (46): 10133 -10140.

Feucht, W. and D. Treutter. (1990). Flavan-3-ols in trichomes pistils and phelloderm of some tree species. *Annals of Botany* 65, 225-230.

Figueiredo, P., Pina, F., Vilas-Boas,L., and Macanita, A.L. (1990). Fluorescence spectra and decays of malvadin 3,5-diglucoside in aqueous solutions. *Journal of Photochemistry and Photobiology*. A: Chem 52: 411-424.

Finnie, J.F. and van Staden, J. (1993). XII. Drosera spp. (sundew): Micropropagation and the in vitro production of plumbagin. In: *Medical and Aromatic Plants*. (Ed. Y.P.S.Bajaj). *Biotechnology in Agriculture and Forestry* (Vol.24).pp. 164-177. Berlin,Heidelberg: Springer-Verlag.

Fischer, N.H. (1991). Sesquiterpene lactones. In *Methods in Plant Biochemistry*. (Eds. B.V. Charlewood and D. Banthorpe). vol. 7, pp.187-212. London: Academic Press.

Fischer, N.H. (1991). Sesquiterpene lactones. In *Methods in Plant Biochemistry*. B.V. Charlewood and D. Banthorpe (eds). Vol. 7, pp. 187-212. Academic Press. London, UK.

Fischer, N.H., Weidenhamer, J.D. and Bradow, J.M. (1989) Inhibition and promotion of germination by several sesquiterpenes. *Journal of Chemical Ecology* 15: 1785-1793.

Franklin, F.C.H., Atwal, K.K., Ride J.P., and Franklin-Tong, V.E. 1994. Towards the elicidation of the mechanisms of pollen tube inhibition during the self-incompatibility response in Papaver rhoeas. In *Molecular and Cellular Aspects of Plant Reproduction*. (Eds. R.J. Scott and A.D. Stead). pp. 173-190. Cambridge: Cambridge University Press.

Franklin-Tong, V.E., Drobak, B.K., Allan, A.C., Watkins, and P.A., Trewavas, A.J. (1996). Growth of pollen tubes of *Papaver rhoeas* is regulated by a slow-moving calcium wave propagated by inositol 1,4,5-triphosphate. *The Plant Cell* 8 (8): 1305-1321.

Franklin-Tong, V.E. (1999). Signaling and the Modulation of Pollen Tube Growth. *The Plant Cell*. 11: 727-738.

Frey-Wyssling, A. and Agthe, C. (1950). Ultraviolet and fluorescence optics of lignified cell. *Verh. Schweiz.Naturforsch. Ges.* 130: 175

Friedman, J., Rushkin, E., and Waller, G.R. (1982). Highly potent germination inhibitors in aqueous eluate of fruits of bishop's weed (*Ammi majus* L.) and avoidance of autoinhibition. *Journal of Chemical Ecology* 8: 55-65.

Frey-Wyssling, A. (1964). Ultraviolet and fluorescence optics of lignified cell. In: *The Formation of Wood in Forest Trees*. M.N. Zimmerman (ed). pp. 153-167. Academic Press. New York, USA.

Frolov, Yu. L., Sapozhnikov, Yu. M., Barer, S.S., Pogodaeva, N.N., and Tyukavkina, N.A. (1974). Luminescence of flavonoid compounds. *Izvestya of USSR Academy of Sciences, Ser. Chemistry* 10: 2364.

Frolov, Yu. L., Sapozhnikov, Yu. M., Chipanina, N.N., Sidorkin, V.F., and Tyukavkina, N.A. (1978). On the correlation of pk values in the ground and electronically excited states. *Izvestya of USSR Academy of Sciences, Ser. Chemistry* 2: 301-304.

Gamaley, I.A. and Klyubin, I.V. (1996). Hydrogen peroxide as signalling molecule. Tsitologia (Cytologia, Russia) 38 (12): 1233-1247.

Gamaley, I.A., and Klybin, I.V. (1999). Roles of ROS: Signaling and Regulation of Cellular Functions. *Int.Rev. Cytol.*, 189: 203-256.

Genders, R. (1994). *Scented Flora of the World.* Robert Hale. London, UK.

Ghisla, S., Massey, V., Lhoste, J.M. and Mayhew, S.G. (1975). Fluorescence characteristics of reduced flavins and flavoproteins. In: *Reactivity of Flavins.* K. Yagi (ed). pp. 15-24. University of Tokyo Press. Tokyo, Japan.

van Gijzel, P. (1961). Autofluorescence and age of some fossil pollen and spores. *Koninkl. Nederl. Acad. Wet.* Proc. Ser B. Physic. Sci. 64 (1): 56-63.

van Gijzel, P. (1967). Autofluorescence of fossil pollen and spores with special reference to age determination and coalification. *Leidse Geologische Mededelingen* 40: 263-317.

van Gijzel, P. (1971). Revie on the ultra-violet-fluorescence microphotometry of fresh and fossill exines and exosporia. In: *Sporopollenin.* J. Brooks, P.R. Grant, M. Muir and van P.R. Gijzel (eds) pp. 659-685. Proc. Symp. at Geology Department, Imperial College, London. 23-25 September 1970. Acad. Press. London, New York.

Gilbert, G.G. (ed) (1990). *Practical Fluorescence.* Second Edition. Marcel Dekker. New York, USA.

Gillispie, G.D. and Lim, E.C. (1976). S_2—S_1 fluorescence of azulene in a Shpol'skii matrix. *J. Chem. Physics.* 65 (10): 4314-4316.

Gimenez-Martin, G., Risueno, M.C., and Lopez-Saez, J.F. (1969). Generative cell envelope in pollen grains as a secretion system, a postulate. *Protoplasma* 67 (2-3): 223-235.

Gitelson, A.A., Buschmann, C. and Lichtenthaler, H.K. (1998). Leaf Chlorophyll Fluorescence Corrected for Re-absorption by Means of Absorption and Reflectance Measurements *Journal of Plant Physiol* 152: 283-296.

Golovkin, B.N., Rudenskaya, R.N., Trofimova, I.A., and Shreter, A.I. (2001). *Biologically Active Substances of Plant Origin.* 3 Volumes Nauka, Moscow, Russia.

Golubinskii, I.N. (1946). About a mutual influence of pollen grains of different species at their mutual germination in the artificial media. *Doklady AN SSSR* 53: 73-76.

Goodbody, K.C. and Lloyd, C.W. (1990) Actin filaments line up across *Tradescantia* epidermal cells, anticipating wound-induced division planes. *Protoplasma* 157 (1-3): 92-101.

Goodwin, T.W. and Mercer, E.I. (1983). *Introduction to Plant Biochemistry*. Second Edition. Pergamon Press. Oxford, New York.

Gordon, Ì. (1952). The azulens. *Chem Rev* 50: 127-200.

Gordon, L.K., Kolesnikov, O.P. and Minibayeva, Ph.V. (1999). Formation of superoxide by redox system of plasmalemma of root cells and its participation in detoxification of xanobiotics. *Doklady of Russian Academy of Sciences* 367 (3): 409-411.

Goryunova, S.V. (1952). The application of the fluorescence microscopy for the assay of living and dead cells of algae. *Trudy of the Institute of Microbiology of USSR* Acad. Sci. (USSR) 2: 64-77.

Gray, J., Janick-Buckner, D., Buckner, B., Close, P.S., and Johal, G.S. (2002). Light-dependent death of maize *lls1* cells is mediated by mature chloroplasts. *Plant Physiol* 130: 1894-1904.

Greenaway, W., English, S. and Whatley, F.R. (1990). Phenolic composition of bud exudates of *Populus deltoides*. *Zeitschrift Naturforschung* 45c: 587-593.

Griesser, H.I. and Wild, U.P. (1980). Energy selection experiments in glassy matrices: The energy selection experiments in glassy matrices: The linewidths of the emissions S_2 - S_0 and S_2 - S_1 of azulene. *J. Chem. Physics* 3 (2): 4715-4719.

Grümmer, G. (1955). *Die gegenseitige Beeinflussung höherer Pflanzen-Allelopathie*. Fischer. Jena, Germany.

Grümmer, G. (1961). The role of toxic substances in the interrelationships between higher plants. *Symp. Soc. Exp. Biol.* 15: 219-228.

Gulvag, B.M. (1968). On the fine structure of the spores of *Equisetum fluviale* var/ *verticillatum* studied in the quiescent, germinated and non-viable state. *Grana Palynologica* 8 (1) 23-69.

Gibson, R.W. (1974). Aphid-trapping glandular hairs on hybrids of *Solanum tuberosum* and *S. berthaultii*. *Potato Res* 17: 152-154.

Gutteridge, J.M. (1995). Signal messenger and trigger molecules from free radical reactions and their control by antioxidants. In: *Signalling Mechanisms - from transcription factors to oxidative stress*. L.P. Packer and K.W.A. Wirtz (eds) H92. pp. 157-164. Springer in cooperation with NATO Sci. Affairs Division. (NATO ASI Ser.). H92. Berlin, Heidelburg, Germany.

Ha, T.N.M., Posterino, G.S., and Fryer, M.W. (1999). Effects of terbutaline on force and intracellular calcium in slow-twitch skeletal muscle fibres of the rat. *British Journal of Pharmacology* 126: 1717-1724.

Hauke, R.L. and Thompson, J.J. (1971). A carotenoid pigment associated with antheridial formation in *Equisetum* gametophytes. *Taiwania* 18 (2): 176-178.

Habuchi, S., Ando, R., Dedecker, P., Verheijen, W., Mizuno, H., Miyawaki, A., and Hofkens, J. (2005). Reversible single-molecule photoswitching in the GFP-like fluorescent protein Dronpa. *Proc. Natl. Acad Sci. USA* 102 (27): 9511-9516.

Haliwell, B. and Gutterige, J.N.C. (1985). *Free Radicals in Biology and Medicine.* Clarendon Press. Oxford, UK.

Halligan, J.P. (1975). Toxic terpenes from *Artemisia californica. Ecology* 56: 999-1003.

Hanson, M.R. and Köhler, R.H. (2001). GFP imaging: methodology and application to investigate cellular compartmentation in plants. *J. Exper. Botany* 52(356): 529-539.

Haraguchi, H., Saito, T., Okamura, N., and Yagi, A. (1995) Inhibition of lipid peroxidation and superoxide generation by diterpenoids from *Rosmarinus officinalis. Planta Medica* 61: 333-336.

Harada, R. and Iwasaki, M. (1982). Volatile components of *Artemisia capillaris. Phytochemistry* 2: 2009-2011.

Harborne, J.B. (1993). *Introduction to Ecological Biochemistry.* Fourth Edition. Academic Press. London, San Diego.

Harrawiji, P., van Oosten, A.M., and Piron, P.G.M. (2001). *Natural Terpenoids as Messengers.* Kluwer Academic. Dordrect, the Netherlands.

Harris, P.J. and Hartley, R.D. (1976). Detection of bounded ferulic acid in cell walls of the Gramineae by ultraviolet fluorescence microscopy. *Nature* 259: 508-510.

Hartley, R.D. (1973). Carbohydrate esters of ferulic acid as components of cell walls of *Lolium multiflorum. Phytochemistry* 12: 661-665.

Hastings, J.W. (1986). Bioluminescence in bacteria and Dinoflagellates. In: *Light Emission by Plants and Bacteria.* (Eds. Govindjee, J, Amesz, and D.C. Fork). pp. 363-398. Orlando: Academic Press.

Haugland, R.P. (1996). *Handbook of Fluorescent Probes and Research Chemicals.* 6 ed. Leiden: Molecular probes, Eugene.

Haugland, R.P. (2000). *Handbook of Fluorescent Probes and Research Chemicals.* 7 ed. Leiden: Molecular probes, Eugen.

Hauke, R.L. and Thompson, J.J. (1973). A carotenoid pigment associated with antheridial formation in *Equisetum* gametophytes. *Taiwania* 18(2) 176-178.

Haupt, W. (1957). Die Induktion der Polaritat bei der Spore von *Equisetum. Planta* 49: 61-90.

Heilbronner, E. (1959). Azulenes. In: *Non-benzenoid Aromatic Compounds.* (Ed. D. Ginsburg). p. 176-278. New York and London: Interscience Publisher.

Heslop-Harrison, Y. (1975). Enzyme release in insectivorous plants. In: *Lysosomes in Biology and Pathology,* G.T. Dingle. and R.T. Dian (eds) Vol. 4, pp. 525-576. North Holland Publ. Amsterdam, the Netherlands.

Heinrich, G., Pfeifhofer, H.W., Stabentheiner, E., and Sawidis, T. (2002). Glandular hairs of *Sigesbeckia jorullensis* Kunth (Asteraceae): Morphology, Histochemistry and Composition of Essential Oil. *Annals of Botany* 89: 459-469.

Hjorth, M., Mondolot, L., Buatois, B., Andary, C., Rapior, S., Kudsk, P., Mathiassen S. K., and Ravn, H.W. (2006). An easy and rapid method using microscopy to determine herbicide effects in Poaceae weed species. *Pest Management Science* (*Pest Manag Sci)* 62: 515–521.

Holz, F.G., Bellman, C., Staudt, S., Schütt, F., and Völcker, H.E. (2001). Fungus autofluorescence and development of geographic atrophy in age-related mucular degeneration. *Investigative Ophthalmology and Visual Science* 42: 1051-1956.

Hoogkamp, T.J.H., Chen, W.-Q., and Niks, R.E. (1998). Specificity of prehaustorial resistance to *Puccinia hordei* and to two inappropriate rust fungi in barley. *Phytopathology* 88: 856-861.

Horner, H.T. and Zindler-Frank, E. (1982). Calcium oxalate crystals and crystal cells in leaves of *Rhynchosia caribaea* (Leguminosae: Papilionoideae). *Protoplasma* 111: 10-18.

Host, P.W.D. (1991). *Quantitation Fuorescence Microscopy*, Vol. 1 and 2, Cambridge Univ. Press. Cambridge, UK.

Hradetzky, D. and Wollenweber, E. (1987). Flavonoids from the leaf resin of snakeweed, *Gutierrezia sarothrae. Z.Naturforsch.* 42 c: 71-76.

Huisen, Y. and Xinfu, X. (2003). A study on fluorescence spectra of Schiff bases. *Chemical Journal of Internet* 5(2): 17-23.

Hulshoff, A. and Lingeman, H. (1985). Fluorescence detection in chromatography. *Molecular Luminescence Spectroscopy. Methods and Applications.* Part 1. pp. 621-715. Wiley - Interscience Publ.New York, Chichester, Brisbane, Toronto, Singapore.

Hurel-Py, G. (1959). Recherches sur le comportement des prothalles de quelques *Equisetum* en milieu aseptique. *Revue Generale deBotanique* 66: 419-449.

Igura, I. (1956). Determination of electric charge and rH by means of the staining methods in prothallia, especially in spermatozoids of ferns. *Bot. Mag.* (Tokyo) 69: 820-821.

Inglett, C.E., Miller, R.R., and Lodge, L.P. (1959). Detection of pollen flavonoids by fluorescence on impregnated papers. *Microchimica Acta* 1: 95-100.

Jabaji-Hare, S.H., Perumalla, C.J., and Kendrick, W.B. (1984). Autofluorescence of vesicles, arbuscules, and intercellular hyphae of a vesicular-arbuscular fungus in leek (*Allium porrum*) roots. *Canadian Journal Botany* 62: 2665-2669.

Jameson, D.M. Reinhart, G.D. (eds) (1989). *Fluorescent Biomolecules.* Plenum Press. New York, London.

Johnson. G.A., Mantha, S.V., and Day, T.A. (2000). A spectrofluorometrie survey of UV-induced blue-green fluorescence in foliage of 35 species. *J. Plant Physiol.* 156: 242-252.

Johri, B.M. and Vasil, I.K. (1961). Physiology of pollen. *Botanical Review* 27: 326-381.

Kaloshina, N.O., Mozul, V.I., Mazulin, O.V., Stoyanovich, S.S., Dmitrieva, S.A., and Burba, O.N. (1992). *Achillea inundata Kondr.* as possible source of azulenes. *Pharmaceutical Zhurnal.* (Ukraine) 2(1): 50-53.

Karnaukhov, V.N. (1972). Spectral methods for investigations of energy regulation in living cells. In: *Biophysics of Living Cell.* G.M. Frank and V.N. Karnaukhov (eds) pp. 100–117. Institute of Biophysics. Pushchino, Russia.

Karnaukhov, V.N. (1973). On the nature and function of yellow ageing pigment lipofuscin. *Exp. Cell Res* 80: 479-483.

Karnaukhov, V.N. (1978). *Luminescent Spectral Analysis of Cell.* Nauka. Moscow, Russia.

Karnaukhov, V.N. (1988a). *Spectral Analysis of Cell in Ecology and Environmental Protection (Cell Biomonitoring).* Biol. Center of USSR Academy of Sciences.Pushchino, Russia

Karnaukhov V.N. (1988b) *Biological Functions of Carotenoids.* Moscow: Nauka.

Karnaukhov, V.N. (1990). Carotenoids: Progress, problems and prospects. *Comp Biochem Physiol* B 95 (1): 1-20.

Karnaukhov, V.N. (2001). *Spectral Analysis in Cell-level Monitoring of Environmental State.* Nauka. Moscow, Russia.

Karnaukhov, V.N., Yashin, V.A., Kulakov, V.I., Vershinin, V.M. and Dudarev, V.V. (1981). Einrichtung zur Untersuchung von Lumineszenzeigenschaften der Microobjekte. *DDR Patent* N147002: 1-32.

Karnaukhov, V.N., Yashin, V.A., Kulakov, V.I., Vershinin, V.M., and Dudarev, V.V. (1982). Apparatus for investigation of fluorescence characteristics of microscopic objects. *US Patent, N4,* 354, 114: 1-14.

Karnaukhov, V.N., Yashin, V.A., Kulakov, V.I., Vershinin, V.M., and Dudarev, V.V. (1983). Apparatus for investigation of fluorescence characteristics of microscopic objects. *Patent of England* 2.039.03 R5R.CHI.

Karnaukhov, V.N., Yashin, V.A., and Krivenko, V.G. (1985). Microspectrofluorimeters. *Proc. of 1-st Sov.-Germany Intern. Symp. Microscop. Fluorimetry and Acoustic Microscopy,* pp.160-164. Nauka. Moscow, Russia.

Karnaukhov, V.N., Yashin, V.A., Kazantsev, A.P., Karnaukhova, N.A., Kulakov, V.I. (1987). Double-wave microfluorimeter-photometer based on standard attachment. *Tsitologia* (*Cytology*, USSR) 29: 113-116.

Kasten, F.H. (1981). Methods for fluorescent microscopy. In: *Staining Procedures.* Fourth Edition. G. Clarck (ed). pp. 39-103. Williams and Wilkins. Baltimore, London.

Katz, M. and Robinson, G. (1986). Nutritional influences on autooxidation, lipofuscin accumulation and aging. In: *Free Radicals, Aging and Degenerative Diseases.* (Ed. R. Alan). pp. 221-259. New York: Allan Liss.

Kawai, H., Cao, L., Dunn, S.M; Dryden ,W.F., and Raftery, M.A. (2000). Interaction of a semirigid agonist with *Torpedo* acetylcholine receptor. *Biochemistry* 39(14): 3867-3876.

Kautsky, H. (1939). Quenching of luminescence by oxygen. *Trans Faraday Soc.* 35: 216-219.

Katoh, Y. (1957). Experimental studies on rhizoid-differentiation of certain ferns. *Phyton* 9(1): 25-40.

Kelsey, R.G. and Shafizadeh, F. (1980). Glandular trichomes and sesquiterpene lactones of Artemisia nova (Asteraceae). *Biochem. Syst. Ecol.* 8: 371-377.

Kendrick, J. and Knox, R.B. (1981). Structure and histochemistry of the stigma and style of some Australian species of *Acacia. Austral. J. Bot.* 29: 733-745.

Kim, S., Mollet, J.C., Dong, J., Zhang, K., Park, S.Y., and Lord, E.M. (2003). Chemocyanin, a small basic protein from the lily stigma, induces pollen tube chemotropism. *Proceedings of Natl. Acad. Sci. USA* 100 (26): 16125-16130.

Knox, R.B. (1979). Pollen and allergy. In: *Studies in Biology*. V. 107. Arnold. London, UK.

Knox, R.B. (1984). Pollen-pistil interactions. In: *Cellular Interactions*. H.F. Linskens and J. Heslop-Harrison Encyclopedia of Plant Physiology., New Ser., Vol. 17., pp. 508-608. Springer, Berlin, Heidelberg, New York.

Knox, R.B., Clarke, A., Harrison, S., Smith, P. and Marchaloni, J.J. (1976). Cell recognition in plants: determinants of the stigma surface and their pollen interactions. *Procedings of National Academy of Sciences of USA*, 73: 2788-2792.

Koga, H. (1994). Hypersensitive death, autofluorescence, and ultrastructural-changes in cells of leaf sheaths of susceptible and resistant near-isogenic lines of rice (PI-Z(T)) in relation to penetration and growth of *Pyricularia oryzae*. *Can J Bot* 72 (10): 1463–1477.

Kohen, E., Thorell, B., Hirschberg, J.G., Wouters, A.W., Kohen, C., Bartick P., Salmon, J.M., Viallet, P., Schachtschabel, D.O., Rabinovitch A., Mintz D., Meda P., Westerhoff, H., Nestor, J., and Ploem, J.S. (1981). Microspectrofluorimetric procedures and their applications in Biological Systems. In: *Modern Fluorescence Spectroscopy*. E.L. Wehry (ed). pp. 295-346. Plenum Press. New York, London.

Kolesnikova, R.D., Chernodubov, A.I. and Deryuzhkin, R.I. (1980). The composition of essential oils of some species belonging to genera *Pinus L.* and *Cedrus L. Rastitelnye Resursy* (Plant Resources, USSR) 16: 108-112.

Konovalov, D.A. (1995). Natural azulenes. *Rastitelnye Resursi* (*Plant Resources*, Russia) 31, 101-132.

Konovalov, D.A. and Starykh, V.V. (1996). Sesquiterpene lactones - phytotoxic substances of plants. In: *Proceedings of 1st All-Russian Conference on Botanical Resurces* A.L. Budantsev (ed) pp. 201-202. Botanical Institute of RAS. Sankt Peterburg.

Konstantinova–Schlesinger, M.A. (ed) (1961). *Luminescent Analysis*. Phys. Mat. Lit. Moscow, Russia.

Kopach, M., Kopac, C., and Klachek, B. (1980). Luminescent properties of some flavonoids. *Zh. Org. Khem* (*J. Organic Chemistry*, USSR). 16 (8): 1721-1725.

Koun, J., Baba, N., Ohni, Y., and Kawano, N. (1988). Triterpenoids from *Agrimonia pilosa*. *Phytochemistry* 27(1): 297-299.

Kovaleva, L.V. and Roshchina, V.V. (1999). Indication of the S-gene (related to a self-incompatibility) expression based on autofluorescence of the system pollen-pistil. In: *Abstracts of 4th All-Russian Congress of Plant Physiology*, Moscow, 4-9 October 1999, p. 598. Inst. of Plant Physiology RAS. Moscow, Russia.

Koteeva, N.K. (2005). New structural type of plant cuticle. *Doklady RAS* 403 (2): 283-285.

Kowalski, S.P., Eannett, N.T., Hirzel, A.T and Steffens, J.C. (1992). Purification and characterization of polyphenol oxidase from glandular trichomes of *Solanum berthaultii*. *Plant Physiol* 100: 677-684.

Kraft, M., Weigel, H.J., Mejer, G.J. and Brandes, F., (1996). Reflectance measurements of leaves for detecting visible and non-visible ozone damage to crops J. Plant Physiol. 148 (1-2): 148-154.

Kubin, A.,Wierrani, F., Burner, U., Alth. G., and Grünberger, W. (2005). Hypericin – the facts about a controversial agent. *Current Pharmaceutical Design* 11: 233-253.

Kumar, S. and Singh, D. (1995). Allelopathy in sustainable agriculture, forestry and environment – a review of an international symposium. *Current Research on Medical and Aromatic Plants* 17: 29-41.

Kupchan, S.M., Cassady, J.M., Kelsey, J.E., Schnces, N.K., Smith, D.H., and Burlingame, A.L. (1966). Structural elucidation and high-resolution mass spectrometry of gaillardin, a new cytotoxic sesquiterpene lactone. *Journal of American Chemical Society* 88: 5292-5302.

Kuznetsova, G.A. (1967). *Natural Coumarins and furanocoumarins*. Nauka. Leningrad, Russia.

Kuzovkina, I.N., Chernysheva, V.V., and Alterman.I.E. (1979). Characteristics of callus tissue strain *Ruta graveolens* produced rutacridone. *Soviet Union Plant Physiology* 26: 492-500.

Kuzovkina, I.N., Ladygina, E.Ya., and Smirnov, A.M. (1975). Root tissue of *Ruta graveolens* studied by luminescence microscopy. *Soviet Union Plant Physiology* 22(3): 598-600.

Kuzovkina, I.N, Roshchina, V.V., Alterman, I.E. and Karnaukhov, V.N. (1999). Study of cultivated cells and pRiT-DNA transformed roots of *Ruta graveolens* by luminescent microscopy. In: *Abstracts of 4th All-Russian Congress of Plant Physiology*, Moscow, 4-9 October 1999, Vol. 2. p. 613. Inst. of Plant Physiology RAS. Moscow, Russia.

Labas, Y. A., Gurskaya, N. G., Yanushevich, Y. G., Fradkov, A. F., Lukyanov, K. A., Lukyanov, S. A., and Matz, M. V. (2002). Diversity and evolution of the green fluorescent protein family. *Proc. Natl. Acad. Sci. USA* 99 (7): 4256-4261.

Labas, Yu. A., Gordeeva, A.V., and Fradkov, A.F. (2003). The light and color of living organisms Fluorescent and Color Proteins. *Priroda* (Nature, Russia) 3: 1051-1061.

Lakowicz, J.R. (1983). *Principles of Fluorescence Spectroscopy*. Plenum Press. New York, USA.

Lakowicz, J.R. (1999) *Principles of Fluorescence Spectroscopy*. Second Edition. Plenum Press. New York and London.

Lang, M., Lichtenthaler, H.K., Sowinska, M., Summ, P., Heisel, F. (1994). Blue, green and red fluorescence signatures and images of tobacco leaves. *Bot. Acta* 107: 230-236.

Leon, J., Lawton, M.A., and Raskin, I. (1995). Hydrogen peroxide stmulates salicylic acid biosynthesis in Tobacco. *Plant Physiol.* 108 (4): 1673-1678.

Lersten, N.R. and Curtis, J.D. (1989). Foliar oil reservoir anatomy and distribution in *Solidago canadensis* (Asteraceae, tribe Astereae). *Nordic Journal of Botany* 9: 281-287.

Lersten, N.R. and Curtis, J.D. (1991). Laminar hydathodes in Urticaceae: survey of tribes and anatomical observations on *Pilea pumila* ans *Urtica dioica*. *Plant Systematics and Evolution* 176: 179-203.

Leuschner, R.M. (1993) Human biometeorology, Part II. *Experientia* 49: 931-942.

Lichtentaler, H.K (Ed) (1988). *Application of Chlorophyll Fluorescence*. Kluwer Academic Publishers. Dordrect, Boston, London.

Lichtenthaler, H.K. and Miehe, J.A. (1997). Fluorescence imaging as a diagnostic tool for plant stress. *Trends Plant Sci.* 2: 316-320.

Lichtenthaler, H.K. and Rinderle, U. (1988). The role of chlorophyll fluorescence in the detection of stress conditions in plants. *CRC crit. Rev. anal. Chem.* 19: S29-S85.

Lichtenthaler, H.K. and Schweiger, J. (1998). Cell wall bounded ferulic acid, the major substance of the blue-green fluorescence emission of plants. *J. Plant Physiol.* 152: 272-282.

London, E. and Feigenson, G.W. (1981). Fluorescence quenching in model membranes. 1. Characterization of quenching caused by a spin-labelled phospholipids. *Biochemistry* 20 (7): 1932-1938.

Lynn, D.Y.C. and Luh, B.S. (1964). (1964). Anthocyanin Pigments in Bing Cherriesa. *Journal of Food Science* 29 (6), 735-743.

Mann, J.L. (1983). Autofluorescence of fungi: an aid to detection in tissue sections. *Am. J. Clin .Pathol.* 79(5): 587-590.

Maquire, Y.P. and Haard, N.F. (1976). Fluorescent product accumulation in ripening fruit. *Nature* (London.) 258 (556): 599-600.

Martin, F.W. (1969). Compounds from the stigmas of ten species. *Amer. J. Bot.* 56(9): 1023-1027.

Mascarenhas, I.P. (1975). The biochemistry of angiosperm pollen development. *Bot Rev.* 41: 259-314.

Mashkovskii, M.D. (2005). *Medicinal Drugs*. 15th Edition. 2 Volumes. Meditsina. Moscow, Russia.

Mathesius, U., Bayliss, C., Weinman, J.J., Schlaman, H.R.M., Spaink, H.P., Rolfe, B.G., McCully, M.E, and Djordjevic, M.A. (1998). Flavonoids synthesized in cortical cells during nodule initiation are early developmental markers in white clover. *Molecular Plant Microbe Interaction* 11 (12): 1223–1232.

Matz, M.V., Fradkov, A.F., Labas, Y.A., Savitsky, A.P., Zaraisky, A.G., and Markelov, S.A. (1999). Fluorescent proteins from nonbioluminescent Anthozoa species. *Nature Biotechnol.* 17. (10): 969-973.

McCormick, S. (1993). Male gametophyte development. *The Plant Cell* 5: 1265-1275.

Meisel, M.N. and Pomoshchinikova, N.A. (1952). The excretory and reductive function of yeast cell. *Trudy of the Institute of Microbiology of USSR* 2: 51-63.

Meisel, M.N. and Gutkina, A.V. (1961). Luminescence microscopy in biology and medicine. In: *Luminescent Analysis.* (Ed. M.A. Konstantinova–Schlesinger). pp. 309-324, Moscow: Phys. Mat. Lit.

Melnikova, E.V., Roshchina, V.V., and Karnaukhov, V.N., (1997a). Microspectrofluorimetry of pollen. *Biophysics* (Russia) 42 (1): 226-233.

Melnikova, E.V., Roshchina, V.V., Spiridonov, N.A., Yashin, V.A., and Karnaukhov, V.N. (1997b). Express-analysis of pollen, perga and propolis by method of microspectrofluorimetry. *Abstracts of 1st Intern. Symposium" Fundamental Sciences and Alternative Medicine"*, 22-25 September 1997, pp. 105-106. Pushchino: Russian Foundation Of Fundamental Studies.

Melnikova, E.V., Roshchina, V.V., Semenova, G.A., Kovaleva, L.V. (2003). Microanalysis of pollen viability. *Biophysics of Living Cell* (in Russian) 7: 145-148.

Merzlyak, M.N., Plakunova, O.V., Gostimsky, S.A., Rumyantseva, V.B. and Kovak, K. (1984). Lipid peroxidation in and photodamage to a light-sensitive chlorescence pea mutant. *Physiologia Plantarum.* 62 (2): 329-334.

Merzlyak, M.N. (1988). Liposoluble fluorescent "ageing pigments" in plants. In: *Lipofuscin - 1987: State of the Art.* I. Nagy (ed). pp. 451-452. Academ. Kiado, Elsevier. Budapest, Amsterdam.

Merzlyak, M.N. (1989). *Active Oxygen and Oxidative Processes in Membranes of Plant Cell.* Itogi Nauki i Tekhniki (Ser. Plant Physiology, Vol. 6.) VINITI, Moscow, Russia.

Merzlyak, M.N. and Pogosyan, S.I. (1988). Oxygen Radicals and lipid peroxidation in plant cell. In: *Oxygen Radicals in Chemistry, Biology and Medicine.* I.B. Afanas'ev (ed). pp. 232-253. Medicinal Institute. Riga, Latvia.

Merzlyak, M.N. and Zhirov, B.K. (1990). Free radical oxidation in chloroplasts at the ageing of plants. Itogi Nauki i Tekhniki, Ser. Biophysics, Vol. 40. pp. 101-125. VINITI, Moscow, Russia.

Mishra, B., Baric, A., Priyadarsini, K.I., and Mohan, H. (2005). Fluorescence spectroscopic studies binding of a flavonoid antioxidant quercetin to serum albumins. *J. Chem Sci.* 117(6): 641-647.

Miyawaki, A. and Tsien, R.Y. (2000). Monitoring protein conformations and interactions by fluorescence resonance energy transfer between mutants of green fluorescent protein. *Methods in Enzymology* 321: 472-500.

Mochalin, V.B. and Porshnev, Yu. N. (1977). Trends in chemistry of azulene. *Usp. Khim. (Trends in Chemistry, USSR)* 44(6): 1002-1040.

Morales, F., Cerovic, Z.G., and Moya, I. (1996). Time-resolved blue-green fluorescence of sugar beet *(Beta vulgaris* L.) leaves. Spectroscopic evidence for the presence of ferulic acid as the main fluorophore of the epidermis. *Biochim. biophys. Acta* 1273: 251-262.

Morales, F., Cerovic, Z.G., Moya, I. (1998). Time-resolved blue-green fluorescence of sugar beet leaves. Temperature-induced changes and consequences for the potential use of blue-green fluorescence as a signature for remote sensing of plants. *Aust. J. Plant Physiol.* 25: 325-334.

Mosebach, G. (1943). Uber die Polarisierung der *Equisetum*-Spore durch das Licht. *Planta.* 49: 61-90.

Mo, Y., Nagel, C., and Taylor, L.P. (1992). Biochemical complementation of chalcone synthase mutants defines a role for flavonoids in functional pollen. Proc. Natl. Acad. Sci. USA 89: 1713-1717.

Murata, S., Iwanaga, C., Toda, T., and Kokobun, H. (1972). Fluorescence and radiationless transitions from the second excited states of azulene derivatives. *Ber. der Bunsen-Gesellschaft.* 76 (10): 1176-1183.

Murav'eva, D.A. (1981). *Pharmacognozie.* Meditsina. Moscow, Russia.

Muravnik, L.E. (2000). The ultrastructure of the secretory cells of glandular hairs in two *Drosera* species as affected by chemical stimulation. *Russian Journal of Plant Physiology* 47 (4): 540-548.

Murphy, S.D. (1992). The determination of the allelopathic potential of pollen and nectar. In: *Plant Toxin Analysis.* H.F. Linskens and I.F. Jackson (eds). pp. 333-357. Springer-Verlag. Berlin, Heidelburg, Germany.

Murphy, S.D. (1999) Is there a role for pollen allelopathy in biological control of weeds. In: *Allelopathy Update,* S.S. Narwal (ed) Vol. 2, pp. 321-332. Science Publishers. Enfield, USA.

Murphy, S.D. (2007). Allelopathic Pollen: Isolating the Allelopathic Effects. In: *Cell Diagnostics.* V.V. Roshchina and S.S. Narwal (eds) pp. 185-198. Science Publisher. Enfield, Plymouth.

Murray, R.D.H., Mendez, J., and Brown, S.A. (1982). *The Natural Coumarins: Occurrence, Chemistry and Biochemistry.* J. Wiley and Sons. Chichester, UK.

Nakagawara, S., Katoh, K., Kusumi, T., Komura, H., Nomoto, K., Konno, H., Huneck, S., and Takeda, R. (1992). Two azulenes produced by the liverwort, *Calypogeia azurea,* during *in vitro* culture. *Phytochemistry* 31 (5): 1667-1670.

Nakazawa, S. (1956). The latent polarity in *Equisetum* spores. *Botanical Magazine* 69 (820-821): 506-509.

Nakazawa, S. (1958). The rhizoid point as the basophilic center of *Equisetum* spores. *Phyton* 10 (1): 1-6.

Nau, W.M. and Wang, X. (2002). Biomolecular and supramolecular kinetics in the submicrosecond time range: the fluorazophore approach. *Chem. Phys. Chem (European Jounal of Chemical Physics and Physical Chemistry* 3 (5): 393-398.

Nau, W.M., Huang, F., Wang, X., Bakirci, H., Gramlich, G., and Marquez, C. (2003). Exploiting long-lived molecular fluorescence. *Chimie* 57: 161-167.

Nienburg, W. (1924). Die Wirkung des Lichtes auf die Keimung der *Equisetum*-Spore. *Ber. Dtsch. Bot. Ges.* 42: 95-99.

Ortega, R.C., Anaya, A.L. and Ramos, L. (1988). Effects of allelopathic compounds of corn pollen on respiration and cell division of watermelon. *J. Chem. Ecol.* 14: 71-86.

Palevitz, B.A., O'Kane, D.J., Kobres, R.E. and Raikhnel, N.V. (1981). The vacuolar system in stomatal cells of *Allium*. Vacuole movements and changes in morphology in differentiating cells as revealed by epifluorescence, video and electron microscopy. *Protoplasma*. 109: 23-55.

Parker, C.A. (1968). *Photoluminescence of solutions*. Elsevier, Amsterdam, the Netherlands

Pawley, J.B. (Ed.). (1990). *Handbook of Biological Confocal Microscopy*. Plenum. New York, USA.

Pawley, J. and Pawley J.B. (2006). *Handbook of Biological Confocal Microscopy*. Springer-Verlag: Berlin, Heidelberg, 985 pp.

Peschek, G.A., Hinterstoisser, B., Pineau, B., and Missbichler, A. (1989). Light-independent NADPH - protochlorophyllide oxidoreductase activity in purified plasma membrane from the cyanobacterium *Anacystis nidulans*. *Biochem. Biophys. Res. Commun.* 162(1): 71-78.

Peumans, W.J., Smets, K., Van Nerum K., Van Leuven F. and Van Damme, E.J.M. 1997. Lectin and alliinase are the predominant proteins in nectar from leek (*Allium porrum* L.) flowers. *Planta* 201: 298-302.

Pheophilov, P.P. (1944). The connection of fluorescence of organic dyes with their chemical structures. *Doklady USSR Acad. Sci*, 45 (9): 587-590.

Phillips, N. J., J. T. Goodwin, et al. (1989). "Characterization of the Fusarium toxin equisetin: the use of phenylboronates in structure assignment." *Journal of the American Chemical Society* 111(21): 8223-31.

Piknova, B., Marsh, D., and Thompson, T.E. (1996). Fluorescence-quenching study of percolation and compartmentalization in two-phase lipid bilayers. *Biophysical Journal* 71 (2): 892-897.

Piszczek, G. (2006). Luminescent metal-ligand complexes as probes of macromolecular interactions and biopolymer dynamics. *Arch Biochem Biophys.* 453 (1): 54-62.

Ploem, I.S., and Tanke, H.J. (1987). *Introduction to Fluorescence Microscopy*. Oxford University Press. Oxford, UK.

Popravko, S.A., Gurevich, A.I. and Kolosov, M.N. (1969). Flavonoid components of propolis. *Chemistry of Natural Compounds* (USSR) 6: 476-482.

Popravko, S.A., Sokolov, I.V., and Torgov, I.V. (1982). New natural phenol triglycerides. *Chemistry of Natural Compounds* (USSR) 2: 169-173.

Posner, G.H., Cumming, J., Ploypradith, P., and Chang, H.O. (1995). Evidence for Fe (IV) = O in the molecular mechanism of action of the trioxane antimalarial artemisinin. *J. Am. Chem. Soc.* 117: 5885-5886.

Pryer, K.M., Schneider, H., Smith, A.R., Cranfill, R., Wolf, P.G., Hunt, J.S. and Sipes, S.D. (2001). Horsetails and ferns are a monophyletic group and the closest living relatives to seed plants. *Nature* 409: 618-622.

Racz-Kotilla, E., Adam, S., and Galin, D. (1968). Protective effect of certain antiulcer medicines associated with azulene on experimental gastric ulcer. *Rev. Med* (Targu-Mures). 14: 331-334.

Radice, S. and Galati, B.G. 2003. Floral nectary ultrastructure of *Prunus persica* (L.) Batch cv. Forastero (Newcomer), an Argentine peach. *Plant Systematics and Evolution* 238: 23-32.

Rao, A.S. (1990). Root flavonoids. *Botanical Review*. 56: 1-90.

Reichling, J and Beiderbeck, R. (1991). X. *Chamomilla recutita* (L.) Rauschert (Camomile) : in vitro culture and the production of secondary metabolites. In: *Biotechnology in Agriculture and Forestry* (Ed. Y.P.S. Bajaj) vol. 15, Medicinal and Aromatic Plants III. pp 156-175. Berlin, Hedelberg: Springer-Verlag.

Reicosky, D.A. and Hanover, J.W. (1978). Physiological effects of surface waxes. Light reflectance for glaucous and nonglaucous *Picea pungens*. *Plant Physiol.* 62 (1): 101-104.

Reigosa Roger, M.J. and Weiss, O. (2001). Fluorescence technique. In: *Handbook of Plant Ecophysiology Technique.*, M.J. Reigosa Roger(ed). pp. 155-171 Kluwer Acad. Publ. Dordrecht, the Netherlands.

Reiling, J. and Beiderbeck, R. (1991). X. Chamomilla recutita (L.) Rauschert (Camomile): in vitro culture and the production of secondary metabolites. In: *Biotechnology in Agriculture and Forestry* Y.P.S. Bajaj (ed) Vol. 15, Medicinal and Aromatic Plants III. pp. 156-175. Springer-Verlag. Berlin, Heidelburg, Germany.

Rekka, E.A., Kourounakis, A.P., and Kourounakis, P.N. (1996). Investigation of the effect of chamazulene on lipid peroxidation and free radical processes. *Research Communications in Molecular Pathology and Pharmacology* 92 (3): 361-364.

Rekka, E.A., Chrysselis, M., Siskou, I. and Kourounakis, A.P. (2002). Synthesis of new azulene derivatives and study of their effect on lipid peroxidation and lipogenase activity. *Chem. Pharm. Bull.* 50 (7): 904-907.

Rice, E.L. (1984). *Allelopathy*. Academic Press. New York, San Francisco, London.

Ribachenko, A.T. and Georgievskii, V.P. (1975). Fluorescence characteristics of oxytransformed flavonoids. *Doklady AN Ukraine SSR*, Ser. B., 11: 1009-1111.

van Riel, M., Hammans, J.K., van den Ven, M., Verwer, W., and Levine, Y.K. (1983). Fluorescence excitation profiles of b-carotene in solution and in lipid/water mixtures *Biochem. Biophys. Res. Commun.* 113(1): 102-107.

Rodriguez, E., Towers, G.H.N., and Mitchell, J.C. (1976). Biological activities of sesquiterpene lactones. *Phytochemistry* 15: 1573-1580.

Rohringer, R., Kim, W.K., Samborski, D.J., and Howes, H.K. (1977). Calcofluor: An optical brightener for fluorescence microscopy of fungal plant parasites in leaves. *Phytopathology* 67: 808-810.

Roshchina, V.D. and Roshchina, V.V. (1989). *The Excretory Function of Higher Plants.* Nauka, Moscow, Russia.

Roshchina, V.V. (1991). *Biomediators in plants. Acetylcholine and biogenic amines.* (Biomediatori v rasteniakh. Atsetilkholin I biogennie amini). Biological Center of USSR Academy of Sciences. Pushchino, Russia.

Roshchina, V.V. (1996). Volatile plant excretions as natural antiozonants and origin of free radicals. In: *Allelopathy. Field Observation* and *Metodology.* S.S. Narwal and P. Tauro (eds). pp. 74-79. Scientific Publishers. Jodhpur, India.

Roshchina, V.V. (1999a). Mechanisms of cell-cell communication. In: *Allelopathy Update.* S.S. Narwal (ed). Vol. 2, pp. 3-25. Science Publishers Enfield, New Hampshire, USA.

Roshchina, V.V. (1999b). Chemosignalization at pollen. *Uspekhi Sovremennoi Biologii (Trends in Modern Biology, Russia)* 119: 557-566.

Roshchina, V.V. (2001a). *Neurotransmitters in Plant Life.* Science Publisher: Einfield New Hampshire, Plymouth.

Roshchina, V.V. (2001b). Molecular-cellular mechanisms in pollen alllelopathy. *Allelopathy Journal.* 8 (1): 11-28.

Roshchina, V.V. (2001c). Autofluorescence of plant cells as a sensor for ozone. In: *Abstracts of 7 th Conference on Methods and Applications of Fluorescence: Spectroscopy, Imaging and Probes.* p. 162 16-19 September 2001. Amsterdam, the Netherlands.

Roshchina, V.V. (2001d). Rutacridone as a fluorescent probe. In: *Abstracts of 7th Conference on Methods and Applications of Fluorescence: Spectroscopy, Imaging and Probes.* p. 161 16-19 September 2001. Amsterdam, the Netherlands.

Roshchina, V.V. (2002). Rutacridone as a fluorescent dye for the study of pollen. *Journal of Fluorescence* 12: 241-243.

Roshchina, V.V. (2003). Autofluorescence of Plant Secreting Cells as a Biosensor and Bioindicator Reaction. *Journal of Fluorescence* 13 (5): 403-420.

Roshchina, V.V. (2004a). Cellular models to study the allelopathic mechanisms. *Allelopathy Journal* 13 (1): 3-16.

Roshchina, V.V. (2004b). Plant microspores as unicellular models for the study of relations between contractile components and chemosignaling. In: *Biological Motility.* Z.A. Podlubnaya (ed) pp. 194-196 International Symposium, Pushchino, 23 May-1 June 2004, pp. 194-196. ONTI. Pushchino.

Roshchina V.V. (2004c) Allelochemicals as possible fluorescent probes // *Abstracts of the Second European Allelopathy Symposium "Allelopathy from understanding to application"*, June 3-5 2004, Pulawy, Poland, p. 160. Pulawy: Polland Acad. Sci.

Roshchina V.V. (2004d) Unicellular plant biosensors for ozone // 8 th World Congress on Biosensors 2004, Elsevier, Granada, Spain. May 25 – 30 2004, Granada, Spain, P. BS12.

Roshchina, V.V. (2005a). Contractile proteins in chemical signal transduction in plant microspores. *Biological Bulleten*, Ser.Biol. 3: 281-286.

Roshchina, V.V. (2005b). Allelochemicals as fluorescent markers, dyes and probes. *Allelopathy Journal* 16 (1): 31-46.

Roshchina, V.V. (2005c). Biosensors for the study of allelopathic mechanisms and testing of natural pesticides. In: *Proceedings of International Workshop on Protocols and Methodologies in Allelopathy*, G.L. Bansal and S.P Sharma (eds) pp 75-87. April 2-4, 2004 Palampur. College of Basic Sciences, Azad Hind Stores. Palampur, India.

Roshchina, V.V. (2005d). Ozone and living organisms. *Science in Russia* 2: 60-63.

Roshchina, V.V. (2005e). Proazulenes, azulene and colchicine as fluorescent dyes for the study of cellular interactions in allelopathy. In: *Proceedings and Selected Papers of the Fourth World Congress on Allelopathy*, 21-26 August 2005. J.D.I. Harper, M. An, H. Wu and J.H. Kent (eds) pp. 521-524 Charles Sturt University, Wagga: Wagga, NSW, Australia.

Roshchina, V.V. (2006a). Plant microspores as biosensors. *Trends in Modern Biology*, 126 (3): 262-274.

Roshchina, V.V. (2006b). Chemosignaling in plant microspore cells. *Biology Bulleten*, 3: 2.

Roshchina, V.V. (2007a). Cellular models as biosensors. In: *Plant Cell Diagnostics*. V.V. Roshchina and S.S. Narwal (eds). pp. 5-22. Science Publisher. Enfield, Plymouth.

Roshchina, V.V. (2007b). Luminescent cell analysis in allelopathy. In: *Plant Cell Diagnostics*. V.V. Roshchina and S.S. Narwal (eds). pp. 103-115. Science Publisher. Enfield, Plymouth.

Roshchina, V.V. and Karnaukhov, V.N. (1999). Changes in pollen autofluorescence induced by ozone. *Biologia Plantarum*. 42 (2): 273-278.

Roshchina, V.V. and Kukushkin, A.K. (1984). Antimycin A fluorescence in chloroplasts and model systems. *Biochemistry* (USSR) 49 (7): 1121-1126.

Roshchina, V.V. and Melnikova, E.V. (1995). Spectral analysis of intact secretory cells and excretions of plants. *Allelopathy J.* 2 (2) 179-188.

Roshchina, V.V. and Melnikova, E.V. (1996). Microspectrofluorimetry: A new technique to study pollen allelopathy. *Allelopathy Journal*. 3 (1): 51-58.

Roshchina, V.V. and Melnikova, E.V. (1998a). Chemosensory reactions at the interaction pollen-pistil. *Biological Bulleten*. 6: 678-685.

Roshchina, V.V. and Melnikova, E.V. (1998b). Allelopathy and plant generative cells. Participation of acetylcholine and histamine in a signalling at the interactions of pollen and pistil. *Allelopathy Journal*. 5: 171-182.

Roshchina, V.V. and Melnikova, E.V. (1999). Microspectrofluorimetry of intact secreting cells, with applications to the study of allelopathy. In: *Principles and Practices in Plant Ecology. Allelochemical Interactions*. Inderjit, K.M.M. Dakshini and C.L. Foy (eds). pp. 99-126. CRC Press.Boca Raton, Fl, USA.

Roshchina, V.V. and Melnikova, E.V. (2000). Contribution of ozone and active oxygen species in the development of cellular systems of plants. In: *Mitochondria, Cells and Active Oxygen Species.* V.P. Skulachev and V.P. Zinchenko (eds). pp. 120-127 Materials of International Conference, 6-9 June 2000, Pushchino. Biological Center of Russian Academy of Sciences. Pushchino, Russia.

Roshchina, V.V. and Melnikova, E.V. (2001). Chemosensitivity of pollen to ozone and peroxides. *Russian Plant Physiology* 48(1): 89-99.

Roshchina, V.V. and Roshchina, V.D. (1993) *The Excretory Function of Higher Plants.* Berlin, Heidelberg: Springer-Verlag. 314 pp.

Roshchina, V.V. and Roshchina, V.D. (2003). *Ozone and Plant Cell.* Kluwer Academ. Publ, . Dordrecht, the Netherlands.

Roshchina, V.V. and Semenova, M.N. (1995). Neurotransmetter system in Plants.Cholinesterase in excreta from flowers and secretory cells of vegetative organs of some species. In: *Proceedings of Plant Growth Regulator Society of America.* D. Green and G. Cutler (eds) pp. 353-357 22 Annual Meeting, 18-20 July 1995, Minneapolis, Fritz C.D. Minneapolis, Minnesota, USA.

Roshchina, V.V., Solomatkin, V.P., and Roshchina, V.D. (1980). Cicutotoxin as an inhibitor of electron transport in photosynthesis. *Soviet Plant Physiol.* 27(4): 704-709.

Roshchina, V.V., Ruzieva, R. Kh., and Mukhin, E.N. (1986). Capsaicin from the fruits of red pepper *Capsicum annuum* L. as a regulator of the photosynthetic electron transport. *Applied Biochemistry and Micobiology* (USSR) 22 (3): 403-409.

Roshchina, V.V., Mel'nikova, E.V., and Kovaleva, L.V. (1996). Autofluorescence in thge pollen -pistil system of *Hippeastrum hybridum. Doklady Biological Sciences* 349: 403-405.

Roshchina, V.V., Melnikova, E.V., Kovaleva, L.V., and Spiridonov, N.A. (1994). Cholinesterase of pollen grains. *Doklady Biological Sciences* 337: 424-427.

Roshchina, V.V., Melnikova, E.V., Spiridonov, N.A., and Kovaleva, L.V. (1995). Azulenes, the blue pigments of pollen. *Doklady. Biol. Sci.* 340 (1): 93-96.

Roshchina, V.V., Mel'nikova, E.V., Karnaukhov, V.N., and Golovkin, B.N. (1997a). Application of microspectrofluorimetry in spectral analysis of plant secretory cells. *Biological Bull* (Russia) 2: 167-171.

Roshchina, V.V., Melnikova, E.V. and Kovaleva, L.V. (1997b). The changes in the fluorescence during the development of male gametophyte. *Russian Plant Physiol.* 47 (1): 45-53.

Roshchina, V.V., Melnikova, E.V., and Spiridonov, N.A. (1997c). Spectral analysis of the pollen, perga and propolis. Azulenes and carotenoids. *Pharmacea* (Russia) 3: 20-23.

Roshchina, V.V., Mel'nikova, E.V, Mit'kovskaya, L.I., and Karnaukhov, V.N. (1998a). Microspectrofluorimetry for the study of intact plant secretory cells. *J. of General Biology* (Russia) 59: 531-554.

Roshchina, V.V., Mel'nikova, E.V., Gordon, R.Ya., Konovalov, D.A. and Kuzin, A.M. (1998b). A study of the radioprotective activity of proazulenes using a chemosensory model of *Hippeastrum hybridum* pollen. *Doklady Biophysics*. 358-360: 20-23.

Roshchina, V.V., Popov, V.I., Novoselov, V.I., Melnikova, E.V., Gordon, R.Ya., Peshenko, I.V., and Fesenko, E.E. (1998c). Transduction of chemosignal in pollen. *Cytologia*. (Russia) 40: 964-971.

Roshchina, V.V., Melnikova, E.V., Popov, V.I., Novoselov, V.I., Peshenko, I.V., Khutsyan, S.S., and Fesenko, E.E. (1998 d). Modelling of chemosignal transduction in pollen. In: *Reception and Intracellular Signalling*. V.P. Zinschenko (ed) pp. 244-247 International Conference, Pushchino, 21-25 September 1998, Institute of Cell Biophysics RAS. Pushchino, Russia.

Roshchina, V.V., Melnikova, E.V., Popov, V.I., Karnaukhov, A.V., Mit'kovskaya, L.I., Gorokhov, A.A., and Karnaukhov, V.N. (1998e). Principles of the bank data creation based on the pollen characteristics. Computer analysis of the fluorescence spectra for the cover composition consideration. In: *Cryoconservation of Genetic Resources*. pp. 221-225 Materials of All-Russian Conference 13-15 October 1998, Pushchino. Biological Center of RAS. Pushchino, Russia.

Roshchina, V.V., Mel'nikova, E.V., and Karnaukhov, V.N. (2000a). Fluorescence of plant cells. *Science in Russia* 6: 53-56.

Roshchina, V.V., Golovkin, B.N., Melnikova, E.V., Novoselov, V.I., and Gordon R.Ya. (2000b). Microanalys of pollen from hothouse (green house) plants. *Bulletin of Central Botanical Garden of Russian Academy of Sciences*. 180: 90-96.

Roshchina, V.V., Mel'nikova, E.V., Yashin, V.A., and Karnaukhov, V.N. (2002). Autofluorescence of intact spores of horsetail *Equisetum arvense* L. during their development. *Biophysics* (Russia) 47 (2): 318-324.

Roshchina, V.V., Miller, A.V., Safronova, V.G., and Karnaukhov, V.N. (2003a). Àctive Oxygen Species and Luminescence of Intact Cells of Microspores. *Biophysics* (Russia) 48: 283-288.

Roshchina, V.V., Yashin, V.A., and Kononov, A.V. (2003b). The study of vegetative microspores of *Equisetum arvense* by confocal microscopy and microspectrofluorimetry. In: *Abstracts of 8 th Conference on Methods and Applications of Fluorescence: Spectroscopy, Imaging and Probes*. p. 207 24-27 August 2003, Prague.

Roshchina, V.V.,Bezuglov, V. V. , Markova , L. N., Sakharova, N. Yu., Buznikov, G.A. (2003c) Fluorescence of Bodipy-neurotransmitters bounded with living cells. *Abstracts of 8th Conference on Methods and Applications of Fluorescence: Spectroscopy, Imaging and Probes*. p. 143, 24-27 August 2003, Prague.

Roshchina, V.V., Bezuglov, V. V., Markova, L. N., Sakharova, N.Yu., Buznikov, G.A., Karnaukhov, V.N., and Chailakhyan, L.M. (2003d). Interaction of living cells with fluorescent derivatives of biogenic amines. *Doklady Russian Academy of Sciences* 393 (6): 832-835.

Roshchina, V.V., Rodionov, A.V. and Bezuglov, V.V. (2003 e). Interactions of biogenic amines with microspores. In: *Materials of International Symposium."Reception and Intracellular Signaling"*. Pushchino, 6-8 June 2003. (Ed. V.P. Zinchenko). pp. 294-296.Pushchino: ONTI.

Roshchina, V.V., Yashin, V.A., and Kononov, A.V. (2004). Autofluorescence of plant microscopes studied by confocal microscopy and microspectrofluorimetry. *Journal of Fluorescence* 14 (6): 745-750.

Roshchina, V.V., Markova, L.N., Bezuglov, V.V., Buznikov, G.A., Shmukler, Yu.B., Yashin, V.A., and Sakharova, N.Yu. (2005). Linkage of fluorescent derivatives of neurotransmitters with plant generative cells and animal embryos. In: *Materials of International Symposium."Reception and Intracellular Signaling"*. Pushchino, 6-8 June 2005. (Ed. V.P. Zinchenko). pp. 399-402. Pushchino: ONTI.

Roshchina, V.V., Yashin, V.A., and Yashina, A.V. (2006). Colchicine as fluorescence marker of contractile elements in unicellular microspores. In: *Biological Motility*, Z.A. Podlubnaya (ed) pp. 101-103. International Symposium, 11-15 May 2006, Biological Center RAS. Pushchino, Russia.

Roshchina, V.V., Yashin, V.A., Kononov, A.V., and Yashina, A.V. (2007a). Laser-scanning confocal microscopy (LSCM) for the study of plant secretory cells. In: *Plant Cell Diagnostics*. V.V. Roshchina and S.S. Narwal (eds). pp. 93-102. Science Publisher. Enfield, Plymouth.

Roshchina, V.V., Yashina, A.V., Yashin, V.A., Prizova, N.K., Vikhlyantsev, I.V., and Vikhlyantseva, E.F. (2007 b). Chemosignaling in plant cell communication: Modelling with unicellular systems in allelopathy. In: *Materials of International Symposium."Reception and Intracellular Signaling"*. Pushchino, 5-7 June 2007. (Ed. V.P.Zinchenko). pp. 282-285. Pushchino: ONTI.

Roshchina, V.V., Kutis, I.S., Kutis, L.S., Gelikonov, V.M., and Kamensky, V.A. (2007c). The study of plant secretory structures by optical coherence microscopy. In: *Plant Cell Diagnostics* (Eds. V.V. Roshchina and S.S. Narwal). pp. 87-92. Science Publisher: Enfield , Jersey (USA), Plymouth.

Ross, A., Laws, W.R., Rousslang, K.W., and Wyssbrod, H.R. (1991). Tyrosine fluorescence and phosphorescence from proteins and polypeptides. In: *Topics in Fluorescence Spectroscopy*. J.R. Lakowicz (ed), Vol. 3, pp. 1-63. Plenum Press. New York and London.

Rost, F.W.D. (1995). *Fluorescence Microscopy*. Cambridge University Press. Cambridge, UK.

Rost, F.W.D. (2000) *Fluorescence Microscopy*. Second Edition: Cambridge University Press. Cambridge, UK.

Rowley, J.R. (1967). Fibrils, microtubules and lamellae in pollen grains. *Review of Paleobotany and Palynology* 3: 213-226.

Rowley, J.R., Mühlethalerm K., and Frey-Wyssling, A. (1959). A route for the transfer of materials through the pollen grain wall. *Journal of Biophysical and Biochemical Cytology* 6 (3): 537-538.

Rubinstein, B. and Luster, D.G. (1993). Plasma membrane redox activity: components and role in plant processes. *Annu Rev Plant Physiol Plant Mol Biol.* 44: 131-155.

Rühland, W. und Wetzel, K. (1924). Der Nachweis von Chloroplasten in den generativen Zellen von Pollenschläuchen. *Ber. Deutsch. Bot. Ges.* 42: 3-14.

Rüngeler, P., Castro, V., Mora, G., Gören, N., Vichnewski, W., Pahl, H.L., Merfort, I., and Schmidt, T.J. (1999). Inhibition of Transcription Factor NF-λB by Sesquiterpene Lactones – A Proposed Molecular Mechanism of Action. *Bioorganic and Medicinal Chemistry* 7: 2343-2352.

Ruzin, S.E. (1999). *Plant Microtechnique and Microscopy.* Oxford Univ. Press. Oxford, UK.

Ryan, J.D., Gregory, P., and Tingey, W.M. (1982). Phenolic oxidation activities in glandular trichomes on hybrids of *Solanum berthaultii. Phytochemistry* 8: 1885-1887.

Rybalko, K.S. (1978). *Natural Sesquiterpene Lactones.* Meditsina. Moscow, Russia.

Saito, Y., Takahashi, K., Nomura, E., Mineuchi, K., Kawahara, T.D., Nomura, A., Kobayashi, S., and Ishii, H. (1997). Visualization of laser-induced fluorescence of plants influenced by environmental stress with a microfluorescence imaging system and a fluorescence imaging lidar system. *SPIE* 3059: 190-198.

Saito, Y., Saito, R., Nomura, A., Kawahara, T.D., Nomura, A., Takaragaki, S., Ida, K., and Takeda, S. (1999). Performance check of vegetation fluorescence imaging lidar through *in vivo* and remote estimation of chlorophyll concentration inside plant leaves. *Opt. Rev.* 6: 155-159.

Sakurovs, R., and Ghiggino, K.P. (1983). Solvent effects on the fluorescence of coumaric acids. *J. Photochem.* 22 (4): 373-377.

Salih, A., Jones, A., Bass, D., and Cox, G. (1997). Confocal imaging of exine for grass pollen analysis. *Grana* 36: 215-224.

Santhanam, M., Houtala, R.R., Sweeny, J.G., and Iacobucci, G.A. (1983). The influence of flavonoid sulfonates on the fluorescence and photochemistry of flavylium cations. *Photochemistry and Photobiology* 38(4): 477-480.

Santhi, A., Kala, U.L., Nedumpara, R.J., Kurian, A., Kurup, M.R.P., Radhakrishnan, P., and Nampoori, V.P.N. (2004). Thermal lens technique to evaluate the fluorescence quantum yield of a Schiff base. *Applied Physics B* 79 (5): 629-633.

Satterwhite, M.B. (1990). Spectral luminescence of plant pollen. In: *Remote Sensing Science for the Nineties, IGARSS'* 90: R. Mills (ed) Vol. 3, pp. 1945-1948 10th Annu Int Geosci and Remote Sensing Symp. 20-24 May 1990, Univ. Maryland, College Park (Maryland): The Institute of Electrical and Electronic Engineers, INC. Washington, D.C., USA.

Sauers, R.R., Zampino, M., Stocki, M., Ferentz, I. and Shams, H. (1983). Synthesis and photophysical properties of some endo-6-substituted norcamphors and pentacyclo [5,4.0.$0^{2.0}$, $0^{3,10}$ $0^{5.9}$] undecan-8-ones. *J. Organic. Chem* 48: 1862-1866.

Sayed, O.H. (2003). Chlorophyll fluorescence as a tool in cereal crop research. *Photosynthetica* 41 (3): 321-330.

Saxena, I., Files, D.J., Rao, S.V., and Costerton, W.J. (2002). Autofluorescence-based bacteria detection using an optical fiber. In: *Optical Diagnostics of Living Cells*, V.D.L. Farkas, C. Robert and V. Leitf (eds) Vol. 4622, pp. 106-111 Proceedings of SPIE.

von Schanz, M. (1962a). Anwendung der Dünnschicht-Chromatographie zur Trennung der Filix-Phloroglucide. *Planta Medica* 10 (1): 22-27.

von Schanz, M. (1962b). Die quantitative Zuammenseitzung der Filix-Phloroglucide von *Dryopteris filix mas* und *Dryopteris austriaca* ssp. *Dilatata*. *Planta Medica* 10 (1): 98-106.

Schildknecht, P.H.P.A., Castro, M. de M., and Vidal, B.C. (2004). Histochemical analysis of the root epidermal mucilage in maize and wheat. *Can J Bot.* 82 (10): 1419-1428.

Schmidt, S.K. and Ley, R.E. (1999). Microbial competition and soil structure limit the expression of allelochemicals in Nature. In: *Principles and Practices in Plant Ecology. Allelochemical Interactions.* Inderjit, K.M.M. Dakshini and C.L. Foy (eds). pp. 339-351. CRC Press. Boca Raton, Fl., USA.

Schmidt, T.J. (1999). Toxic Activities of Sesquiterpene Lactones–Structural and Biochemical Aspects. *Current Organic Chemistry* 3: 577-605.

Schnabl, H., Weissenböck, G., and Scharf, H. (1986). In vivo-microspectrophotometric characterization of flavonol glycosides in *Vicia faba* guard and epidermal cells. *Journal of Experimental Botany* 37: 61-72.

Schopfer, P. (1996). Hydrogen peroxide mediated cell-wall stiffening *in vitro* in maize coleoptiles. *Planta* 199 (1): 43-49.

Schreck, R., Rieber, P., and Baeurle, P.A. (1991). Reactive oxygen intermediates as apparently widely used messengers in the activation of the NF-Lambda B transcription factor and HIV-1. *EMBO Journal* 10 (8): 2247-2258.

Schulman, S.G. (1985). Luminescent spectroscopy: an overview. In: *Molecular Luminescence Spectroscopy. Methods and Applications* S.G. Schulman (ed). pp. 1-28. Wiley.New York, Chichester, Brisbane, Toronto, Singapore.

Schur, J.M. (1991). Fluorescence studies of nucleic acids: Dynamics, rigidities and structures. In: *Topics in Fluorescence Spectroscopy*. J.R. Lakowicz (ed), Vol. 3, pp. 137-229. Plenum Press. New York and London.

Scott, R.J. (1994). Pollen exine - the sporopollenin enigma and the physics of pattern. In: *Molecular and Cellular Aspects of Plant Reproduction* R.J. Scott and M.A. Stead (eds). pp. 49-81. Cambridge University Press. Cambridge, UK.

Sejalon-Delmas, N., Magnier, A., Douds, D.D. Jr., and Becard, G. (1998). Cytoplasmic autofluorescence of an arbuscular mycorrhizal fungus *Gigaspora gigantea* and nondestructive fungal observations in planta. *Mycologia* 90 (50): 921-926.

Sestak, Z. (1999). Chlorophyll fluorescence kinetic depends on age of leaves and plants. - In: *The Chloroplast: From Molecular Biology to Biotechnology*. J.H. Argyroudi-Akoyunoglou and H. Senger (eds). pp. 291-296. Kluwer Academic Pub. Dordrect, Boston, London.

Sestak, Z. and Siffel, P. (1997). Leaf-age related differences in chlorophyll fluorescence. *Photosynthetica* 33: 347-369.

Shaner, N.C., Campbell., R.E., Steinbach, P.A., Giepmans, B.N.G., Palmer, A.E., and Tsien, R.Y. (2004). Improved monomeric red orange and yellow. Fluorescent proteins derived from *Discosoma* sp. red fluorescent protein. *Nature Biotechnology* 22(12): 1567-1572.

Shapovalov, A.A. (1973). Possibility and perspectives in application of fluorescent method of allelopathic investigations. In: *Physiologo-biochemical Base of the Plant Interactions in Phytocenosis*. A.M. Grodzinskii (ed) Vol. 4, pp. 128-130. Naukova Dumka. Kiev.

Shwartsburd, P.M. and Aslanidi, K.B. (1991). Spectrokinetic characteristics of two types of fluorescence of refractive granules in native individual cells from ascitic tumours. *Biomedical Science* 2: 391-397.

Siegel, U., Mues, R., Doenig, R., Eicher,T., Blechschmidt, M., and Becker, H. (1992). Ten azulenes from *Plagiochila longispina* and *Calypogeia azurea*. *Phytochemistry*., 31(5): 1671-1678.

Singh, H.P., Batish, D.R., and Kohli, R.K. (2001). Allelopathy in agrosystems and overview. In: *Allelopathy in Agrosystems*. R.K. Kohli, H.P. Singh, and D.R. Batish (eds). pp. 1-41. The Hanworth Press. New York, London.

Smekal, E. (1982). Quantum yield and lifetime measurements of some protoberberine alkaloids and their complexes with DNA. *Studia Biophysica* 87(2/3): 211-212.

Smekal, E. (1982). Fluorescence polarization of complexes of nucleic acids with alkaloids berberine and coralyne. *Studia Biophysica* 87(2/3): 213-214.

Smekal, E. and Pavelka, S. (1977). Interaction of berberine and its derivatives with biopolymers. Fluorescence properties of some berberine derivatives and related compounds with berbine structure. *Studia Biophysica* 64 (3): 183-192.

Smekal, E., Koudelka, J., and Hung, M.A. (1980). Fluorescence investigation of berberine-nucleic acid complexes. *Studia Biophysica* 81: 89-90.

Southon, I.W. and Buckingham, J. (1989). *Dictionary of Alkaloids*. Chapman and Hall, London, UK.

Sperling Pagni, P.G., Walne, P.L., and Wehry, E.L. (1976). Fluorimetric evidence for flavins in isolated eyespots of *Euglena gracilis* var. *Bacillaris*. *Photochemistry and Photobiology* 24 (4): 373-375.

Stahl, E. (1885). Einfluss der Beleuchtungsrichtung auf die Theilung der Equisetum-Sporen. *Ber. Deutsch. Bot. Ges.* 3: 334-340.

Staiger C.J. (2000) Signaling to the actin cytoskeleton in plant. *Annu. Rev. Plant Physiol. Plant Mol. Biol.* 51: 257-288.

Stanley, R.G. and Linskens, H.F. (1974). *Pollen. Biology, Biochemistry, Managements*. Springer, Berlin, Germany.

Stevens, K.L. and Merrill, G.B. (1984). Sesquiterpene lactones and allelochemicals from *Centaurea* specvies. In: *The Chemistry of Allelopathy. Biochemical Interactions Among Plants*. A.C. Thompson (ed) ACS Symposium Ser., Vol, 268. pp. 83-98. Amer Chem Soc. Washington D.C., USA.

Stewart, C.N. (2006). Go with the glow: fluorescent proteins to light transgenic organisms. *Trends in Biotechnology* 24 (4): 155-162.

Stober, F. and Lichtenthaler, H.K. (1992). Changes of the laser-induced blue, green and red fluorescence signatures during greening of etiolated leaves of wheat. *J. Plant Physiol.* 140: 673-680.

Stober, F. and Lichtenthaler, H.K. (1993). Characterization of the laser-induced blue, green and red fluorescence signatures of leaves of wheat and soybean grown under different irradiance. *Physiologia Plantarum* 88: 696-704.

Stober, F., Lang, M., and Lichtenthaler, H.K. (1994). Studies on the blue, green and red fluorescence signatures of green, etiolated and white leaves. *Remote Sens. Environ.* 47: 65-71.

Sukhada, K.D. and Jayachandra (1980). Pollen allelopathy - a new phenomenon. *New Phytologist* 84: 739-746.

Swain, T. (1965). The Tannins. In: *Plant Biochemistry*, (Eds. J.Bonner and J.E. Varner). pp. 552-586. New York: Academic Press.

Szöllosi, J., Matyus, L., Tron, L., Balazs, M., Ember, I., Fulwyler, M.J., and Dajanovich, S. (1987). Flow cytometric measurements of fluorescence energy transfer using single lase excitation. *Cytometry* 8 (2): 120-128.

Tan, X, Xie, M., Kim,Y.J., Zhou, J., Klessig, D.F., and Martin, G.B. (1999). Overexpression of Pto activates defense responses and confers broad resistance. *Plant Cell* 11: 15-30.

Taylor, D.L. and Salmon, E.D. (1989). Basic fluorescence microscopy. In: *Methods in Cell Biology. Living Cell in Culture.* J.L. Wang and D.L. Taylor (eds). pp. 207-237. Acad. Press. San Diego, New York.

Terenin, A.N. (1945). The effect of the medium on the photoluminescence of aromatic compounds. *Bulleten of USSR Acad. Sci.*, Ser. Phisical, 9(4-5): 305-316.

Thomson, I.D., Andrews, B.I., and R.C. Plowright. (1982). The effect of a foreign pollen on ovule development in *Dervilla lonicera (Caprifoliaceae) New Phytologist* 90: 777-783.

Thompson, R.B. (1991). Fluorescence-based fiber-optic sensors. In: *Topics in Fluorescence Spectroscopy.* J.R. Lakowicz (ed). Vol. 2, pp. 345-365. Plenum Press. New York and London.

Thorp, R.W., Briggs, D.L., Esters, J.R., and Erickson, E.H. (1975). Nectar fluorescence under ultraviolet irradiation. *Science* 189: 476-478.

Thompson, R.B. (Ed.) (2006). *Fluorescence Sensors and Biosensors.* Boca Raton: CRC Press, 416pp.

Towers,G.H.N.(1987a).Comparative antibacteriophage activity of naturally occurring photosensitizers. *Planta medica* 53: 536-539.

Towers, G.H.N. (1987 b). Fungicidal activity of naturally occurring photosensitizers. *American Chemical Society Symposium Seria* 339: 231-240.

Traidi-Hoffmann, C., Kasche, A., Jacob, T., und Behrendt, H. (2005). Pollen-associated lipid mediators (PALMs): Messenger substances from pollen with proinflammatory and immunomodulating effects on cells involved in allergic inflammation. *GSF-Forschungszentrum für Umwelt und Gesundheit GmbH in der Helmholtz- Gemeinschaft Annual Reports*. pp. 61-66.

Tretyak, T.M. and Arkhipova, L.V. (1992). Intracellular activity of neuromediators. Uspekhi Sovremennoi Biologii (Trends in Modern Biology, Russia) 112(2): 265-272.

Tsien, R.Y. (1994). Fluorescence Imaging creaties a window on the cell. *Chem. Eng. New* 72: 34-37.

Tsien, R.Y. (1998). The green fluorescent protein. *Annu. Rev.Biochem.* 67: 509-544.

Tsien, R.Y. (2004). Improved monomeric red, orange and yellow. Fluorescent proteins derived from *Discosoma* sp. red fluorescent protein. *Nature Biotechnology* 22 (12): 1567-1572.

Tswett, M. (1911). Über Reicherts Fluoreszenzmikroskop und einige damit angestellte Beobachtungen über Chlorophyll und Cyanophyll. *Ber. Dtsch. Bot. Ges.* 29: 744.

Tyukavkina, N.A., Pogodaeva, N.N., Brodskaya, E.I., and Sapozhnokov, Yu.M. (1975). Ultraviolet absorbance of flavonoids. V. The structure of 3- and 5-oxyflavons. *Chemistry of Natural Compounds (USSR)* 5: 583-587.

Uehara, K. and Murakami, S. (1995). Arrangement of microtubules during spore formation in *Equisetum arvense* (Equisetaceae). *American Journal of Botany* 82 (1): 75-80.

Uma, L., Balasubramanian, D., and Sharma, Y. (1994). In situ fluorescence spectroscopic studies on bovine cornea. *Photochemistry and Photobiology* 59 (5): 557-561.

Valeur, B. (2002). *Molecular Fluorescence.* Weinheim: Wiley-VCH.

Vandenkooi, J.M. (1991). Tryptophan phosphorescence from proteins at room temperature. In: *Topics in Fluorescence Spectroscopy.* J.R. Lakowicz (ed), Vol. 3, pp. 113-136. Plenum Press. New York and London.

Vasilyev, A.E. (1977). *Functional Morphology of Plant Secretory Cells.* Nauka. Leningrad, Russia.

Vasilyev, A.E. and Muravnik, L.E. (1988). The ultrastructure of the digestive glands in *Pinguicula vulgaris* L. (Lentibulariaceae) relative to their function. 1. The changes during maturation. *Annals of Botany* 62: 329-341.

Vaughan, M.A. and Vaughan, K.C. (1988). Mitotic disrupters from higher plants and their potential uses as herbicides. *Weed technology.* 2: 533-539.

Veeger, C., Visser, T., Eweg, J.K., Grande, H., de Abreu, R., de Graaf-Hess, A., and Müller, F. (1980). Fluorescence of oxidzed and reduced flavins and flavoproteins. In: *Flavins and Flavoproteins.* K. Yagi and T. Yamano (eds). pp. 349-357. Japan Scientific Societies Press and University Park Press Baltimore. Tokyo, Baltimore Vidali, L., McKenna, S.T., and Hepler, P.K. (2001). Actin polymerization is essential for pollen tube growth. *Mol Biol Cell* 12 (8): 2534-2545.

Viswanath, G. and Kasha, M. (1956). Confirmation of the anomalous fluorscence of azulene. *J. Chem. Phys.* 24(3): 574-577.

Vogelmann, T.C. (1993). Plant tissue optics. *Ann. Rev. Plant Physiol and Plant Mol. Biol.* 44: 231-251.

Vogt, T., Pollak, P., Tarlin, N., and Taylor, L.P. (1994). Pollination – or wound-induced kaempferol accumulation in *Petunia* stigmas enhances seed production. *The Plant Cell* 6 (1): 11-2.

Wagner, H. and Wolff, P. Eds. (1977) *New Natural Products with Pharmacological Biological or Therapeutical Activity.* Berlin: Springer Verlag,.

Wagner, G.J. (1990). Secreting glandular trichomes: more than just hairs. *Plant Physiol* 96: 675-679.

Wang, W.W., Gorsuch, J.W., and Hughes, J. (eds.) (1997). *Plants for Environmental Studies.* CRC Press. Boca Raton, Fl., USA.

Wang, X.F., and Herman, B. (1996). (eds) *Fluorescence Imaging Spectroscopy and Microscopy.* John Wiley. London, UK.

Wardlow, C.W. (1965). Physiology of embryonic development in cormophytes. In: *Handbuch der Pflanzenphysiologie. (Encyclopedia of Plant Physiology)*, Bd. 15, teil part 1. pp. 844-965. Springer-Verlag.Berlin, Heidelburg, New York.

Wehling, K., Niester, Ch., Boon, J.J., Willemse, M.T.M., and Wiermann, R. (1989). *p*-Coumaric acid - monomer in the sporopollenin skeleton. *Planta* 179(3): 376-380.

Weissenböck, G.W., Schnabl, H., Scharf, H., and Sachs, G. (1987). On the properties of fluorescing compounds in guard and epidermal cells of *Allium cepa* L. *Planta* 171(1): 88-95.

Weiss, O. and Reigosa Roger, M.J. (2001). Modulated fluorescence. In: *Handbook of Plant Ecophysiology Technique.* M.J. Reigosa Roger (ed). pp. 173-183. Kluwer Acad. Publ. Dordrecht, the Netherlands.

Werker, E. (2000). Trichome diversity and development. *Advances in Botanical Research* 31: 1-35.

Wettstein, D. (1965). Die Induction und experimentelle Beeinflussung der Polaritat bei Pflanzen. In: *Handbuch der Pflanzenphysiologie (Encyclopedia of Plant Physiology).* Bd.15, teil Part 1, pp. 275-330 Springer-Verlag, Berlin, Heidelburg, New York.

Whitaker, M. (1995). Fluorescence imaging in living cells.In: *Cell Biology. A Laboratory Handbook.* J.E. Celis (ed). Vol. 2. pp. 37-43. Academic Press. San Diego, New York.

White, N.S., Errington, R.J., Wood, J.L., and Fricker, M.D. (1996). Quantative measurements in multidimensional botanical fluorescent images. *Journal of Microscopy* 181(2): 99-116.

White, R.A., Kutz, K.J., and Wampler, J.E. (1991). Fundamental of Fluorescence Microscopy. In: *Topics in Fluorescence Spectroscopy.* J.R. Lakowicz (ed) Vol. 1, pp. 379-410. Plenum Press. New York and London.

Wilhelm, J. and Wilhelmova, N. (1981). Accumulation of lipofuscin-like pigments in chloroplasts from scenescing leaves of *Phaseolus vulgaris. Photosynthetica.* 15 (1): 55-60.

Willemse, M.T.M. (1971). Morphological and fluorescence microscopical investigation on sporopollenin formation at *Pinus sylvestris* and *Gasteria verrucosa*. In: *Sporopollenin*. J, Brooks, P.R. Grant, M. Muir and P.R. Gijzel van (eds) pp. 68-91 Proc. Symp. at Geology Department, Imperial College, London, 1970. 23-25 September. Acad. Press. London, New York.

Williams, S. (1938). Experimental Morphology. In: *Manual of Pteridology*. Fr. Verdoorn (ed). pp. 105-140. Martinus Nijhoff. The Hague.

Willingham, M.C. and Pasan, I. (1978). The visualization of fluorescent proteins in living cells by video intensification microscopy (VIM). *Cell* 13 (3): 501-507.

Wittstock, U., Hadacek, F., Wurz, G., Teusscher, E., and Greger. H. (1995). Polyacetylenes from water hemlock, *Cicuta virosa*. *Planta medica* 61(5): 439-445.

Wollenweber, E. (1984), The systematic implication of flavonoids secreted by plants. In: *Biology and Chemistry of Plant Trichomes*. (Eds. E.Rodriquez, P.L.Healey, and I. Mehta). Pp. 53-69. New York: Plenum Press.

Wolfbeis, O.S. (1985). The fluorescence of organic natural products In: *Molecular Luminescence Spectroscopy. Methods and Applications* S.G. Schulman (ed). pp. 167-370. Wiley. New York, Chichester, Brisbane, Toronto, Singapore.

Wollenweber, E., Asakawa, Y., Schillo, D., Lehmann, U., and Weigel, H. (1987). A novel caffeic acid derivative and other constituents in *Populus* bud excretion and propolis (bee-glue). *Z. Naturforsch.* 42 C (6): 1030-1034.

Wollenweber, E., Mann, K., and Roitman, J.N. (1991). Flavonoid aglycones from the bud exudates of three Betulaceae. *Z. Naturforsch.* 46 C (3): 495-497.

Wu, W., Nelson, P.E., Cook, M.E. and Smalley, E.B. (1990). Fusarochromanone production by *Fusarium* isolates. *Applied and Environmental Microbiology* 56 (10): 2989-2993.

Yakovlev, G.P. and Blinova, K.E. (2002). *Encyclopedia Dictionary of Medicinal Plants and Products of Animals*. SpetsLit. CPKhFA.Sankt Peterburg, Russia.

Yashina, A.V., Yashin V.A., and Roshchina V.V., (2007). Laser-scanning confocal microscopyfor the study of pollen state and communication. In: *Materials of International Symposium."Reception and Intracellular Signaling"*. Pushchino, 5-7 June 2007. (Ed. V.P. Zinchenko). pp. 285-287. Pushchino: ONTI.

Yasuda, H. and Shinoda, H. (1985). The studies of the spherical bodies containg anthocyanins in plant cells. 1. Cytological and cytochemical observations on the bodies appearing in the seedlings hypocotyls of radish plants. *Cytologia* 50: 397-403.

Yeloff, D. and Hunt, C. (2005). Fluorescence microscopy of pollen and spores: a tool for investigating environmental change. *Review of Palaeobotany and Palynology:* 133(3-4): 203-219.

Young, A.J. and Low, G.M (2001). Antioxidant and prooxidant properties of carotenoids. *Arch Biochem Biophys.* 385 (1): 20-27.

Yurin, P.V., Shef, R.P., and Chernysheva, V.I. (1972). Role of free radicals at the physiologo-biochemical plant interactions in agrophytocenosis.In: *Physiologo-Biochemical Base of Plants Interactions in Phytocenosis*. A.M. Grodsinskii (ed). Vol. 3, pp. 127-134. Naukova Dumka. Kiev, Ukraine.

Zander, M. (1981). Fluorimetrie. Springer-Verlag Berlin, Heidelburg, New York.

Zanoni, D.G. (1930). Antagonismo pollinico. *Revista di Biologia* 12: 126-133.

Zeiger, E. and Hepler, P.K. (1979). Blue light-induced, intristic vacuolar fluorescence in onion guard cells. *Journal of Cell Science* 37 (1): 1-10.

Zeiger, E. (1980). The blue-light response of stomata and the green vacuolar fluorescence of guard cells. In: The *Blue Light Syndrome*. (Eds. H. Senger). pp.629-636. Berlin Heidelberg New York: Springer-Verlag.

Zeiger, E. (1981). Novel approaches to the biology of stomatal guard cells: protoplast and fluorescence studies. In: *Stomatal Physiology*.(Eds. P.G.Jarvis and T.A. Mansfield). pp.71-85. Campbridge: Cambridge University Press.

Zeiger, E. (1983). The biology of stomatal guard cells. *Annu. Rev. Plant. Physiol.* 34: 441-475.

Zelinskii, V.V. (1947). One of possible causes of the internal quenching of fluorescence of complex organic dyes. *Dokl USSR Acad. Sci* 56 (4): 383-385.

Zeng, H., MacAulay, C., McLean, D.I., and Palcic, B. (1995). Spectroscopic and microscopic characteristics of human skin autofluorescence emission. *Photochemistry and Photobiology* 61 (6): 639-645.

Zerback, R., Bokel, M. Geiger, H. and Hess, D. (1989). A kaempferol 3-glucosylgalactoside and further flavonoids from pollen *of Petunia hybrida. Phytochemistry*, 8: 897-899.

Zhang, Y., Aslan, K., Previte, M.J.R. and Geddes, C.D. (2006). Metal-enhanced S_2 fluorescence from azulene. *Chemical Physics Letters* 432: 526-532.

Zhou, Y.C. and Zheng, R.L. (1991). Phenolic compound and analogy as superoxide anion scavengers and antioxidants. *Biochem. Pharmacol.* 42: 1177-1179.

Zobel, A.M. and Brown, S.A. (1989). Histological localization of furanocoumarins in Ruta graveolens shoots. *Can J. Bot* 67 (3): 915-921.

Zobel, A.M. and Brown, S.A. (1990). Dermatitus-inducing furanocoumarins on the leaf surfaces eight species of rutaceous and umbelliferous plants. *Journal of Chemical Ecology* 16: 693-700.

Zobel, A.M. and March, R.E. (1993). Autofluorescence reveals different histological localization of furanocoumarins in fruits of some *Umbelliferae and Leguminosae. Annals of Botany* 71: 251-255.

Zobel, A.M. and Brown, S.A. (1995). Coumarins in the interaction between plant and its environment. *Allelopathy Journal* 2: 9-20.

Latin Index

Subject Index